CUSTOM COMPONENTS IN ARCHITECTURE

This book offers architects strategies in the design and manufacturing of custom, repetitively manufactured building components.

A total of 36 case studies from around the globe demonstrate the diversity of CRM in architecture and are contributed by architecture firms, including Diller Scofidio + Renfro, Kengo Kuma & Associates, Abin Design Studio, Behnisch Architekten, Belzberg Architects, and many more. The book is organized by manufacturing process and covers the use of various types of glass, clay, plastic, metal, wood, plaster, and concrete. Each process is described with diagrams and text and expanded with one or more examples of customized building components. Projects included are of buildings of various types, sizes, and clients, and many deviate from the typical manufacturing process as they include a secondary process (e.g. casting glass, then slumping it), special tooling modifications (e.g. dams used

to subdivide a mold), post-production processes, or other notable manufacturing features. Each case study includes a project overview, behind-the-scenes images of the component manufacturing, and original diagrams that illustrate how those components were customized.

Custom Components in Architecture will be essential reading for any architect interested in building design details and keeping up-to-speed on material advances. Upper-level students of digital architecture, fabrication, and building technology will also find this a useful tool.

Dana K. Gulling is a professor at North Carolina State University, School of Architecture and is a licensed architect. Her work is at the intersection of building technology and design. She has won two national book awards and a national teaching award.

W0113175

CUSTOM COMPONENTS IN ARCHITECTURE

Strategies for Customizing Repetitive Manufacturing

Dana K. Gulling

Routledge
Taylor & Francis Group

LONDON AND NEW YORK

Designed cover image: Chipster Blister House by AUM with Corian screen panels. Photo by Studio Erick Saillet.

First published 2024
by Routledge
4 Park Square, Milton Park, Abingdon, Oxon OX14 4RN

and by Routledge
605 Third Avenue, New York, NY 10158

Routledge is an imprint of the Taylor & Francis Group, an informa business

© 2024 Dana K. Gulling

The right of Dana K. Gulling to be identified as author of this work has been asserted in accordance with sections 77 and 78 of the Copyright, Designs and Patents Act 1988.

All rights reserved. No part of this book may be reprinted or reproduced or utilised in any form or by any electronic, mechanical, or other means, now known or hereafter invented, including photocopying and recording, or in any information storage or retrieval system, without permission in writing from the publishers.

Trademark notice: Product or corporate names may be trademarks or registered trademarks, and are used only for identification and explanation without intent to infringe.

British Library Cataloguing-in-Publication Data
A catalogue record for this book is available from the British Library

Library of Congress Cataloging-in-Publication Data
Names: Gulling, Dana K., author.
Title: Custom components in architecture : strategies for customizing repetitive manufacturing / Dana K. Gulling.
Description: Abingdon, Oxon ; New York, NY : Routledge, 2024. | Includes bibliographical references and index. |
Identifiers: LCCN 2023025686 | ISBN 9781032289335 (hbk) | ISBN 9781032289328 (pbk) | ISBN 9781003299196 (ebk)
Subjects: LCSH: Building fittings--Design and construction. | Manufacturing processes. | Repetitive manufacturing systems. | Architecture--Details.
Classification: LCC TH6010 .G85 2024 | DDC 670--dc23/eng/20230927
LC record available at https://lccn.loc.gov/2023025686

ISBN: 978-1-032-28933-5 (hbk)
ISBN: 978-1-032-28932-8 (pbk)
ISBN: 978-1-003-29919-6 (ebk)

DOI: 10.4324/9781003299196

Typeset in Univers LT Std
by KnowledgeWorks Global Ltd.

Contents

List of Figures

Abbreviations

hr	hour or hours
min	minute or minutes
sec	second or seconds

km	kilometer or kilometers
kg	kilogram or kilograms
°C	degrees Celsius

Imperial Units

in	Inch or inches
fl oz	fluid ounce
ft	Foot or feet
ft^2	square foot or square feet
gal	galleon or galleons
ksi	kip or kips per square inch (equals 1000 lb/in^2)
mi	Mile or miles
lb	pound or pounds
lb/in^2	pound or pounds per square inch
k/in^2	kip or kips per square inch (1 kip = 1000 lb)
lb/ft^2	pound or pounds per square foot
qt	quart or quarts
°F	degrees Fahrenheit

SI Units

kPa	kilo Pascal or Pascals (equals 1000 Pa)
mm	millimeter or millimeters
cm	centimeter or centimeters
m	meter or meters
m^2	square meter or meters
mL	milliliter or milliliters
L	liter or liter
Pa	Pascals

Acronyms and Abbreviations

AAMA	Asia America Multi-technology Association
ABS	acrylonitrile butadiene styrene
AEC	architecture, engineering, construction
APA	American Plywood Association
ASTM	American Society of Testing and Materials
BIM	building information modeling
BS	British Standards
CAM	computer-aided manufacturing
CAD	computer-aided design/computer-aided drafting
CNC	computer numeric controlled
CLT	cross-laminated timber
CRM	customized repetitive manufacturing
DIY	do it yourself
EDM	electronic discharge machine
EXW	explosive welding
FRP	fiber reinforced plastic
glulam	glue-laminated
LED	light-emitting diode
LEED	Leadership in Energy and Environmental Design
low-E	low emissivity
MDF	medium density fiberboard
OSB	oriented strand board
PET	polyethylene terephthalate polymer (plastic)

PET-G	polyethylene terephthalate glycol-modified polymer	UL	Underwriters Laboratory
PS	polystyrene	US	United States (of America)
PVB	polyvinyl butyral	USD	United States dollar
PVC	polyvinyl chloride	UV	ultraviolet light
SPF	super-plastic forming	VOC	volatile organic compounds

Prologue

I am fascinated by how things are made and as an architect interested in building construction and assemblies this is a great characteristic to have. Many years ago, early in my days of architecture practice, I worked on the design development for a large university dormitory. The lead team designer wanted all the dorm room entry doors to be offset by 6 in from the room's interior corner. As he explained, 6 in would allow for the 2-in wide hollow-metal door frame and 4 in of wall that would easily fit a painter's brush and hand. I am not certain that his notion of construction was fully accurate—it would be more efficient for the sub-contractor to spray the rooms with paint rather than use a brush and roller—but nevertheless it imprinted onto me the idea that architects can design from an understanding of how a building would be constructed.

Through my subsequent experiences in practice, teaching, and research, I gained an appreciation for making, recognizing that there is a deep beauty to buildings that are both well-designed and well-crafted. As Richard Sennett articulated through his book, *The Craftsman*, beauty and poetics can come from the act of making. For me, I link design and making in this way: *if we know how something is made, then we can design it.*

Custom Components in Architecture comes out of my love of making and is for those architects who also love making. The book highlights architectural projects that have customized repetitive manufacturing (CRM) processes to produce building components that are unique to a particular building project. The manufacturing processes in *Custom Components* extend beyond processes enabled by computer numerical controlled (CNC) equipment, as CRM processes often require direct human interaction for production. For many CRM projects, CNC equipment is used to make custom tooling (e.g. molds, patterns, or jigs) that are then used repeatedly to produce custom components. Despite our assumptions that repetitive manufacturing relies on assembly lines and automated machines, *customized* repetitive manufacturing is done in small batches and therefore demands human involvement in its making.

Custom Components builds upon another book of mine, *Manufacturing Architecture: An Architect's Guide to Custom Processes, Materials, and Applications*. *Manufacturing Architecture* is a guide for architects who are interested in customizing repetitive manufacturing processes and includes case studies to illustrate potential architectural applications. Since *Manufacturing Architecture* was published, I have continued to gather case studies of CRM in architecture. As a result of my gathering, I believed that another book was needed, one that primarily focused on the case studies, to tell the stories of this type of architectural work. Through this book's research, I now have amassed a list of over 460 examples of CRM in architecture.

Custom Components is organized by manufacturing processes, with similar manufacturing processes grouped together. Each chapter includes a brief visual and written overview of the typical manufacturing process to provide context. The chapters' focus is on the case studies in architecture that make use of that process. Each case study includes a project overview, describes the custom components, and illustrates any modifications of the typical manufacturing process for production. It is through the new case studies that we learn how each manufacturing process has been—and by extension, could be—used in the making of custom components.

By sharing these case studies with you, I share my joy in learning about making and I promote the idea that designing custom components for a project is something that every architect can do. Design and making custom components are not just for large projects with big budgets, or just for high-profile architecture firms with features in architecture magazines; instead, all architects can consider designing custom components for their buildings. Use this book. Learn from the case studies. Be inspired. Because, once we know how something is made, then we can design it.

Dana K. Gulling

https://orcid.org/0000-0001-5554-4505

Introduction

Figure 0.1
Tooling is a manufacturing term that refers to molds, patterns, dies, or
jigs that are used to shape the produced units. (a) Wood and metal molds
used in metal spinning. (b) Tool steel dies used for extruding aluminum.

DOI: 10.4324/9781003299196-1

Figure 0.2
Zaha Hadid Architects' Guangzhou Opera House. Photographs by Terri Meyer Boake.

Figure 0.3
San Francisco Museum of Modern Art Expansion by Snohetta used a single-use,
CNC-milled foam mold to form each one of its individual fiber-reinforced plastic facade
panels. Photograph by Henrik Kam, courtesy of Snohetta and Kreysler & Associates.

In architecture, computer-aided manufacturing (CAM) has revolutionized the relationship between design and fabrication, using computer numeric controlled (CNC) machines to make individual and unique architecture components that are not prohibitively expensive. However, the benefits of CAM are not straightforward, and architects should recognize the differences between directly and indirectly using CNC equipment to produce architecture components. Directly, CNC equipment can cut, carve, and shape materials to fabricate architecture components; indirectly, CNC equipment can cut, carve, and shape custom tooling (e.g. molds, patterns, and jigs) to manufacture architecture components. An example of direct CNC use is the granite stones for Zaha Hadid Architects' Guangzhou Opera House that were custom cut to clad the building's curving skin. An example of indirect CNC use is Snohetta's San Francisco Museum of Modern Art, for which a five-axis CNC mill fabricated the single-use Styrofoam molds for the facade's glass fiber-reinforced plastic (GFRP) panels.[1] Instead of fabricating a single-use mold that then is either disposed or recycled, this book focuses on custom-building components that are made using durable, reusable tooling—otherwise known as *customized repetitive manufacturing*.

Today, repetitive manufacturers use CNC equipment to fabricate tooling. Contact fiberglass molders and plastic thermoformers use CNC-milled, high-density foam to fabricate their tooling. CNC millers and electrical discharge machines (EDM) fabricate hardened-steel molds for injection molding and dies for extrusion. In addition, manufacturing research has investigated the use of rapid prototyping (RP) equipment, often known as 3D printers, to make tooling for repetitive manufacturing. For example, sand-casters can use 3D-printed patterns for small production

runs,[2] and manufacturing production researchers have investigated the use of metal laser sintering to make molds for injection molding plastic.[3] With its indirect use, CNC equipment has made it affordable and practical to customize repetitive manufacturing (CRM) on a per-building basis, making CRM an option in producing custom architecture components. Architectural examples include COOKFOX Architects' One South First (Chapter 4.1) that used three-dimensional printed and CNC-milled, carbon-fiber-reinforced plastic mold inserts to form its exterior precast concrete panels and Mei Architects and Planners' Gnome Parking Garage (Chapter 1.5) that used a CNC-milled, mated steel die to stamp its stainless-steel panels.

CRM has notable advantages over the direct use of CNC equipment to produce custom architecture components. First, there is a wide range of forms, materials, and finishes available in CRM. Processes such as slumping glass and clay (Chapter 1.1), rotational molding plastic (Chapter 3.4), and casting glass (Chapter 4.3) are done with a mold and cannot be replicated with the direct use of CNC equipment. Second, most CRM processes only use as much materials as the mold, pattern, or jig needs. By reusing tooling and reducing raw material requirements, repetitive manufacturing can have little-to-no production waste. For many of the remaining CRM processes, the production waste can be recycled directly onsite. For example, excess clay from extruding bricks is sent directly back into the extruder to be re-extruded. Third, manufacturing tolerances for most of these processes are high and have the potential to rival the tolerances of CNC equipment. Fourth, because of CNC technology, designers can customize the molds, patterns, or jigs, with limited additional costs, and the added cost is amortized over the number of units produced. Fifth, CNC equipment can have longer production times

Figure 0.4
One South First Kent by COOKFOX Architects. Precast concrete panels
were made with molds that included both 3D-printed and hand-fabricated
"plugs" for window openings.

and thus higher fabrication costs than CRM.[4] Finally, soft costs associated with design fees,[5] machine programming, and construction labor will likely be lower with CRM than CAM.

CAM, CRM, and the Human Element

The relationship between CAM and CRM is complex, and although CRM can overlap with CAM, it is distinct from it (see Figure 0.5). CRM necessitates

human interactions that CAM does not. CAM, by definition, *must* use CNC equipment or robots in its manufacturing processes; in contrast, CRM *can* use CNC equipment to make the tooling needed, but it is not necessary, as tooling can also be made by hand. CRM requires that its tooling be durable and reusable; therefore, CRM tooling can be made by hand, by CNC machine, or some combination of the two. One South First by COOKFOX Architects combined the 3D-printed carbon fiber molds that were CNC-milled post-printing in combination with hand-built plywood molds, which are typical in the precast

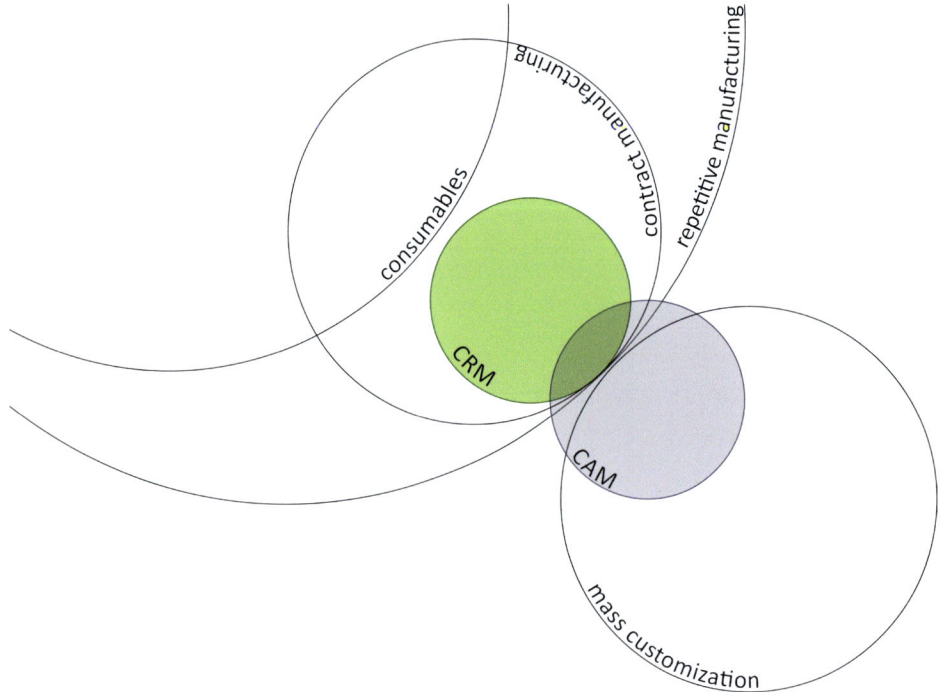

Figure 0.5
Diagram of the relationship between CAM and CRM.

Figure 0.6
Paul Smith Store Expansion by 6a Architects. New storefront made from sand-casted iron panels. The complex pattern of interlocking circles was made by forming sand molds on CNC-milled foam patterns. Included in a few panels are Paul Smith sketches that were carved by hand directly into the sand molds, prior to casting. Photos by 6a Architects.

concrete industry. Similarly, the Paul Smith Store by 6a Architects combined CNC-milled foam patterns to form sand molds for the project's cast iron panels along with Paul Smith's sketches that were hand carved directly into the mold surface (see Figure 0.6). CRM must accommodate smaller production runs than repetitive manufacturing and often relies on manual labor for its flexibility in production. In addition, some CRM processes invite the added value of craft in the production of the custom components in a way that CAM cannot.

Although the cost of CNC equipment continues to decrease, many manufacturing industries subcontract their tooling fabrication to fabricators as many are reluctant to take on the cost of purchasing, operating, and maintaining CNC equipment.

For some manufacturing processes, this is because their tooling is complex, with multiple moving parts, and mold-makers are specialized. Examples include pultrusion (Chapter 2.2), pressing (Chapter 4.6), injection molding (Chapter 4.7), extruding plastic, and plastic blow molding. For some manufacturing processes, metal tooling must be made with thin walls so that it can release heat during its process. Examples include rotational molding (Chapter 3.4), spin or centrifugal casting (Chapter 3.5), and casting concrete (Chapter 4.1). Shaping thin metal walls for tooling cannot be cost effectively exclusively formed by current CNC processes and instead will be built up by steel fabricators.[6]

If purchasing CNC-fabricated molds from toolmakers, then manufacturers will want to ensure the project can support production runs to offset the tooling costs. With One S First, Oak Ridge National Laboratories developed the 3D-printed molds, while Gate Precast carpenters hand fabricated the plywood portions of the mold. Both 3D-printed and hand-made plywood molds formed the project's window "plugs" that could be rearranged and flipped between concrete pours to create different panels. To meet the project schedule, One S First precast production was completed by multiple Gate plants. The Gate plant in Winchester, KY used the 3D-printed plugs with plywood mold edges, while their facility in Oxford, NC only used hand-built plywood molds for both the plugs and the mold edges. The form of the mold was relatively simple and could be constructed cost effectively by hand; however, the 3D-printed molds were more durable than the plywood molds. The 3D-printed plugs could be used for 200 or more casts, while the plywood molds only lasted for 10 casts before degrading too much. Even though the cost of the 3D-printed molds was 50 times greater than the cost of the plywood molds, they were cost effective because of the production run.

Repetitive manufacturing may conjure visions of big assembly lines, producing millions of the same unit, and for highly consumable items like drink bottle caps, Legos, and glass bottles, this is the reality. Instead, *customized* repetitive manufacturing requires smaller production runs than repetitive manufacturing of consumables, because it requires flexibility by its manufacturers in their production processes. Most CRM manufacturers are contract manufacturers, meaning that they are contracted to specifically produce small production runs of items. Occasionally contract manufacturers may make some of their own products, but they have dedicated machines or production schedules that are dedicated to fulfilling contract work.

Generally, contract manufacturing requires human engagement between the maker and material. For metal stamping, people load the metal blanks into and out of the press. For making bent plywood, people pass the wood veneers through a glue curtain before layering them into the mold and turning the press cycle on. For casting metal, people are moving the crucible from the furnace and pouring it into the mold. For plastic blow molding, people are removing the blown items from the press and trimming off the flashing between machine cycles. The benefit of using people, rather than robots and assembly lines, is to maximize the contract manufacturer's flexibility to produce small batches of different components. This reduces capital costs associated with purchasing and maintaining the robots and soft costs of labor associated with their programming.

Additionally, some CRM processes have not been mechanized by their industries and therefore necessitate a direct interaction between maker and mold to manufacture a component. Processes such as contact molding (Chapter 3.1), vibration tamping (Chapter 4.5), casting (Chapters 4.1–4.3), and slip casting (Chapter 3.6) are primarily done by people, with some machine assistance.[7] A familiar architectural example is precasting architectural concrete. It is people that prepare the mold, place the reinforcing steel and embeds, operate the gantry crane and concrete pour bucket, spread the concrete, screed the concrete, strip the mold, inspect the component, and do any post-finishing that is required. Currently and in the foreseeable future, the manufacturing of precast concrete relies almost solely on people in its process.

Some contract manufacturers are what we would consider more craftspeople than manufacturer, as they operate small workshops or studios rather than manufacturing facilities. This removes CRM even

Figure 0.7
Workers at Davis Plywood, a contract manufacturer, manipulating wood veneers before they place them in the press.

Figure 0.8
Workers at Gate Precast placing concrete into a mold. An overhead gantry crane carries the concrete bucket, and the worker operates an opening lever to place the concrete in the panel where needed.

Figure 0.9
Bill Ganz demonstrating metal spinning at his workshop in Midvale,
UT. All his work is done by hand on a lathe.

Figure 0.10
Corian sunshade panels made by a French cabinet maker for
AUM's Chipster Blister House. Photo by Studio Erick Saillet.

further from our preconceptions of assembly lines and machine automation. Just outside of Salt Lake City, Utah, is located Bill Ganz & Company, a metal spinner, making axisymmetric components on a lathe out of metal. Bill Ganz is the sole proprietor and single worker in his metal spinning company, which operates out of his neighborhood workshop. He manufactures a little bit of everything, including custom light fixture parts for a lighting company to components for NASA (United States National Aeronautics and Space Administration). Similarly, Image, a French cabinet maker known for making custom furniture, thermoformed the Corian sunshade panels for AUM's Chipster Blister House (Chapter 1.2).

Contract manufacturing is often locally based, meaning that there are certain types of manufacturing in some areas but not others. In North Carolina, because of our large clay deposits, we have a high concentration of brick manufacturers, and because of our history in boat building, we also have several contact molders that work with

Figure 0.11
The Compton, 30 Lodge Road by E8 Architecture. The building's custom, sand-cased
aluminum panels were made approximately 50 mi (80 km) from the building.
(a) Photograph by E8 Architecture at AATi. (b) Image courtesy of Regal London.

fiberglass-reinforced plastic (FRP). Conversely, North Carolina has no manufacturers that do metal spinning, whereas there are 12 metal spinners in Illinois. Typically, items that are produced locally, by small businesses are valued higher than those from far-flung places. This can be attributed to a growing awareness to reduce the energy footprint of imported goods for sustainable living, and a desire to spend money that supports the local economy and workforce. Generally, contract manufacturers are those small businesses that operate locally and provide this additional value.

Architectural examples of locally made CRM components can be found throughout our case studies. These include the custom bent, glue-laminated, Monterey pine structural members made by Techlam for the Wellington International Airport by Warren Mahoney (Chapter 1.4), which were made approximately 60 mi (100 km) from the project; the custom-cast aluminum panels made by AATi Architectural Limited for The Compton (Chapter 4.2), which were made approximately 50 mi (80 km) from the site; and the custom concrete masonry units for Phra Pradeang House by all(zone) (Chapter 4.4), which were made by Tangrungjareon Engineering that has a facility only 43 mi (71 km) from the house's neighborhood. Through my interviews and discussions, the architecture firms of the projects placed additional value onto the CRM component because of its local connection.

Architects institutionalized the value of collaborations between designers (architects) and makers (craftspeople) through our modern education practices with the formation of the Bauhaus in the 1920s. The intent of that curriculum was to erode the distinction between designer and maker, and

this principle continues to inform architects today. Architects speak to design intention, aesthetics, performance, and form, whereas craftspeople speak to making, materials, shaping, and craft. The expertise of each discipline is appreciated and balanced by the other. Architects value the craftsperson's experience with materials and processes of making, and craftspeople value the architect's ability to expand what their processes can produce. Similarly, CRM bridges between designers and makers in a way that CAM does not. Unlike CAM which relies heavily—and maybe exclusively—on CNC equipment and robots for production, CRM *may* use CNC-fabricated tooling but relies heavily on humans in its manufacturing processes for production. Even in more industrial processes such as extruding bricks, humans can interact with the assembly line. For Lantern House (Chapter 2.1), workers at Taylor Clay Products hand-threw minerals onto the bricks' surface before the bricks were cut into shape.

Many architects appreciate building design at multiple scales—from a building's massing to the execution of its details. Similarly, those architects respect the processes needed to get a building built; it is not only the artifact, or building, that they value but also the means and methods used to produce the building. In other words, buildings are enhanced if they are well executed and well crafted. These characteristics may not visually be apparent in our buildings, but as Juhani Pallasmaa argues in *The Thinking Hand*, the human condition can be imprinted into an object through its making.[8] Similarly, Peter Zumthor writes in *Thinking Architecture* that he is tempted to think that the effort and skill put into making things are inherent

Figure 0.12
Worker at Taylor Clay Products hand-throwing minerals on the brick surface for Lantern House.

parts of the things themselves.[9] In other words, the architectural value of a building is not just the artifact itself but also how it was made.

This argument should extend beyond buildings to include the things that make up our buildings— our building components. About The Compton, E8 Architecture's website promotes the craft and provenance of the project's custom-cast aluminum panels. They wrote, "We were able to mold the aluminum and carve bespoke patterns and textured shapes, producing a structure that is finely crafted; made from recycled materials; locally sourced from Essex, and manufactured by a dedicated team of highly skilled tradespeople."[10] Similarly, the Eastside Townhouse by Michael K. Chen Architecture (MKCA) (Chapter 3.6) has four different CRM components made by Boston Valley Terra Cotta, using four different processes—hand pressing, hydraulic ram pressing, extrusion, and slip casting. The project has been publicized and awarded for its attention to "material research, craft, fabrication, and collaboration."[11] In

these examples, the CRM processes used to make the components have a human element that gives additional worth to both the components and the projects.

One of the compelling things about CRM is that there are many possibilities for making. In Richard Sennet's *The Craftsman*, he wrote, "The enlightened way to use a machine is to judge is powers, fashion its uses, in light of our own limits rather than the machines potential."[12] Unlike CAM, which necessitates using CNC equipment or robots, CRM offers a potential balance between human and machine. Through CRM's nuances, there is the opportunity to ask how should the component be produced? Can the component be made by craftspeople, artists, or contract manufacturers that have tacit knowledge in forming materials and can be found locally? Would the tooling be more cost effective if it were made by hand or by machine? What advantages might be afforded by either?

Figure 0.13
Eastside Townhouse by Michael K. Chen Architecture (MKCA) incorporates four different CRM components into this highly crafted house. (a) Photograph of the back facade that includes custom extruded tiles and custom slip-cast planters. (b) Photograph of the ground-level interior that includes custom hydraulic ram-pressed tiles as a feature wall for the back stairs. The slip-cast planter can be seen outside the back door. Photography by Alan Tansey.

CRM in Architecture

Through research, we have collected over 460 examples of CRM in architecture. The CRM examples have been mined from publicly accessed architecture and design databases (e.g. *ArchDaily*, *Architizer*, and *DesignBoom*), architecture firm websites, manufacturer websites, manufacturing association websites (e.g. Precast/Prestressed Concrete Institute (PCI) and Tile of Spain Awards), books, and architecture publications (e.g. *Architect Magazine*, *Architectural Record*, *The Architect's Newspaper*). Sources must state that the CRM component was custom made for the project, and they must indicate that either multiples of the custom component were made or that durable and reusable tooling was used to manufacture the component, as another component *could* be made using the same tooling.

The collection of case studies is organized by component, with each CRM component getting its own entry as several buildings have more than one CRM component. An example is MKCA's Eastside Townhouse that has four CRM components—interior ram-pressed wall tiles, exterior hand-pressed molding for the historic front facade, extruded exterior rainscreen tiles for the rear facade, and slip-cast planter boxes for its vertical planter. Some projects have CRM components that use more than one manufacturing process. When this happens, if the manufacturing process is distinct, then it has its own entry; if the manufacturing process is an extension of the first, then it does not. An example of the former is Hariri Pontarini Architects' Baha'i Temple of South America that has custom kiln-cast glass panels for its exterior skin with approximately 10–15% of those glass panels slumped by another manufacturer. An example of the latter is

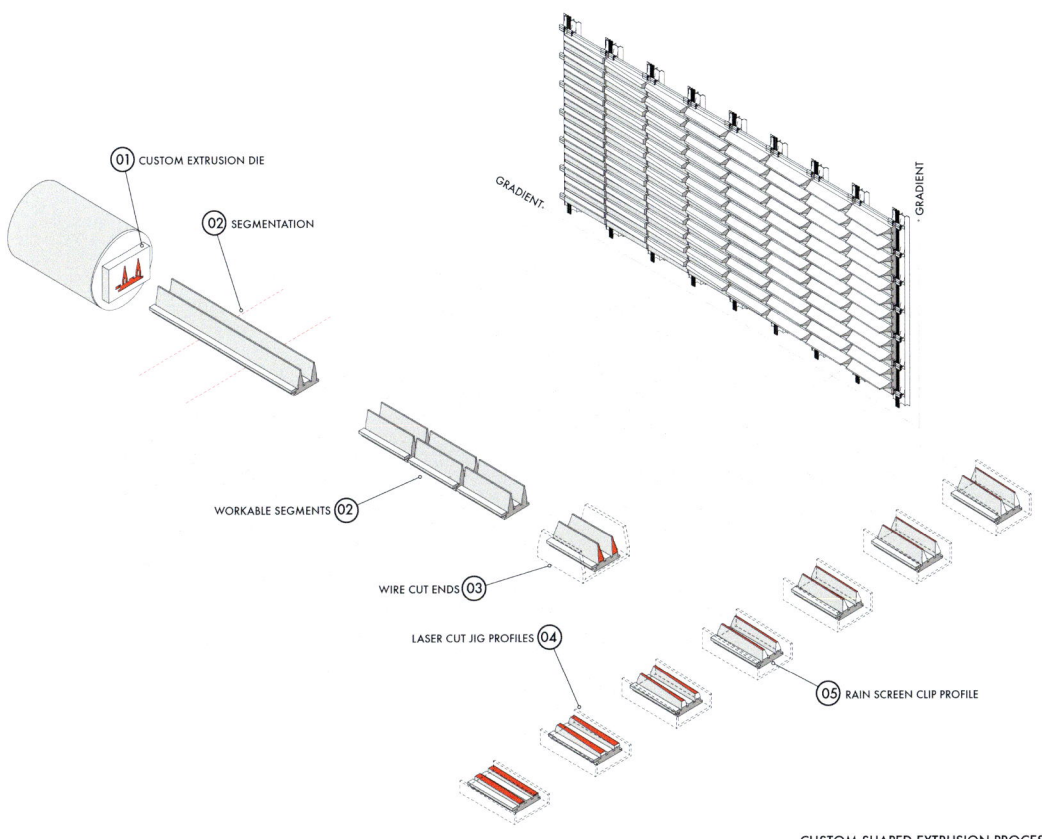

01 CUSTOM EXTRUSION DIE

02 SEGMENTATION

GRADIENT.

· GRADIENT

WORKABLE SEGMENTS 02

WIRE CUT ENDS 03

LASER CUT JIG PROFILES 04

05 RAIN SCREEN CLIP PROFILE

CUSTOM SHAPED EXTRUSION PROCESS

Figure 0.14
Diagram by MKCA, illustrating how the custom extruded cross sections were modified.
Since there was only one custom die and all the components' use were the same, this
component has one entry in the database. Drawings/Diagrams courtesy MKCA.

in Eastside Townhouse, after extruding, Boston Valley passed some of the rainscreen tiles under a laser cutter to trim off portions of their profile to modify the extrudate's cross section (see Figure 0.15). If the CRM components have the same manufacturing process, but are distinct in their use, then each has its own entry. An example is Rudy Ricciotti Architecte's Museum of European and Mediterranean Civilizations (MuCEM) that has four separate custom cast concrete CRM components—the horizontal floor decks, the exterior columns, the exterior screen panels, and the U-shaped, pedestrian bridge elements.

Each component entry includes project information such as architect, location, completion year, and building type; and component information such as component use and manufacturing processes. The CRM manufacturing process names and groupings are consistent with *Custom Component*'s chapter

organization. All included CRM examples are for components in structures that enclose space, like buildings, pavilions, 3D installations, or 3D architectural studies; the collection does not include two-dimensional, art-like installations or wall-only assembly studies. Most of the CRM examples are of buildings, with only a handful of the cases being temporary pavilions, or 3D installations or studies. To be included in the collection, a CRM component needs to have played a critical role in the overall design of the building and impact the overall aesthetics of the project. Generally, the CRM components are found in the building exterior envelope, such as cladding materials, louvers, or screen components. In a few examples, the CRM components may be structural, but if they are structural, then the customization has not solely been done for a utilitarian purpose; instead, the CRM is used

Figure 0.15
Kolumba Museum by Peter Zumthor. Photograph by Timothy Brown.[13]

to improve the structural component's aesthetics. In almost all cases, a CRM component was chosen over available mass-market building products because the CRM component was necessary to the overall vision for the building design.

For each CRM example included in the collection, the custom component was manufactured for the particular architectural project and was not conceptualized nor manufactured commercially for purchase. Occasionally, the custom component was made available to the mass market after the building was completed. This is seen in the Kolumba Museum by Peter Zumthor, and while the building was under construction, its custom brick was commercially developed by Petersen Tegl for purchase. In addition, the CRM examples highlighted are integrated into the building's construction and, as such, are not considered building products—such as light fixtures and doorknobs—that could be easily relocated to another project.

The CRM examples include a few historic projects, such as Frank Lloyd Wright's Textile Block houses in Southern California (e.g. Ennis and Freeman Houses) which used custom, onsite manufactured concrete blocks,[14] and Harrison & Abramovitz's high-rise projects in Northeastern United States (e.g. Socony-Mobil and Alcoa Buildings) which used custom stamped aluminum panels. While this sample of historic CRM projects helps us better understand CRM's changing role in architecture, this book maintains a focus on current architectural examples of CRM. The included case studies in *Custom Components* are found in recently completed projects that are both contemporary and aspirational in their designs.

Analysis of CRM Case Studies

The CRM examples are located around the world and on every continent—excluding Antarctica; this demonstrates a global application of CRM processes for custom architectural components in architecture[15] (see Figure 0.17). Most of the collected projects

Figure 0.16
World map of all CRM component locations. Map includes manufacturing process and select project identification tags. Visualization by Author, using the software Tableau.

Figure 0.17
Europe has the highest density of CRM components. Visualization by Author, using the software Tableau.

Figure 0.18
New York City metropolitan area has the highest number (41) of CRM components.
Visualization by Author, using the software Tableau.

are in the United States, particularly along the East Coast; in Europe, primarily in the United Kingdom, the Iberian Peninsula, and Central Europe from The Netherlands through to northern Italy; and southern Asia, primarily in India, China, South Korea, and Japan. Overall, Europe has the highest density of CRM projects. The highest urban density of CRM is the New York metropolitan area with 42 examples; followed by London with 21; Paris and Tokyo with 8 each; and Seoul, Beijing, and Los Angeles with 7 each (see Figures 0.17–0.20).

Globally, precast architectural concrete is found evenly distributed across the five continents, whereas extruding is found on four, with South America being excluded. In Europe and in populous global cities, there are high concentrations of manufacturing processes that manipulate sheet goods (e.g. stamping, hydroforming, and slumping) and contact molding. These processes form thin components from materials such as sheet metal, glass, and fiber-reinforced plastic, which are generally thin, light, and can be nested for ease of transport to reduce shipping

Figure 0.19
London metropolitan area has the second highest number (21) of CRM components.
Visualization by Author, using the software Tableau.

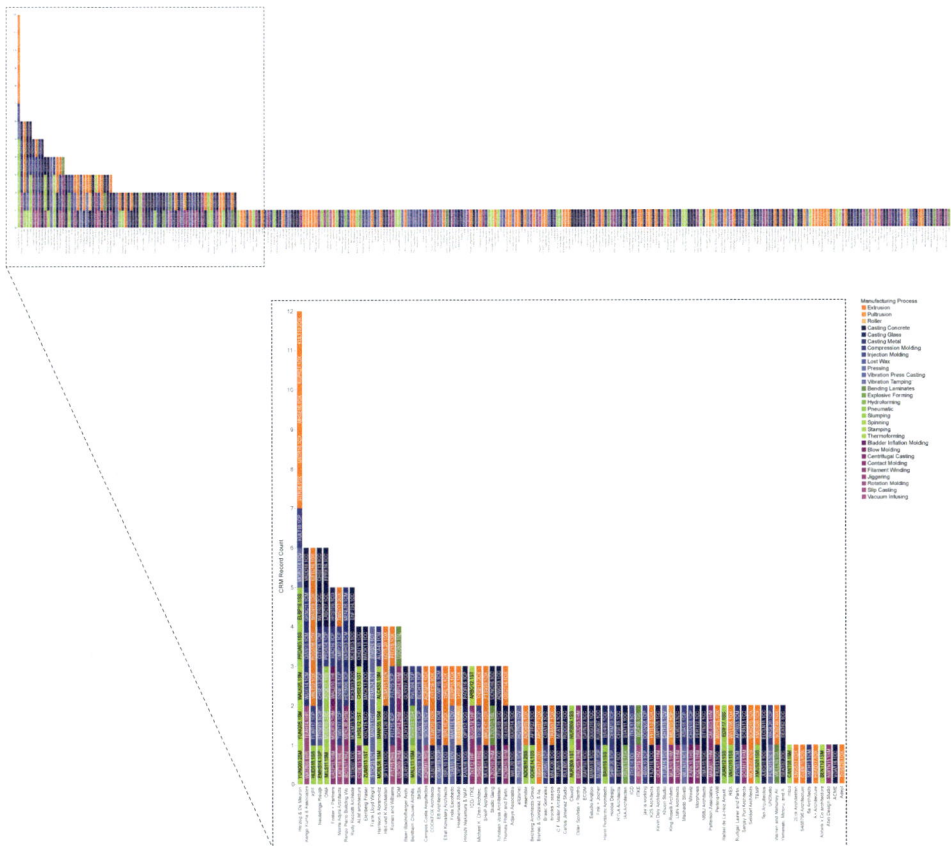

Figure 0.20
Number of CRM components by architecture firm. Includes manufacturing process
and identification tag. Visualization by Author, using the software Tableau.

costs. In Asia, Central America, and South America, we find an increase in casting processes, such as casting glass and making CMU, that make heavy components with a limited shipping radius. Generally, this necessitates that component manufacturing should be relatively local to the construction site.

A wide range of architecture practices have embraced CRM for their designs, including large global offices such as Foster and Partners, Kohn Pedersen Fox Associates (KPF), and SOM, as well as globally recognized design firms like Kengo Kuma and Associates, OMA, and Neutelings Riedijk. It is compelling when internationally recognized architecture practices—such as Herzog and deMeuron, Morphosis, University of Stuttgart Institute for Computational Design and Construction (ICD), and the Institute for Building Structures and Structural Design (ITKE)—that are well known for CAM processes appear with multiple CRM projects on our list. This demonstrates that architecture firms, known for exploring direct CAM processes, are exploring the more nuanced opportunities afforded by CRM.

Smaller architecture firms are also adopting these processes. The case study collection includes mid-sized practices such as Brooks + Scarpa, Belzberg Architects, and Hild und K Architekten, along with small, experimental, and award-winning practices such as Antistatics, E8 Architecture, and all(zone). Additional scales of architectural practices include self-build explorations, often but not exclusively, by university-led design-build courses and research projects. Examples include the University of Stuttgart ICD, ITKE, and their collaborative pavilions and a UK-based collective, Assemble, which manufactured and hung custom clay tiles on a renovated building in Brooklyn, New York.

Approximately three-quarters of the firms included in the case study list have only one example of CRM; the remaining one quarter is "repeat offenders," having two or more examples. [16] Many of the repeat offenders are large offices, with the presumable resources to explore different CRM processes for their projects. However, several small, award-winning, local practices also have multiple CRM projects and processes, including Brisac Gonzalez in London, Studio Marco Vermeulen in the Netherlands, Hiroshi Nakamura in Japan, and Kevin Daly Architects in California. Studio Marco Vermeulen used vacuum infusion process (VIP) for the fiber-reinforced plastic (FRP) panels for the Gas Receiving Station (Chapter 3.2) and will be using compression molding for FRP panels for the Netherlands ProRail's new prototype stations. By using more than one CRM process for more than one project, these small practices illustrate the capacity of differently sized firms to work with CRM.

Our CRM projects also include a wide range of building types (see Figure 0.22). The most frequent application of CRM is for museums, followed by single residential units, commercial spaces or buildings, mixed-use buildings, and multi-family residential buildings, in descending order. Although building type is not the same as project budget, they are linked; generally, large institutional projects like museums have larger budgets than smaller sized projects, as institutional projects can absorb what are seen as additional costs. However, it is interesting to note although the most dominant building type is museums, the remaining four of the top five building types are not known for having large construction budgets. This demonstrates that CRM components are appropriate for small projects and do not require institutional budgets or support.

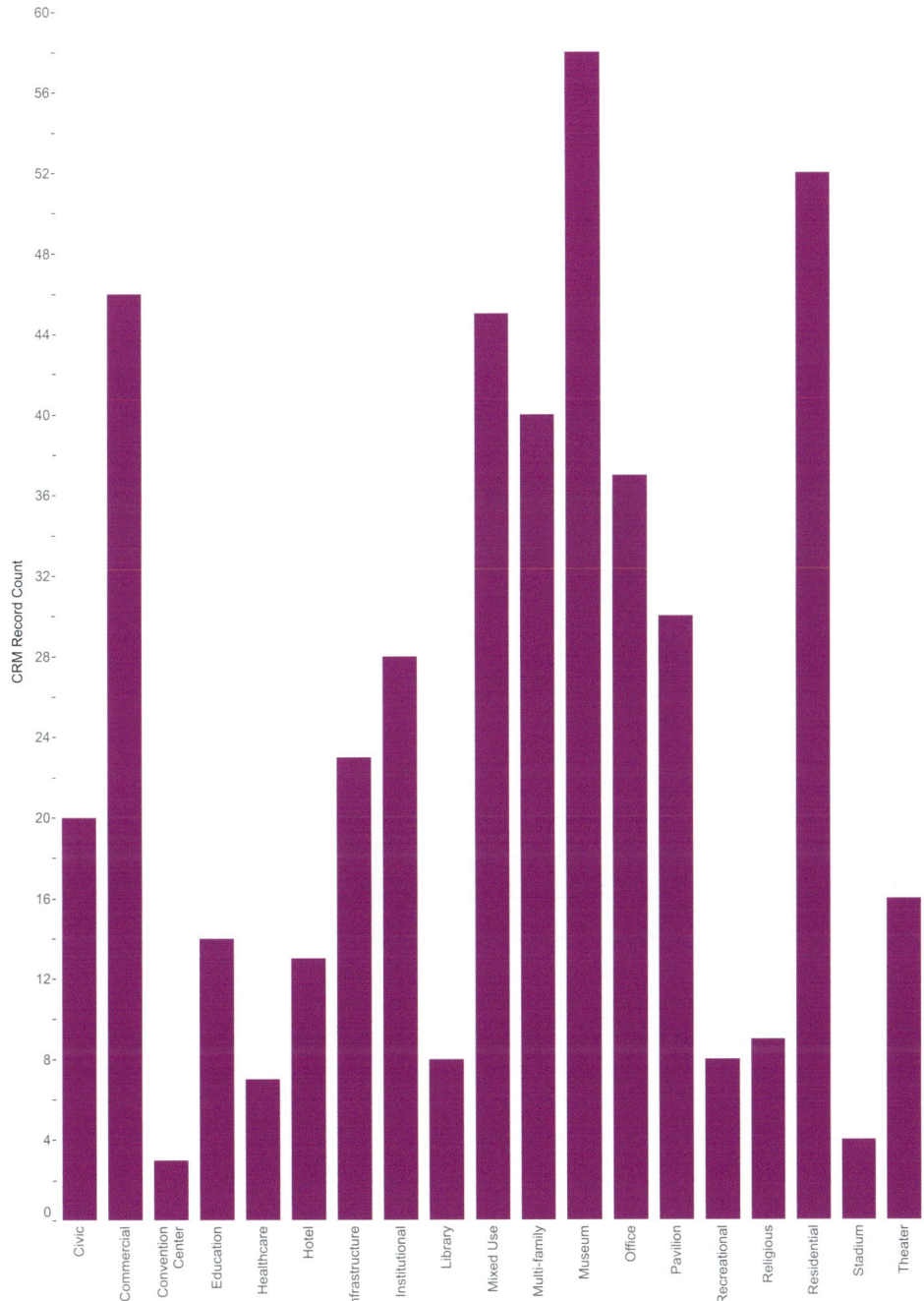

Figure 0.21
Number of CRM components by building type. Visualization by Author, using the
software Tableau.

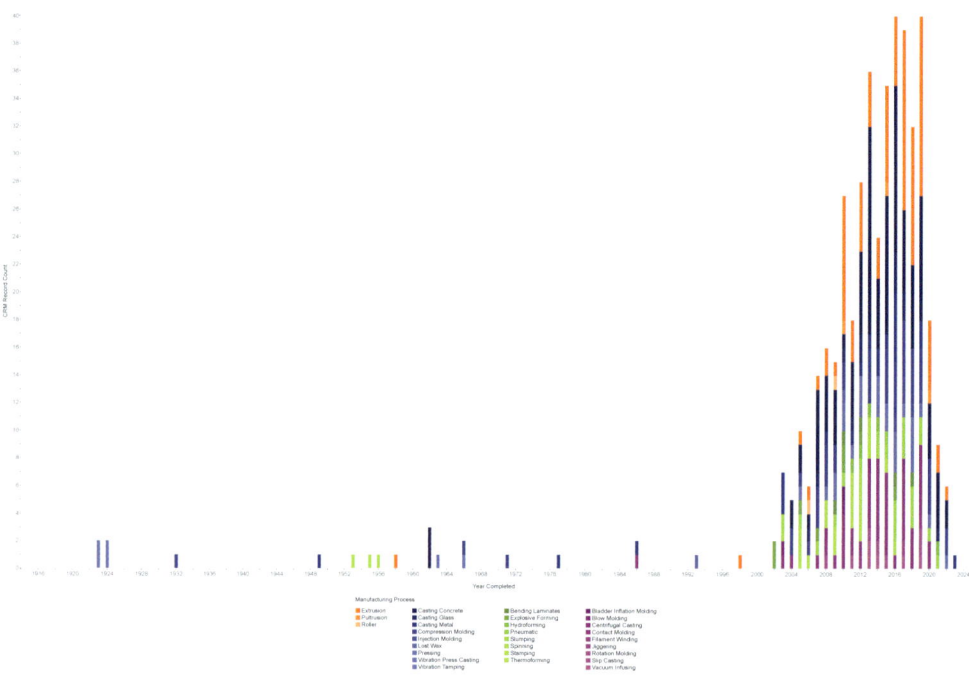

Figure 0.22
Timeline of CRM project completions, including manufacturing processes. Visualization by Author, using the software Tableau.

A snapshot of our current collection of CRM examples indicates a project completion year range from 1923 to the present (see Figure 0.23). The collection's acquisition sources favor contemporary projects over historic, and therefore it is most likely that there are more CRM examples completed prior to 2002 than our list indicates. Additionally, recently completed projects are continually added to the collection, as projects are published and shared with the public; therefore, it is likely that the peak of the adoption of these processes has yet to be reached. Despite the uncertainties, it is important to note the substantial increase in the number of CRM components completed since 2002. This is compelling as the rise of CRM components correlates with the widespread adoption of CAM and CNC technology for fabrication and manufacturing. Although distinct from CAM, CRM often relies on CNC technology to lower tooling costs to make it cost effective to manufacture custom-building components. At the same time, CRM is a competitor of CAM, as CRM provides an alternative to CAM's production waste, long production schedules, and added costs.

In the Timeline of CRM Projects, a shift appears in the prevalence of certain manufacturing categories and processes over time. The manufacturing category of Making Solid, with the specific manufacturing process of casting concrete, dominates the CRM processes since 2003. Manipulating Sheet is popular from 2002 to 2018 and significantly decreases after 2018. In addition, the manufacturing processes within Manipulating Sheet transition from processes that shape metal (e.g. stamping, explosive forming, and hydroforming) to slumping glass. In 2010, we see a rise in the prevalence of extrusion and contact molding with the exception of a few dips; those processes maintain a consistent popularity to the present. Our gathered information does not indicate why the application of certain CRM processes shifts over time. It could be any combination of reasons, including the high costs of tooling for manipulating sheet metal, changes in the availability of CRM manufacturers, or changes in architecture's design values.

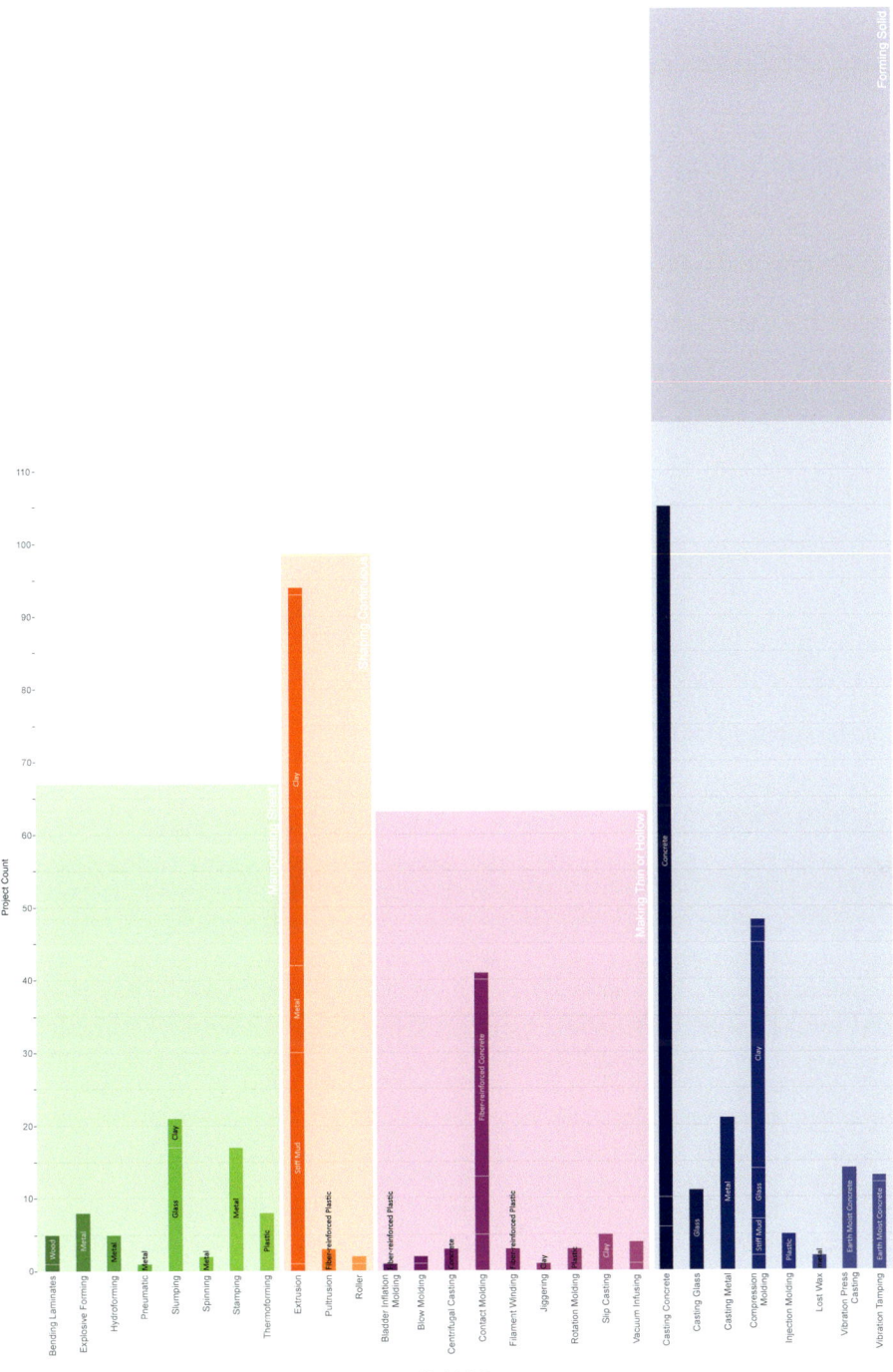

Figure 0.23
CRM processes used in component manufacturing. Visualization by Author, using the
software Tableau.

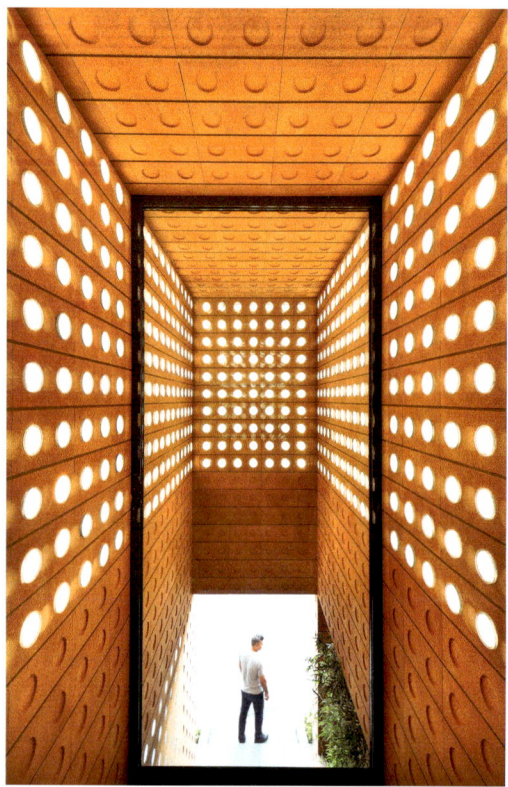

Figure 0.24
Kohan Ceram Building by Hooba Design with its two different, custom-pressed bricks. Photograph by Parham Taghioff.

Figure 0.25
Lantern House by Heatherwick Studio with its custom extruded bullnose bricks used for its iconic bay window jambs. Credit Kevin Scott.

Sorting the case studies by CRM manufacturing processes, we see which processes are most prevalent in architecture (see Figure 0.24). The manufacturing categories, in descending order of frequency, are Forming Solid, Continuous Shaping, Manipulating Sheet, and Making Thin or Hollow. In Forming Solid, most are made from casting concrete—pouring a wet concrete mixture into a mold. This process primarily includes traditional, Portland cement-based concrete mixes but can also include fiber-reinforced concrete (FRC), ultra-high-performance concrete (UHPC), and polymer-based concrete mixes. Given the value of architectural precast concrete as an exterior building face material and that architectural precast concrete manufacturers are equipped to produce custom architectural components, it is not a surprise that casting concrete dominates the CRM processes.

Forming Solid (Part 4) are the manufacturing processes that produce solid components. In the CRM case studies, the second most prevalent manufacturing process in Forming Solid is compression molding, also known as pressing[17] (Chapter 4.7). In pressing, the manufacturing media is placed into a mold and the mold is compressed until the media fills all the mold. For architectural applications, most pressing is done with clay or stiff mud but can include glass, plastic, for FRP; and pressing be done by mechanical or hydraulic presses, or by hand. Examples include the screen units made from pressed stiff mud for Kohan Ceram Building by Hooba Design or glass for Le Prisme by Brisac Gonzalez (Chapter 4.6). Other processes in the category include casting metal (Chapter 4.2); casting glass (Chapter 4.3); vibration press casting, which is the process used to make concrete masonry units (Chapter 4.4); vibration tamping, which is the process used to make cast stone components (Chapter 4.5); injection molding, which is injecting plastic into a closed mold (Chapter 4.6); and lost wax casting, which is making a sacrificial mold around a wax pattern, done for complex metal castings with high surface quality requirements.

Figure 0.26
Gnome Parking Garage by Mei Architects
and Planners with its custom-stamped
stainless steel perforated screen panels.
Photo by © Jeroen Musch.

Second, Continuous Shaping (Part 2) are the processes that produce a continuous cross section along the length of the manufactured component. This category is dominated by extrusion, which is pushing a medium through a die with stiff mud or clay materials being the most prevalent (Chapter 2.1). This book's examples include the terracotta rainscreen units for London Wall Place by Make Architects and bullnose bricks for Lantern House by Heatherwick Studio. The other processes in this category include pultrusion, which is pulling a composite medium—typically FRP—through a die (Chapter 2.2); and rolling, which is passing a soften medium under a continuous roller to imprint a surface texture.

Third, Manipulating Sheet (Part 1) are processes that use preformed, flat sheet material, and deform them into their final shape. This category most often uses metal sheets but can include wood laminates or veneers, glass panels, plastic sheets, or clay. In comparison to the other manufacturing categories,

the CRM processes in this category are relatively evenly distributed with slumping used most frequently, followed closely by stamping. In slumping, gravity slowly deforms either kiln-heated glass or a clay slab over a mold (Chapter 1.1). In metal stamping, a metal sheet is placed between two matched dies and struck one or more times (Chapter 1.5). Examples include the exterior slumped glass windows at the Bibliothèque in Caen by OMA and the stamped metal screen panels for the Gnome Parking Garage by Mei Architects and Planners. Other processes in this category include thermoforming, which used a vacuum to form a heated sheet over an open mold (Chapter 1.2); explosive forming, which uses the force from an explosive charge to deform sheet metal onto an open mold (Chapter 1.3); bending plies, which laminates multiple thin layers over a shaped mold (Chapter 1.4); hydroforming, which deforms sheet metal using a hydrostatic pressure (Chapter 1.6); pneumatic forming, which uses

Figure 0.27
The Broad by Diller Scofidio + Renfro with its custom contact molded glass
fiber-reinforced concrete panels used for the museum's brise soleil.

positive air pressure to force a softened medium against and open mold; and metal spinning, which uses mechanical leverage to slowly deform room-temperature metal against a spinning mold.

Finally, Making Thin or Hollow (Part 3) are processes that make components with relatively thin cross sections and their shape may be open or closed. This category is dominated by contact molding, in which a thin medium is applied onto an open mold (Chapter 3.1). In architecture, most contact-molded components are made from FRC, with fiber-reinforced plastic (FRP) and fiber-reinforced gypsum (FRG) plaster a distant second and third, respectively. This book's examples include the FRC veil of The Broad by Diller Scofidio + Renfro, the FRG feature wall panels for the Tobin Center of the Performing Arts by LMN Architects, and FRP screen panes for the Newtown School by Abin Design Studio. The other processes in this category include VIP, which uses a vacuum to pull resin through dry fibers (Chapter 3.2);

filament winding, which wraps FRP around a mold or a frame (Chapter 3.3); rotational molding, which uses gravity, a slowly turning mold, and a liquified medium to coat the inside of a mold (Chapter 3.4); spin or centrifugal casting, which uses centrifugal force to press a medium against a mold (Chapter 3.5); slip casting, which uses slip (liquid clay) poured inside of a plaster mold to solidify (Chapter 3.6); jiggering which uses a profile to shape spinning clay against a mold; bladder inflation molding (BIM), which uses an inflated bladder to make hollow items from FRP; and blow molding, which blows air into a heated medium to press it against a mold.

Conclusion

CRM balances an architect's desire for custom-building components, with the economics associated with repetitive manufacturing. CRM processes can use CNC equipment to fabricate custom tooling, but tooling can also be made by hand or a combination of hand and CNC equipment. Although repetitive manufacturing conjures images of assembly lines and automation, CRM is done by contract manufacturers or small workshops that are flexible and able to fulfill small to mid-sized production runs. Additionally, there are CRM processes that have not been mechanized by their industries and therefore necessitate human labor for production. The collaborative relationship between designer (architect) and maker (manufacturer) is valued by the architectural profession and can be considered an added value for CRM-made components.

By analyzing our collection of over 460 case studies, we can see trends of CRM in architecture. First, there is a high concentration of CRM projects in the United States, Europe, and South Asia. Projects are located across six continents and demonstrate a global application of CRM. Second, this type of work is done by high-profile design firms, global multi-office firms, as well as small, local practices. About a quarter of all firms are repeat offenders, with multiple CRM projects on the list. Third, CRM is not limited to large, institutional projects, and most often can be found in single-family residential, commercial, mixed-use, and multi-family projects. Fourth, the timeline indicates that the number of CRM projects has rapidly increased since 2003. While Forming Solid appears consistently, we see a shift in popularity from manipulating sheets to extrusion. Finally, casting concrete is the most dominant manufacturing process; however, other CRM processes such as extruding, pressing, stamping, slumping, and contact molding appear frequently.

The first step to understanding the possibilities afforded by CRM is to study the built examples included in *Custom Components*. As both the analysis of the CRM projects and the case studies themselves indicate, no project or firm is too small to incorporate CRM in the making of custom components. Learn the basics of how each CRM process works, how the case studies applied that process to their components, and, possibly, how the case studies may have modified the manufacturing process. Seek out contract manufacturers that are willing to collaborate with architects for custom productions. Hopefully, the projects presented here serve as a guide for architects interested in expanding their design scope to include the design of custom components for their buildings.

Notes

1. The molds for the San Francisco Museum of Art were recycled after their single use; however, it is important to note that recycling is not cost neutral. Recycled plastic is a fluctuating commodity that is exported by many countries, the United States included, and its acceptance depends on trade politics between countries. Next, post-consumable recycled plastic is generally of lower quality than virgin plastic and therefore limited in its application after being recycled. Finally, for the museum, energy was spent running the CNC equipment, transporting the Styrofoam for recycling, and processing the recycled plastic.
2. Both fused deposition modeling (FDM) and stereolithography (SLA) printed plastic patterns can be used in sand casting. Plastic patterns in sand-casting tend to wear out faster than wood or cast aluminum patterns and can be used for production runs under 50 units.

3. Kuo, Chil-Chyuan, et al. "Characterizations of Polymer Injection Molding Tools with Conformal Cooling Channels Fabricated by Direct and Indirect Rapid Tooling Technologies." *International Journal of Advanced Manufacturing Technology*, vol. 117, no. 1–2, 2021, pp. 343–360. Lan, Hongbo. "Web-Based Rapid Prototyping and Manufacturing Systems: A Review." *Computers in Industry*, vol. 60, no. 9, 2009, pp. 643–656. Combrinck, J., et al. "Limited Run Production using Alumide Tooling for the Plastic Injection Moulding Process." *South African Journal of Industrial Engineering*, vol. 23, no. 2, 2012, pp. 131–146.

4. Gulling, Dana K. *Manufacturing Architecture: An Architect's Guide to Custom Processes, Materials, and Applications* (London: Laurence King Publishing, 2018). Minutillo, Josephine "Sculptural Skins: Digital Fabrication Comes into its Own for Creating Precisely Crafted, Complex Building Envelopes, Even on Large Projects" *Architectural Record*, September 2014.

5. Thomas, Douglas. "Costs, Benefits, and Adoption of Additive Manufacturing: A Supply Chain Perspective." *International Journal of Advanced Manufacturing Technology*, vol. 85, no. 5–8, 2015. pp. 1857–1876.

6. Subtractive CNC processes (e.g. milling and EDM) would be wasteful and time intensive, additive processes in metal are limited in their size, and incremental sheet forming is time intensive. Mold parts may be sheet CNC-cut by laser, water, or plasma but are then assembled by hand.

7. These processes *can* be mechanized, using assembly lines, CNC machines, or robots; however, in my visits and discussions with manufacturers, these processes have not been mechanized.

8. Pallasmaa, Juhani. *The Thinking Hand: Existential and Embodied Wisdom in Architecture*. Chichester, West Sussex, John Wiley & Sons, 2009.

9. Zumthor, Peter. *Thinking Architecture*. Basel, Switzerland, Birkhauser GmbH, 2010.

10. E8 Architecture. *30 Lodge Road*. https://www.e8architecture.co.uk/home#/thirty-lodge-road/. Accessed 26 January 2023.

11. Raskin, Laura. "Machine for Entertaining." *Metropolis*, May 2018, pp. 138–145. Almost identical language appears in Keegan, Edward. "2017 Residential Architect Design Awards." *Architect*, December 2017, pp. 132–167.

12. Sennet, Richard. *The Craftsman*. New Haven, CT, Yale University Press, 2008.

13. Brown, Timothy. *Flickr*, 3 June 2008.

14. The textile blocks were manufactured in lifts rather than the full mold filled at once. Technically speaking, this means that the textile blocks are cast stone, in small masonry-sized blocks, not concrete blocks as typically referenced.

15. We believe it likely that there are more CRM examples in Africa, but due to our collection sources, we acknowledge that this continent is underrepresented.

16. We hypothesize that the percentage of repeat offenders will continue to increase as we add additional projects to our list and will track this in future analyses.

17. Typically, the manufacturing industry uses the term "compression molding" to refer to a type of plastic manufacturing, whereas clay and glass are "pressed." Both processes use the same methods, regardless of the medium.

References

Combrinck, J., et al. "Limited Run Production using Alumide Tooling for the Plastic Injection Moulding Process." *South African Journal of Industrial Engineering*, vol. 23, no. 2, 2012, pp. 131–146.

Gulling, Dana K. *Manufacturing Architecture: An Architect's Guide to Custom Processes, Materials, and Applications*. London, Laurence King Publishing, 2018.

Kuo, Chil-Chyuan, et al. "Characterizations of Polymer Injection Molding Tools with Conformal Cooling Channels Fabricated by Direct and Indirect Rapid Tooling Technologies." *International Journal of Advanced Manufacturing Technology*, vol. 117, no. 1–2, 2021, pp. 343–360.

Lan, Hongbo. "Web-Based Rapid Prototyping and Manufacturing Systems: A Review." *Computers in Industry*, vol. 60, no. 9, 2009, pp. 643–656.

Minutillo, Josephine. "Sculptural Skins: Digital Fabrication Comes into its Own for Creating Precisely Crafted, Complex Building Envelopes, Even on Large Projects." *Architectural Record*, September 2014.

Pallasmaa, Juhani. *The Thinking Hand: Existential and Embodied Wisdom in Architecture*. Chichester, West Sussex, John Wiley & Sons, 2009.

Scanlon, Jessie. "Frank Gehry for the Rest of Us." *Wired Magazine,* http://www.wired.com/wired/archive/12.11/gehry.html. Accessed November 2004.

Sennet, Richard. *The Craftsman*. New Haven, Connecticut, Yale University Press, 2008.

Thomas, Douglas. "Costs, Benefits, and Adoption of Additive Manufacturing: A Supply Chain Perspective." *International Journal of Advanced Manufacturing Technology*, vol. 85, no. 5–8, 2015. pp. 1857–1876.

Zumthor, Peter. *Thinking Architecture*. Basel, Switzerland, Birkhauser GmbH, 2010.

Manipulating Sheet

1

This part includes those manufacturing processes that deform existing sheet goods (e.g. plastic sheets and glass panes) into their intended shape. Manufacturing processes include Chapter 1.1, Slumping; Chapter 1.2, Thermoforming; Chapter 1.3, Explosive Forming; Chapter 1.4, Bending Plies; Chapter 1.5, Stamping; and Chapter 1.6, Hydroforming. Materials in this part are varied and include metal, plastic, glass, clay, and wood veneers or laminates. The surface areas of components formed with these processes are significantly greater than their cross-sectional areas. In most cases, the deformed materials will be stiffer than the material in its original flat condition.

DOI: 10.4324/9781003299196-2

CHAPTER

1.1

Slumping

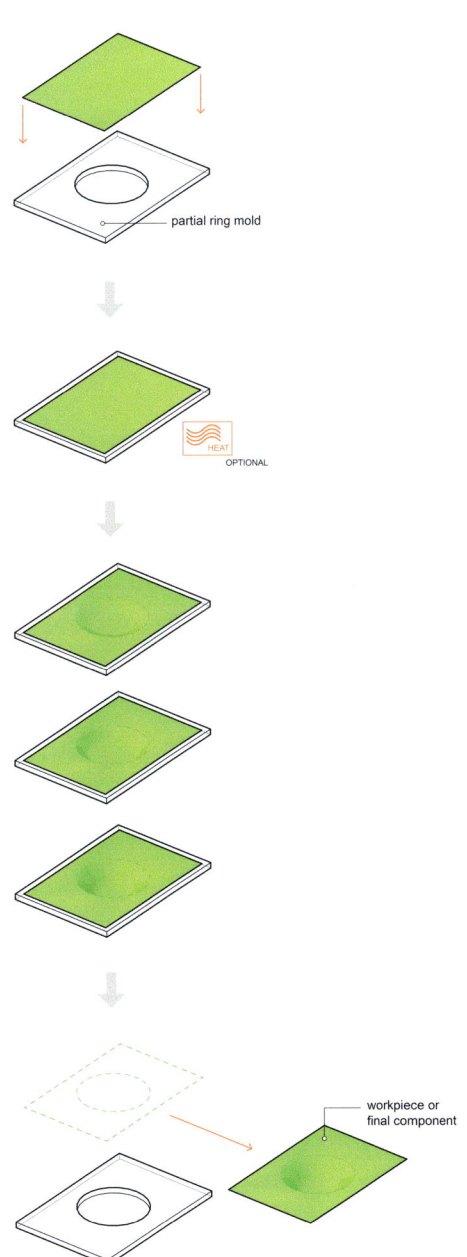

Figure 1.1.1
Slumping process diagram. Drawing by author.
A version of this diagram originally appeared in
Manufacturing Architecture (Laurence King, 2018).

DOI: 10.4324/9781003299196-3

Slumping is the manufacturing process that uses gravity to pull a softened medium into its desired shape. For architectural applications, slumping uses clay or glass, but the process can use plastic. Slumping is generally used to produce gentle complex curves but can be used to make surfaces with fine details and crisp edges. Unlike the other manufacturing processes in Part 1: Manipulating Sheet, slumping is not an active manufacturing process; there are no large presses or high-intensity forces needed to deform the medium. Since slumping uses gravity for deformation, little pressure is placed on the mold and therefore tooling can be made of a wide range of inexpensive materials and capital costs are low. The depth of slump achievable with this process depends on the medium, sheet thickness, material viscosity, and the cycle time. Generally, slumping is time intensive and often multiple molds are used for parallel productions to reduce the overall production schedule.

In slumping, the sheet material is placed onto the mold. If the material is glass or plastic, heat will be used to soften the sheet; if the medium is a clay slab, then the clay is kept moist with water or flexible with additives. The softened sheet will slowly deform due to the pull of gravity. After the slump is complete, the medium is hardened through cooling or drying. Once it is stiff enough, the slumped medium will be removed from the mold for optional post-production processes.

Slumping uses full or partial, open molds. In a full mold, the medium slumps until it is in full contact with the mold surface. Full molds are used to create complex shapes, sharp details, or to imprint a surface texture, and they are used when the medium behavior or cycle times are inconsistent or unpredictable. For a partial mold, such as a ring mold, the mold only supports a portion of the slumping medium. The portion of the sheet that is not supported forms into a complex, catenary curve. In glass slumping, partial molds are best for optical quality and to be closest to vision-glass quality.

Clay

Slumped clay may also be known as *draped* or *humped* clay. Slumped clay can either be done with clay sheets, typically referred to as *slabs*, or with extruded clay components to transform them with a gentle curve (see case study). Typically, clay slumping is done with full molds, because compared to glass, the flow of the clay is often less predictable. High- and low-volume production molds are made of fired clay, plaster, wood and wood products, and Styrofoam.[1] Prototype molds can be made of almost anything, including draped canvas or wadded newspaper.[2]

Glass

Slumped glass may also be known as *bent*, *curved*, *sagged*, *draped*, or *kiln-formed* glass. Glass slumping can be done with a single sheet, multiple stacked sheets, or to make laminated glass. If slumping stacked sheets, then the heat needed for slumping will likely fuse the glass sheets together, but a chemical barrier can be used to keep the sheets separate. Slumped glass is generally stronger than a comparable thickness of flat float glass; however, slumped glass can be further strengthened by tempering if needed. High-volume production molds may be made from steel or cast iron and low-volume production molds may be made from refractory plaster, refractory concrete, clay, or sand.

University of Arizona Health Services Innovation Building (HSIB) in Tucson, AZ
By CO Architects

The University of Arizona (UA) Health Services Innovation Building (HSIB) is a medical and health education training, research, and simulated practice facility that serves the Colleges of Medicine

Figure 1.1.2
University of Arizona Health Services Innovation Building (HSIB) by CO Architects. Facade facing campus pedestrian mall, clad in custom and standard extruded profiles, and custom slumped curved louvers. Photography by Bill Timmerman, courtesy of CO Architects.

Figure 1.1.3
South-facing facade, showing the terracotta-clad porch and the glazed loft.
Photography by Bill Timmerman, courtesy of CO Architects.

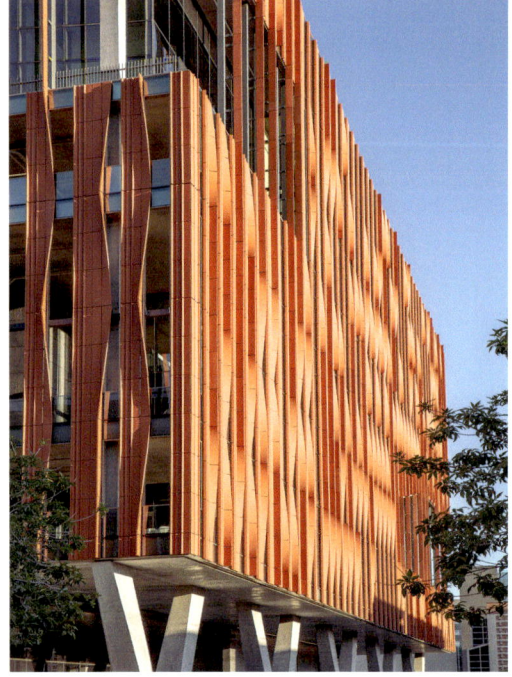

Figure 1.1.4
Oblique photograph showing the curved louvers.
Photography by Bill Timmerman, courtesy of CO
Architects.

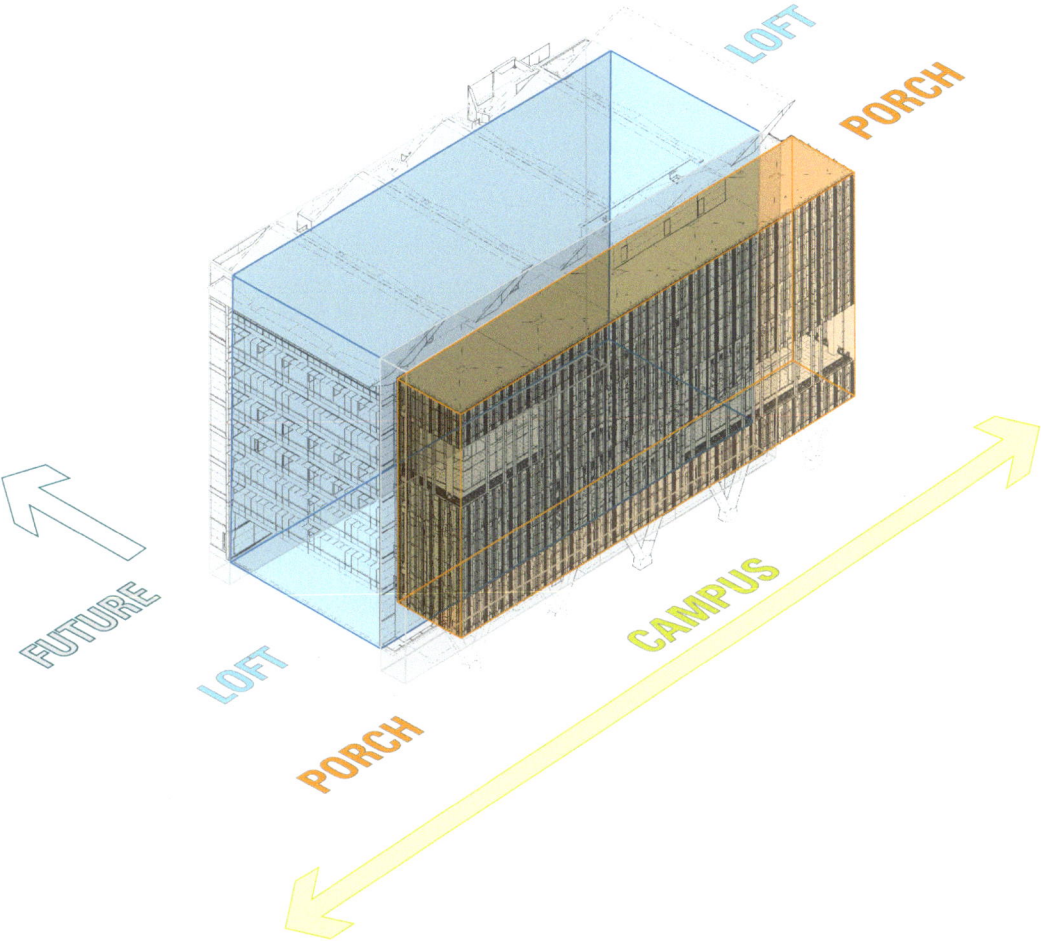

Figure 1.1.5
Building parti diagram. Courtesy of CO Architects.

(both UA's Tucson and Phoenix campuses), Nursing, Pharmacy, and Public Health, and supports the Arizona Simulation Technology and Education Center (ASTEC). The HSIB was a fast-track construction delivery method, with design and construction taking 36 months to complete. The 230,000 ft² (21,368 m²) building is nine stories tall and includes clinical skills and simulation rooms, classrooms, flexible team learning areas, research labs, and open social and collaborative spaces. The building

parti is divided into two parts. The *porch* houses the student-centered spaces and terraces; it is the entry facade and faces east toward a campus pedestrian mall. The *loft* has the learning spaces, skills and simulation rooms, and research labs. This building is the third building that CO Architects has completed at the UA. HSIB is LEED Gold certified and CO Architects designed the building envelope to be self-shading to reduce heat gain from the hot desert sun.

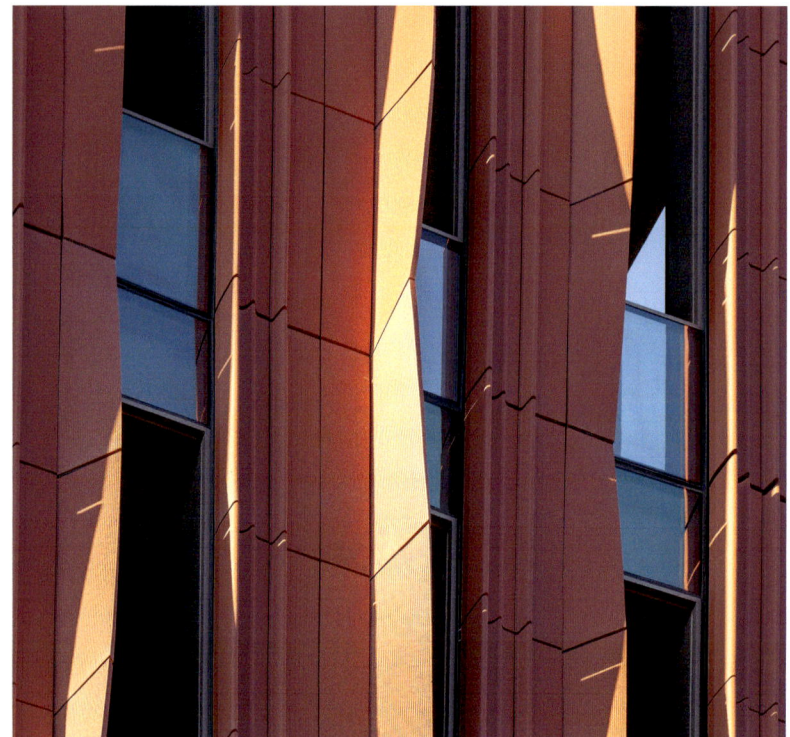

Figure 1.1.6
The deep louvers shade the glass from the sun. Photography by Bill Timmerman, courtesy of CO Architects.

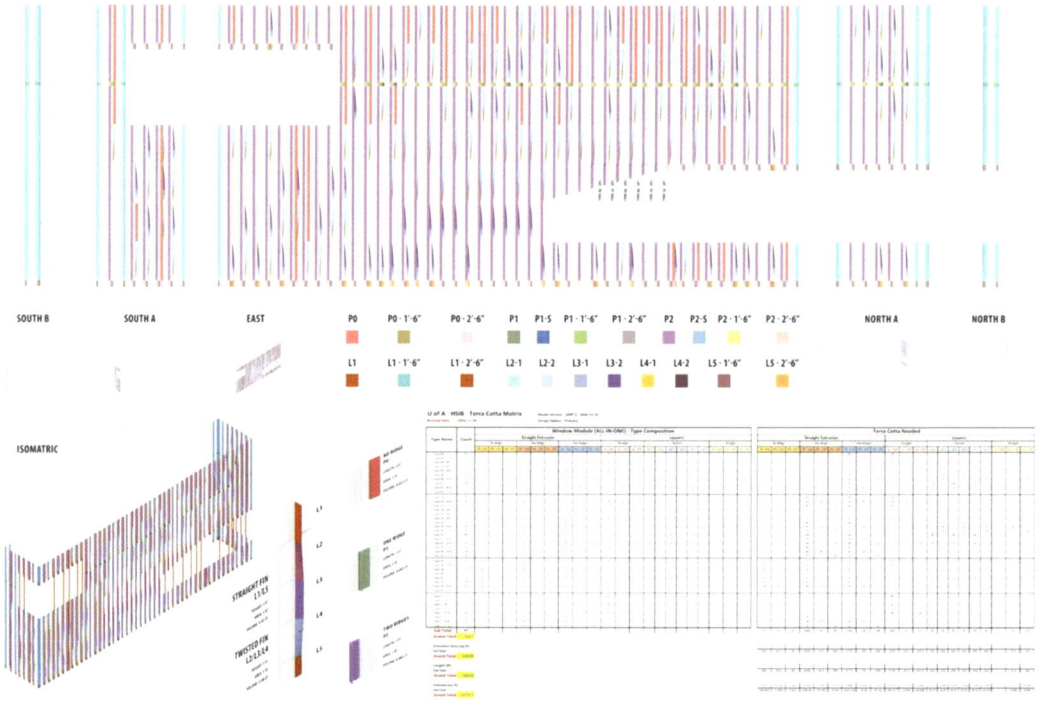

Figure 1.1.7
A final diagram of the different profiles, the amount of twist, and their length as distributed across the building. Courtesy of CO Architects.

Inspired by Arizona's saguaro cactus, which has deep ribs that shade parts of the cactus surface, CO Architects designed the porch envelope with terracotta cladding tiles that have deep vertical ribs that self-shade the terracotta surface and vertical terracotta louvers that shade the porch's east-facing windows. The dark red color of the terracotta panels references the deep red color of the neighboring building bricks and is a through-body color that uses clays, minerals, and stains to produce the color. The terracotta louvers are 60 in (1525 cm) long and are stacked three panels high per floor with the three forming up to a 90-degree twist. Using a combination of program analysis and parametric modeling, CO Architects determined the locations and degrees of the twist to meet programmatic and performance needs. The terracotta louvers are perpendicular to the glass surface, where interior views and daylight are important and parallel to the glass surface in areas that require more sun control or privacy.

According to CO Architects Principal Alex Korter, CO Architects knew they wanted to use terracotta at the beginning of the project's design. Being familiar with Boston Valley Terra Cotta through various facade conferences, CO Architects went to Boston Valley with their design intent and had a 2- to 3-hr brainstorming session to learn about what the family-owned manufacturer could produce.[3] Boston Valley shared with CO Architects their manufacturing parameters. These included the parameters for extruding and slumping clay, as well as dimension limits so that the maximum number of units would fit on the rolling racks that

Boston Valley uses to dry, move, and fire their terracotta components. As a publicly bid project, the specifications for the project were written in such a way that Boston Valley was the only terracotta manufacturer that could make the components for this job. Kovach Building Enclosures was awarded the bid to be the project's facade subcontractor and they contracted with Boston Valley to supply the terracotta.

Kovach's curtain wall system for the porch included vertical strips of glass curtain wall with louvers alternating with insulated rainscreen panels clad with terracotta extrusions. Both the glass and terracotta panels were unitized, and the terracotta extrusions were pre-mounted to their insulated panel before installation to lessen onsite construction labor. The slumped louvers were installed onsite with custom brackets that integrated with Kovach's unitized curtain walls. The brackets cantilevered from the mullions, matching the angles of the louvers. On each bracket were pins (i.e. male connectors) that would fit into the internal (i.e. female) cavities of the louvers, supporting the louvers at their top and bottom edges.

The project has two custom extruded profiles—a two-ridge profile (P2) and a one-ridge profile (P1)—and a flat standard profile (P0). Immediately after extruding, Boston Valley added raked lines to the surfaces of profiles P1 and P0. Profiles P2 and P1 are used for the rainscreen portions of the facade, while P0 is used both in the rainscreen and for the vertical louvers. To produce the curved louvers, after the raked lines were added, Boston Valley placed the still-malleable clay onto a three-part,

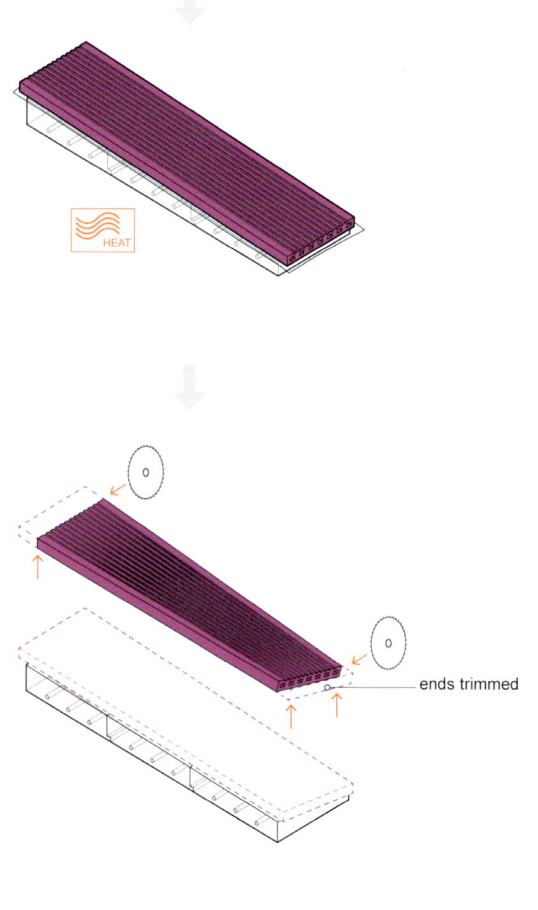

Figure 1.1.8
A diagram depicting the steps used to slump the louvers and their placement on the building. Diagram by author.

custom clay
extrusion

propriety paper

cnc - milled
plaster mold

HEAT

ends trimmed

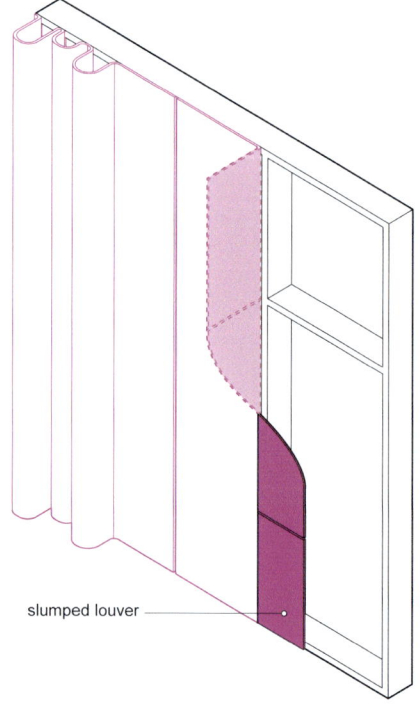

slumped louver

Figure 1.1.9
A drawing of the standard cladding profiles. On the left, P2 and P1, mounted on a rainscreen. The louver is made from a stand profile (P0) with raked lines and slumped to make the 90-degree twist. Courtesy CO Architects.

Figure 1.1.10
Three slumped louvers ganged together to form the 90-degree twist. Courtesy CO Architects.

paper-lined, CNC-milled plaster mold. The malleable clay deformed to the shape of the mold and the mold, paper, and slumped component would all be placed in the kiln for firing.[4] Korter recalled that the specialty paper protected the mold and louver surfaces, kept the clay from sticking to the mold, and burned off during firing.

According to Korter, this was the first completed project that Boston Valley had used slumped clay, although Boston Valley had done some tests with the process prior to this project. Boston Valley had been concerned about being able to maintain tight tolerances with the louvers, but CO Architects were accepting of a more crafted appearance. Korter acknowledged that louvers did not always align from one panel to the next, and as CO Architects has written of the project, they placed a value on its crafted appearance.[5]

Figure 1.1.11
The slumped louvers at Boston Valley, before packing for shipping. Courtesy CO Architects.

Bibliotheque in Caen, France
By OMA

OMA won the commission to design the Caen public library as part of a 2010 architecture competition. The library is in Presqu'ile de Caen, a former industrial area near docklands and the city's railway station, and it sits at the intersection of Orne River and the Canal de Caen. The library's

Figure 1.1.12
Bibliotheque Alexis de Tocqueville, public library in Caen, France. View from entry side.
Photograph by Delfino Sisto Legnani and Marco Cappelletti, Courtesy OMA.

Figure 1.1.13
Photograph of Main Reading Room, looking through the slumped glass windows. Note the column-free space. Photograph by Delfino Sisto Legnani and Marco Cappelletti, Courtesy OMA.

design needed to respect both French urban planner Philippe Panerai's 2001 redevelopment masterplan[6] and a 2011 masterplan developed by MVRDV. The library has 129,167 ft² (12,000 m²) of enclosed space. The library is three stories and 59 ft (18 m) tall with a below grade level that houses most of the library's collection in compact shelving. In addition to the library's large reading rooms and book stacks, the program includes a lobby, a canal-side restaurant with outdoor terrace, an exhibit space, and a 150-seat auditorium. The building is X-shaped in plan, with each of the four arms pointing to Caen monuments. The interior of the building is open, while its ends house the building cores, support offices, and study rooms. The top level of the building primarily houses the children's reading room and administrative offices. The upper level was designed as a full-story high truss, allowing the building's main reading room on the second

Figure 1.1.14
Photograph of slumped windows, looking outside. © VS-A.

Figure 1.1.15
Photograph of the IGUs during construction. The glass is slumped almost 15 in
(381 mm) out of plane. © VS-A.

floor to be column free with floor-to-ceiling windows providing uninterrupted views of the city and its waterways.

To maintain the main reading room's openness, the design team wanted its exterior glazing to be supported at the top and bottom, without vertical mullions. This means that the glass would need to be strong enough to span approximately 20 ft (6 m) from floor to ceiling; however, with a publicly funded project, the design team needed a solution that would work within the constraints of the budget. So, OMA reached out to facade consultant VS-A Design for solutions. According to VS-A Founder Robert-Jan Van Santen, the solution came from a misinterpreted sketch that Van Santen made for OMA's design team.[7] In beginning discussions, Van Santen sketched a metal reinforcement element adhered to the glazing as a stiffener; whereas OMA interpreted the sketch as complex curved glass without anything adhered

to it. Once the misinterpretation was clarified, Van Santen realized that curved glass was a smart solution to the problem.

The library's custom slumped glass is the outermost layers of the insulated glass unit (IGU) and carries much of the units' wind load. VS-A calculated that the exterior layer was to be made of 9/16 and 3/8 in (15 and 10 mm) laminated glass and the inner layer made of two plies of 5/16 in (8 mm) laminated glass. The two plies were slumped and fused together at the same time. The curved sheets were slumped with a double curve in the middle that is approximately 15 in (381 mm) out of plane and creates an inner cavity in the IGU with more than 35 ft^3 (1 m^3) of air. A thermal protection coating was added to the air space face (i.e. number 5 face) of inside, flat laminated glass, as it could not be applied to the slumped glass surface.

Sunglass, an Italian glass manufacturer, slumped the glass and made the IGUs. Sunglass used a

Figure 1.1.16
Photograph of the slumped prototype at Sunglass. © OMA.

Figure 1.1.17
Photograph of Sunglass maneuvering the large, slumped glass for evaluation and strength testing. © VS-A.

partial mold and gravity and controlled the amount of slump with temperature and time. Initially, SV-A designed a double curve that they believed would be the most optimum in meeting the project requirements; however, Sunglass could not form their intended shape. Instead, Sunglass could respect two regulating lines—one vertical and one horizontal—at the center of the glass and could ensure that the glass edges would be flat to form the IGU. Any of the points between the center regulating lines and edges would be formed by the sagging, catenary curve of the glass, deforming under its self-weight, and could not meet the curve profile that SV-A designed. The design team accepted this change.

Sunglass did an initial test for the slump and measured the amount of distortion at the edges that the glass would undergo during the slumping process. Based on the test, Sunglass calculated the blank shape required that would result in straight edges so that no post-production trimming would be required (see Figure 1.1.18). During their visit to Sunglass, the design team brought a prototype outside to see the effects that the curve would have on the glass's transparency. A prototype was also statically tested with bags of sand to see how the glass would perform under loading conditions and to measure its deflection. According to Van Santen, no additional tests were required. The IGUs were supported with standard mullion and support systems and, despite the slumped glazing, the IGUs were warrantied for ten years. Overall, Van Staten stated, "All the components, in fact, are traditional."

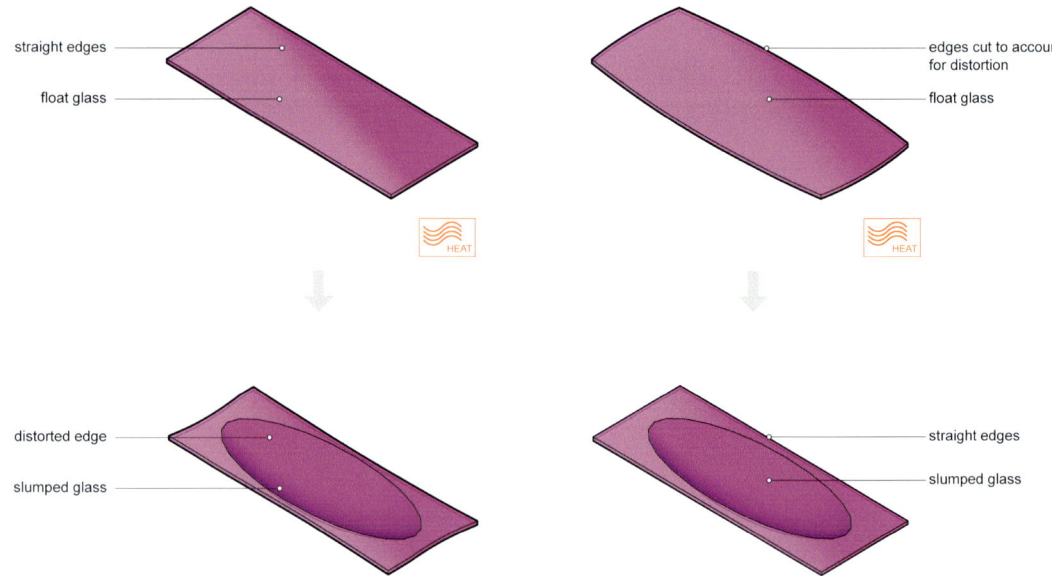

Figure 1.1.18
Through calculations, Sunglass could determine the amount of deformation the glass would undergo during slumping and would calculate the blank sheet shape to ensure straight edges after slumping.

Figure 1.1.19
Photograph of visual distortion due to slumping. © OMA

Notes

1. Steel molds can be used, but often the cost of the material is too high, and the stresses imparted by the process are low that steel is not cost effective.
2. Gulling, Dana. *Manufacturing Architecture: An Architect's Guide to Custom Materials, Processes, and Applications.* London, Laurence King Publishing, Ltd., 2018.
3. Korter, Alex. *Personal Interview.* 17 June 2022.
4. *Ibid.*
5. CO Architects. *Health Sciences Innovation Building Fact Sheet.*
6. Ayers, Andrew. "X Marks the Spot." *Architectural Review*, vol. 1440, 2017, pp. 92–99. According to Ayers, the Panerai plan required that the new library be built up to the four corners of the site. In the competition brief, the librarians had thought that each of the four corners would house its own separate library—arts, science and technology, literature, and human sciences—and be linked by a footbridge. The OMA proposal maintains the idea of separate libraries at the ends of the X-shaped legs but uses the reading rooms to connect them.
7. Van Santen, Robert-Jan. *Personal Interview.* 7 December 2022.

References

Ayers, Andrew. "X Marks the Spot." *Architectural Review*, vol. 1440, 2017, pp. 92–99.

Gulling, Dana. *Manufacturing Architecture: An Architect's Guide to Custom Materials, Processes, and Applications*. London, Laurence King Publishing, Ltd., 2018.

Korter, Alex. *Personal Interview*. 17 June 2022.

Van Santen, Robert-Jan. *Personal Interview*. 7 December 2022.

Thermoforming

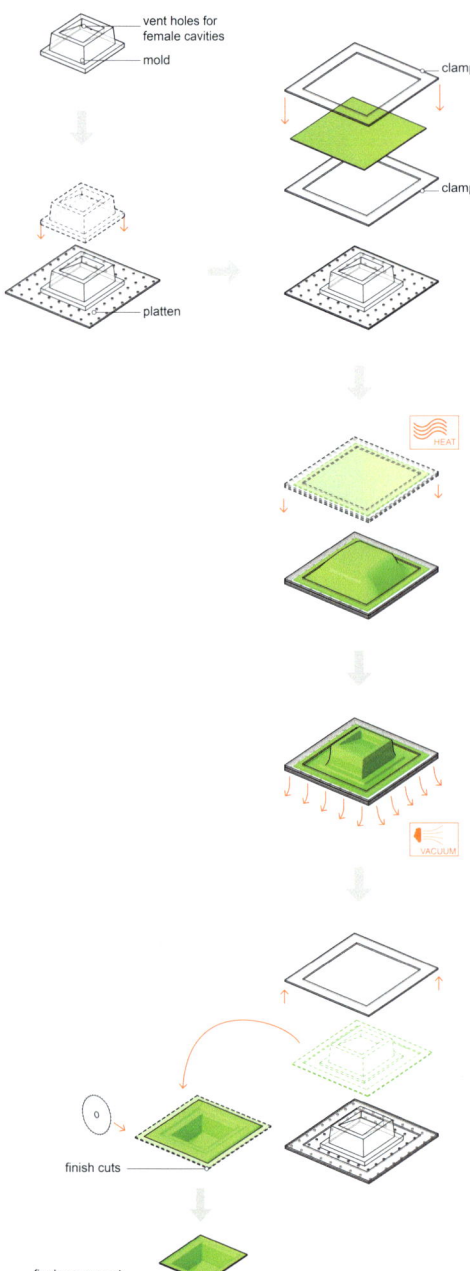

Figure 1.2.1

Thermoforming process diagram. Note that through-mold vent holes must be placed in any of the mold's female cavities. Drawing by author. A version of this diagram originally appeared in *Manufacturing Architecture* (Laurence King, 2018).

DOI: 10.4324/9781003299196-4

Thermoforming is the manufacturing process that uses air pressure to force a medium, made pliable through heating, against a mold. Once the medium has been cooled, it is removed from the mold for optional post-production processes. Thermoforming is most commonly done with thermoplastic but can also be done with super-plastic metal alloys[1] that have been specifically developed for this process; however, due to high material and manufacturing costs, the architectural application for super-plastic alloys is limited. Thermoforming can produce complex curves, ribs, surface textures, and deep draws. The air pressures used for this process are low and place little stress on the mold; however, the mold material needs to withstand stresses caused by heating and cooling during each cycle. Molds can be made of a range of different materials and therefore tooling costs can be kept low if needed for small-production runs. Generally, cycle times for this process are longer and labor costs are higher compared to other plastic manufacturing, but because of low capital costs, thermoforming can be cost effective for CRM.

In thermoforming, molds can be made through a variety of processes, with CNC being the most prevalent. The thermoforming medium is placed inside a clamping frame and is heated until it is ductile enough to stretch without tearing. The frame drops over the mold and air pressure forces the medium to the mold's surface. Typically, the air pressure is applied by a vacuum beneath the mold (as shown in Figure 2.2.1) but can be applied by positive air pressure from above the workpiece or a combination of positive air pressure above and a vacuum below. The thermoformed component is then cooled enough so that

it retains its shape and workers do not make marks on the workpiece surface during handling. The workpiece is then demolded and prepared for optional post-production trimming and other processes.

Molds for thermoforming are open, full molds, and may be male or female. Female mold cavities will need through-mold vent holes so that the air pressure inside of the cavity can be reduced, allowing the medium to fully form against the mold cavity. Thermoforming plastic molds can be made of high-density foam, wood, MDF, thermoset plastics, FRP, plaster, or cast or machined metals. Metal molds are the most durable, are often embedded with cooling lines to actively cool the medium after thermoforming, and are appropriate for high-volume productions. The other mold materials are less costly than metal molds and are appropriate for small- to mid-size production runs.

Thermoplastic

Thermoformed thermoplastic may also be known as *vacuum forming* or *pressure forming*, depending on the specifics of the air pressures used in the process. All thermoplastics can be thermoformed, but not all thermoplastics are suited for repetitive manufacturing. The heating temperature and the cycle times depend on the plastic used, the sheet size and thickness, and ambient humidity. Manufacturers prefer PS, ABS, polycarbonate, PVC, and PVC/Acrylic blends as they have a good temperature range for heat softening and are more forgiving than other types of thermoplastics.

Chipster Blister House in Lyon, France
By AUM

The Chipster Blister House is a single-family home with a rectangular footprint and cantilevers to the west and east that provide cover for the owner's cars, sun shading, and weather protection for an

Figure 1.2.2
Chipster Blister House. South-facing elevation. Photo by Studio Erick Saillet.

Figure 1.2.3
Chipster Blister House. Photo by Studio Erick Saillet.

entry door. The house is in a suburban neighbor-hood with a primary, open view to the south, but with neighbors all around. Founding AUM Principal Pierre Minassian's intent was to design a house that provided privacy for the client on a site that is not private, while still allowing for expansive views to the south.[2]

In many of Minassian's houses, he uses a con-temporary form of an Islamic architectural fea-ture, called the *mashrabiya*. In the Chipster Blister House, the mashrabiya is the black Corian, thermo-formed panels suspended on tension cables out-side of the bedroom windows. True mashrabiyas provide sun protection, ventilation, and privacy to the interior spaces, while AUM's mashrabiyas pro-vide sun protection, privacy, and a unifying design element. AUM put their mashrabiya elements close to the glass to provide privacy during the day as outsiders cannot see into the large bedroom

windows. The large sizes of the Corian panels for the Chipster Blister House provide enough cover-ing while allowing the occupants full views to the south.

AUM specified and designed the black Corian screen panels to contrast with the white cement of the house's site-cast concrete walls and floor, and their organic shape contrasts with the build-ing's rectilinear form and straight planes of concrete and glass. Mounted on cables that are tensioned from the roof to the second-floor slab, the panels do move as breezes pass over them. Minassian compared the panels' movement to "leaves on a tree."[3] Image, a contemporary cabinet maker with expertise in wood and Corian, thermoformed and trimmed the approximately 15 in tall and 56 in wide (37 cm × 143 cm) panels.

There were two things that led to AUM using custom thermoformed panels for the Chipster

Figure 1.2.4
Chipster Blister House. Detail image of Corian screen panels. Photo by Studio Erick Saillet.

Figure 1.2.5
Photograph of the wood screen for AUM's Biscuit House (2008). Photo by Studio Erick Saillet.

Blister house. First was a previously completed AUM project, the Biscuit House, in which the floating elements of the mashrabiya were made of wood; however, for the Chipster Blister House, Minassian wanted a material that would be longer lasting and require less maintenance than the wood. Second, AUM had collaborated with Image for custom furniture on another project. During their visit to Image's fabrication facilities, Minassian saw their thermoforming capabilities and thought it would be interesting to use them on a project as it is easy to make multiples of the same shape and the Corian was light and durable. Minassian saw the opportunity to use the material and equipment when working on the Chipster Blister House.

AUM designed the component, making prototypes in its office to help determine the shape. Image CNC milled MDF to make the mold. In Image's thermoforming process, they use a moldable sheet over the heated Corian workpiece to add even pressure to help form the plastic against the mold. For post-production trimming, the thermoformed panel was placed on a trim mold and the final shape was trimmed by CNC. Minassian estimated that production of all 60 panels took approximately one month.

The panel installation was done by AUM, with a team of four or five people from the office. AUM prides itself on their ability to do both the intellectual and manual labor involved in its projects. In addition, Minassian finds that it is difficult to find contractors who are willing to do the specialty work of installation for a cost that is reasonable. To ensure that the panels were consistently spaced and leveled properly, AUM detailed and fabricated a support ladder made from dimensional lumber to hold the panels in proper position as the tension cables were strung, tensioned, and the compression rings set. Installing the panels at the project's end is a good time for the office; as Minassian phrased it, "it is like the cherry on top of their relationship with their clients" and is a signature of the office's projects.

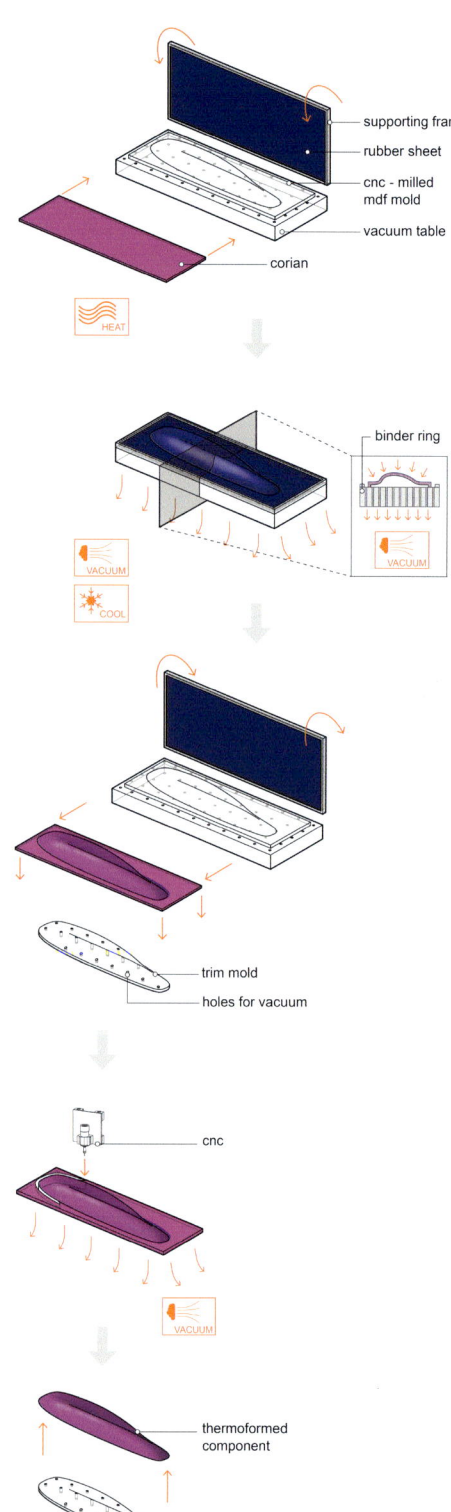

supporting frame
rubber sheet
cnc - milled
mdf mold
vacuum table
corian

HEAT

binder ring

VACUUM

VACUUM

COOL

trim mold
holes for vacuum

cnc

VACUUM

thermoformed
component

Figure 1.2.6
Diagram of the thermoforming of the
Chipster Blister screen panels.

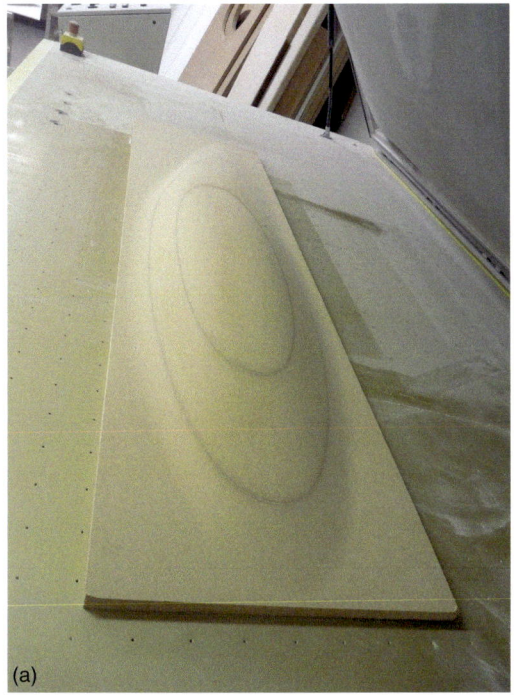

(a)

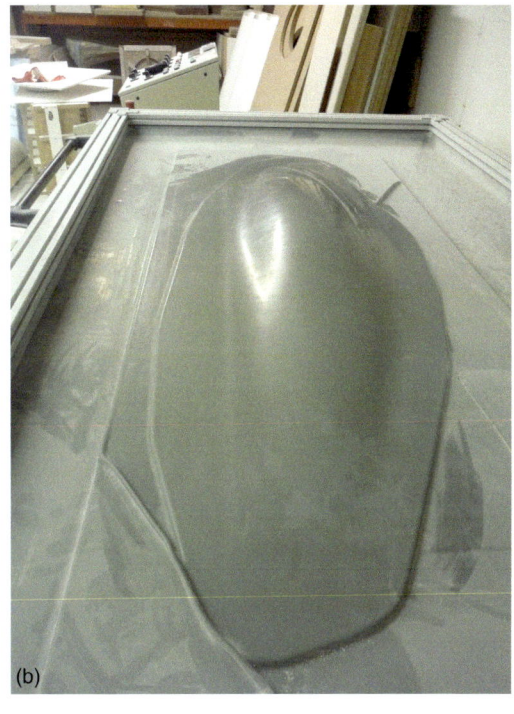

(b)

(c)

Figure 1.2.7
(a)–(c) Photographs of Image's thermoforming process for the Chipster Blister Corian panels. (a) CNC-milled MDF mold on vacuum table. (b) Mold and panel thermo-formed, under sheet for added even pressure. (c) Panel ready for demolding. Photos by AUM Pierre Minassian.

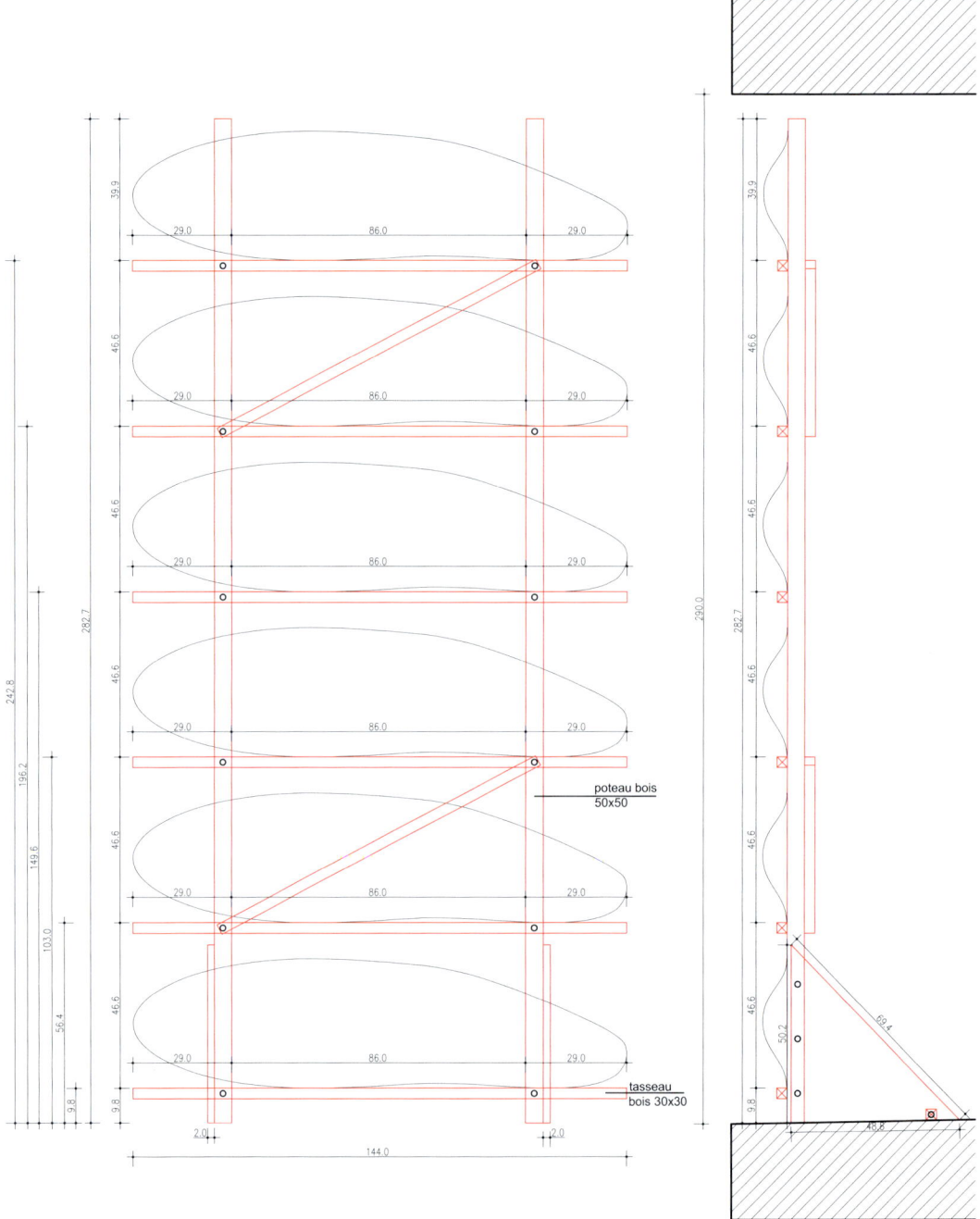

Figure 1.2.8
Detail drawing of panels and support ladder, used by AUM for panel installation.
Photo by AUM Pierre Minassian.

Figure 1.2.9
Photograph of the white Corian screen for AUM's House by the Lake (2015).
Photo by AUM Pierre Minassian.

Since Chipster Blister, AUM used thermoformed Corian on the House by the Lake but changed the panels' mounting details for added stability. In the Chipster Blister House, AUM adhered small tabs on the panels inside face at the top and bottom edges. The tension cables were strung through the tabs, and crimp rings were added to keep the panels in place. After the panels were mounted on the Chipster Blister House, a bad storm came through, the tabs broke off, and several panels came loose. AUM researched a better adhesive for the tabs and found a solution that was more akin to plastic soldering than a traditional adhesive. AUM took off all the panels off the Chipster Blister house to repair them with this new technique. For House by the Lake, AUM designed the panels with a continuous tube adhered to the back of the panel, in which the tension cable was strung; the tube provided more contact between the connectors and the panels than the tabs.

Figure 1.2.10
Back of the panels, showing the adhered tabs that connect the panels to
the tension cables. Photo by Studio Erick Saillet.

Notes

1. Super-plastic alloys are highly ductile and deform with little necking under tension. The super-plastic alloys are aluminum, magnesium, or titanium based and include AA2004, AA7075, AA8090, SUPRAL, and ICONEL alloy 718SPF. SUPRAL and ICONEL are both propriety alloys and may not be available to third-party manufacturers.

2. Minassian, Pierre. *Personal interview*. 18 April 2022.
3. *Ibid*.

Reference

Minassian, Pierre. *Personal interview*. 18 April 2022.

Explosive Forming

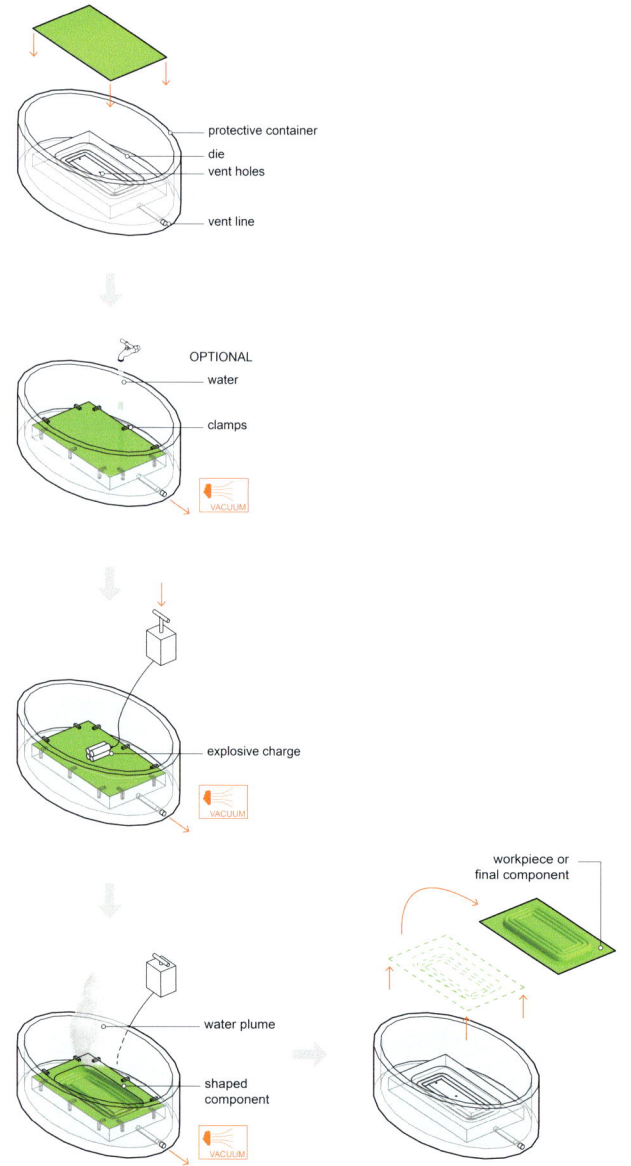

protective container
die
vent holes

vent line

OPTIONAL
water

clamps

VACUUM

explosive charge

VACUUM

workpiece or
final component

water plume

shaped
component

VACUUM

Figure 1.3.1
Explosive forming process diagram.
Drawing by author. A version of this
diagram originally appeared in
Manufacturing Architecture
(Laurence King, 2018).

 DOI: 10.4324/9781003299196-5

Explosive forming is the manufacturing process that uses the energy of an explosion to deform a sheet of metal against a die to form a desired shape. Aluminum, titanium, steel, and stainless steel can be formed with this process and are well suited for architectural cladding applications. Explosive forming is best for large sheets of metal that might be too large or thick for stamping (Chapter 1.5) or hydroforming (Chapter 1.6). Explosive forming forms deep draws, smooth curves, and imprint surface textures. Manufacturers can use a range of different explosive materials, configurations, and placements relative to the metal blank, and each affects the types of metal, shape, size, thickness, and draw depth that can be formed. Explosive-forming dies are made of a range of materials and material costs; however, the safety regulations, transportation, and storage associated with using explosives can be costly. Explosive forming requires long set-up times between cycles and so this process is best suited for small productions.

In explosive forming, the metal blank is clamped onto a mold and inserted into an underground tank or an above-ground container. A vacuum is formed between the blank and the die to eliminate air from being trapped so that the blank can fully deform against the die face. Water or air can be used to transfer the force from the explosive to the workpiece. When the explosive detonates, the force and subsequent shock waves travel through the medium, pushing the blank against the die. Then the workpiece is removed from the die and explosive forming tank for post-production.

Dies for explosive forming can be open or closed, and open dies can be full or partial. Full molds are generally female so that the manufacturer can form a vacuum between the blank and die for proper deformation. Partial dies, such as ring molds, can be used to produce conical or hemispherical shapes with the exact shape being determined by the explosive shape, amount, distance to the blank, manufacturing medium, and transfer medium. Dies can be made of precast concrete, fiberglass reinforced plastic (FRP), or cast, machined, or sheet metal.

Explosive forming may also be known as *explosion forming*, *explosive working*, *high-energy rate forming (HERF)*, *high-energy hydroforming (HEH)*,[1] or *high-velocity metal forming (HVMF)*.

Metal

Explosive forming can be done not only on ductile metals or metal alloys but also works well on tough metals that may be difficult to form with other sheet metal forming processes, such as stamping, hydroforming, or spinning. There is substantially less spring back with explosive forming than stamping; therefore, dies can be designed to near net dimensions without allowances for overbending.[2] The explosive force will permanently bond dissimilar metals together during deformation in a process called *explosive welding* (EXW).

Lightrailstation in the Hague, the Netherlands
By ZJA

Figure 1.3.2
Photograph of HSE, looking across the Bus Depot.

Figure 1.3.3
Photograph of HSE, looking across the Bus Depot at its connection to Central Station.

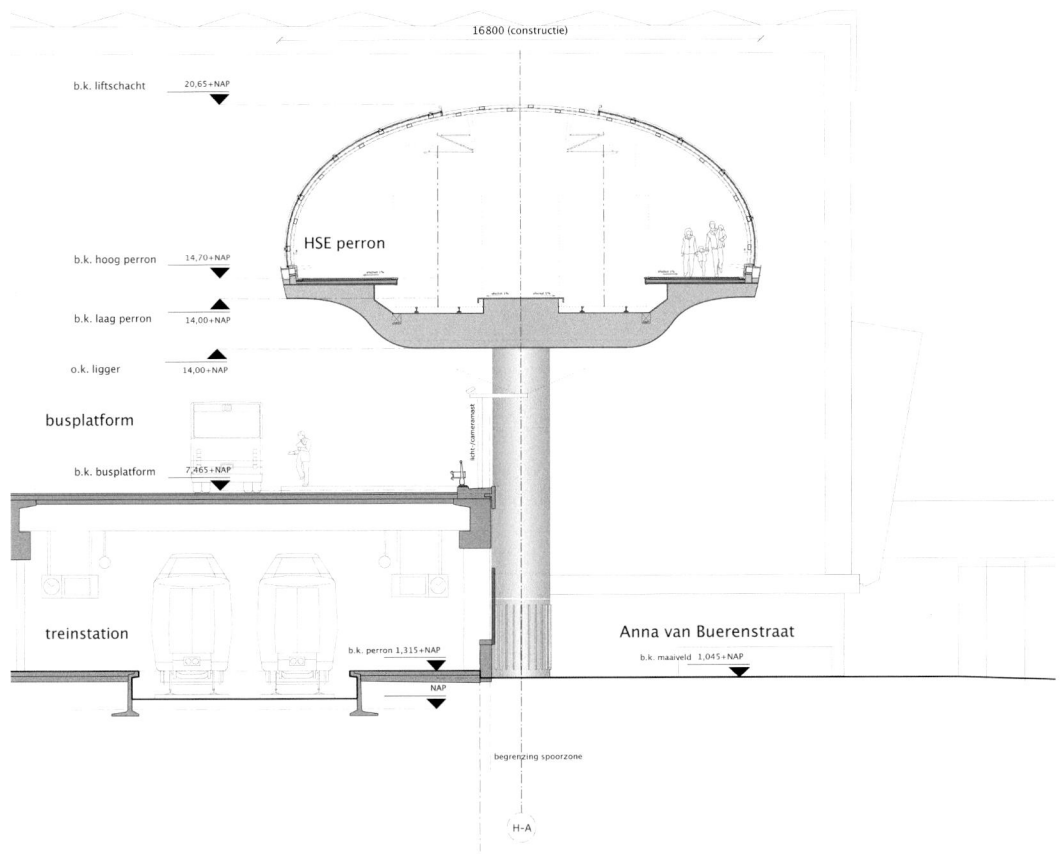

Figure 1.3.4
Transverse Section of HSE, showing at-grade passenger train tracks, second-level
Bus Depot, and upper-level light rail. Outline of Central Station, beyond.

The Hague's Lightrailstation, or *Haag Startation E-line* (HSE), is part of Netherlands' RandstadRail, a network of light rail systems that connects The Hague, Zoetermeer, Rotterdam, and the regions in between. The Lightrailstation physically connects to The Hague's Central Station and forms the city's *Openbaar Vervoer Terminal*, or Public Transport Terminal, connecting the light rail network, the Central Station's heavier passenger trains, and a bus depot. The HSE is organized linearly along the Central Station passenger train tracks and is raised, approximately 50 ft (15 m), above the existing grade and tracks. The HSE provides access to the Erasmus Line terminal on its upper most level and, one level below provides passenger access to the bus

platform and connects to the second floor of the Central Station. Despite being tucked into the urban fabric and hidden behind Central Station, the municipality of The Hague asked ZJA to design the station with its own architectural identity and they wanted it to be distinct from that of Central Station.[3]

Prior to the HSE, ZJA designed another light rail station for The Hague, called the Beatrixlaan. According to ZJA Partner Ralph Kieft, The Hague asked ZJA to design the HSE to resemble the Beatrixlaan.[4] For the HSE, the design team used a material palette similar to the Beatrixlaan while distinguishing the HSE by using more fluid and organic shapes than Beatrixlaan. Kieft believes the HSE is more futuristic than the Beatrixlaan and an outstanding object

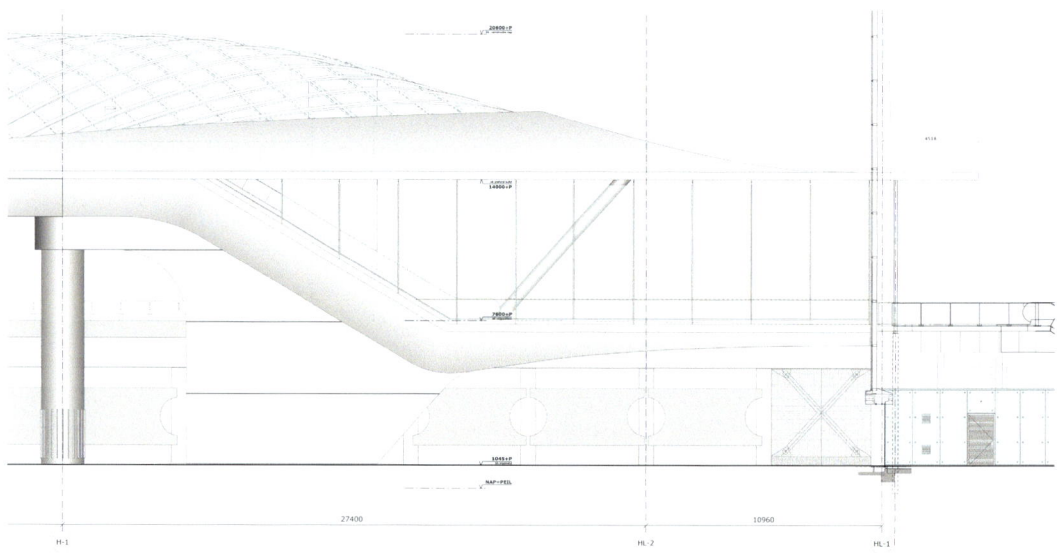

Figure 1.3.5
Elevation of HSE, as it meets Central Station's curtain wall.

in its tight site. Most publications on the HSE have focused on its column-free, slender, shallow curved glass and rolled structural steel canopy that covers the upper level light rail platform.[5] As passengers leave the platform, the glass and steel canopy transitions to solid metal panel and splits along the escalator, forming both a new, lower canopy and the lower level passenger path to Central Station (see Figure 1.3.5). The solid metal panels are ¼ in (6 mm) thick, stainless steel and are painted. The canopy pierces the Central Station's glass curtain wall and end inside the station.

According to ZJA Technical Contractor René Sjoerdstra, the project was done under a time pressure that had an impact on its design and delivery approach. ZJA designed the steel canopy and escalator parametrically, using Rhinoceros and Grasshopper, to maximize the number of single-curved panels and minimize the number of double-curved panels, while meeting the original design intent. 3D Metal Forming,[6] a Netherland-based explosive former located in Lelystad, approximately 68 mi (110 km) from The Hague, formed the double curved panels. The HSE contractor selected 3D Metal Forming and its explosive forming techniques to form the double-curved panels as it was the least expensive option. 3D Metal Forming uses water

as its explosive medium and can shape metal sheet thicknesses ranging from .01 in. (.3 mm) to approximately 6 in. (150 mm), and the blanks can be assembled and welded prior to forming. There was little direct collaboration between ZJA and 3D Metal Forming; instead ZJA shared their digital files with the HSE steel contractor and 3D Metal Forming. According to Kieft, this project was one of ZJA's first projects in which they shared their digital files directly with the contractor and manufacturer.

Figure 1.3.6
Photograph of single curve panels being installed on the lower level, against Central Station's curtain wall.

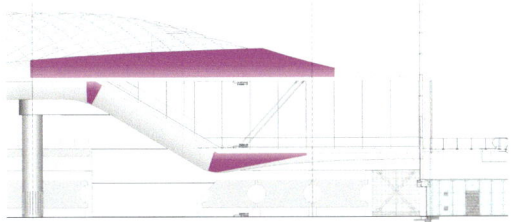

Figure 1.3.7
Highlighted areas depict where explosive-formed panels are located on the station. Drawing by ZJA, highlights by author.

Beneath the metal cladding panels is a painted steel substructure with bolted, stainless-steel flanges to receive the panels. The panels are mechanically fastened with counter-sunk, self-tapping screws to the flanges (see Figure 1.3.6). The joints and fastener heads then are filled with epoxy and finished with multiple onsite coats ending in a high-gloss, white polyurethane finish. This

Figure 1.3.8
Photograph of the double-curved, explosive-formed panels being installed under the escalator.

reduced the number of visible joints.[7] At the top of the canopy, where the geometry transitions from the steel and glass to solid panels there are 72 double-curved panels that were formed by explosive forming. Under the escalator are 26 double-curved

Figure 1.3.9
Photograph of the HSE during construction and the solid metal canopy installed.

Figure 1.3.10
Photograph of the completed HSE from ground level.

panels that were formed by explosive forming (see Figure 1.3.7a and b). The panel sizes ranged from 69 × 83 in. to 83 × 156 in. (1750 × 2100 mm to 2120 × 3960 mm) on the canopy and 26 × 65 in. to 54 × 177 in. (650 × 1650 mm to 1360 × 4500 mm) under the escalator. 3D Metal Forming assembled the canopy panels and welded stiffeners to the panels off-site while the escalator panels were installed on site. Sjoerdstra estimates that the production of the explosive-formed panels took approximately four months.

Notes

1. This term is only used if a fluid is used as the explosion transfer medium, rather than air.
2. Gulling, Dana K. *Manufacturing Architecture: An Architect's Guide to Custom Processes, Materials, and Applications.* London, Laurence King Publishing Ltd., 2018.
3. Helbig, T. et al. "Double-Layer Curved Steel-Structure with Bent Glazing; New Departure Station Erasmuline, The Hague (NL)". *Challenging Glass 5 – Conference on Architectural and Structural Applications of Glass, Ghent University*, June 2016, edited by Belis, Bos & Louter.
4. Kieft, Ralph and René Sjoerdstra. *Personal Interview.* 10 October 2022.
5. Printed project publications include Helbig, T. et al. "Double-Layer Curved Steel-Structure with Bent Glazing; New Departure Station Erasmuline, The Hague (NL)". *Challenging Glass 5 – Conference on Architectural and Structural Applications of Glass, Ghent University*, June 2016, edited by Belis, Bos & Louter. Helbig, Thorsten. "Curved Directly Glazed Steel Structure: New Departure Station, E-Line, The Hague." *Steel Construction Design and Research*, November 2016. pp. 363–368.
6. Company formerly called Exploform B.V.
7. There are expansion joints between some of the metal panels to allow for movement.

References

Gulling, Dana K. *Manufacturing Architecture: An Architect's Guide to Custom Processes, Materials, and Applications*. London, Laurence King Publishing Ltd., 2018.

Helbig, Thorsten. "Curved Directly Glazed Steel Structure: New Departure Station, E-Line, The Hague." *Steel Construction Design and Research*, November 2016, pp. 363–368.

Helbig, Thorsten, et al. "Double-Layer Curved Steel-Structure with Bent Glazing; New Departure Station Erasmuline, The Hague (NL)." *Challenging Glass 5 – Conference on Architectural and Structural Applications of Glass*, Ghent University, June 2016, edited by Belis, Bos & Louter.

Kieft, Ralph and René Sjoerdstra. *Personal Interview*. 10 October 2022.

Bending Plies

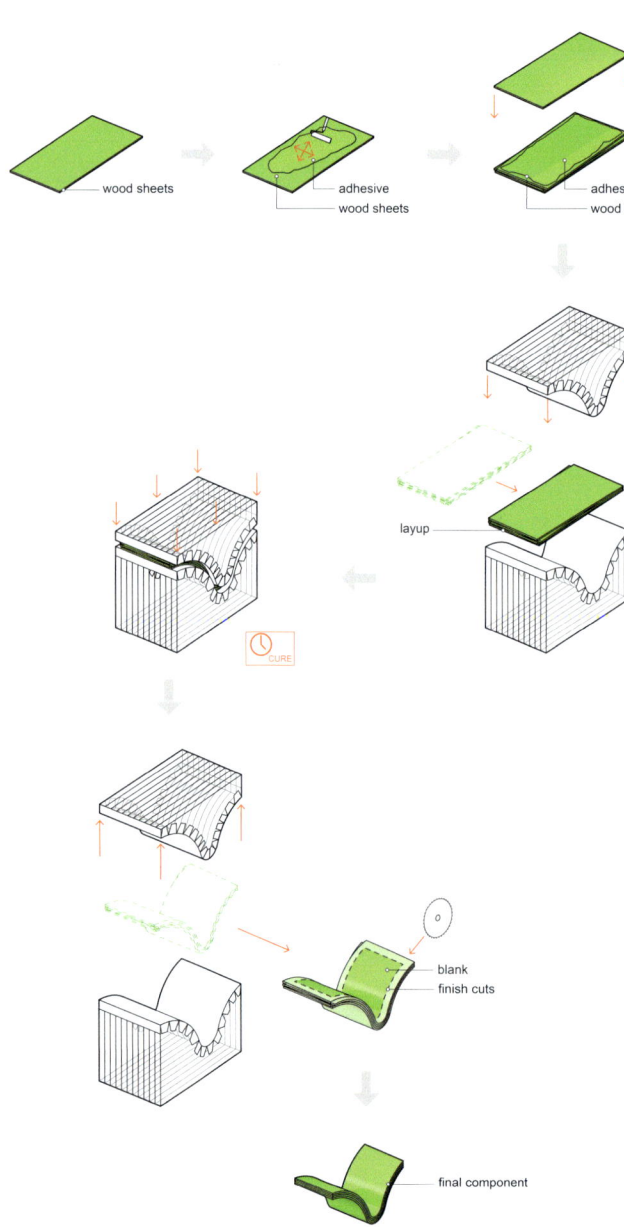

wood sheets

adhesive
wood sheets

adhesive
wood sheets

layup

CURE

blank
finish cuts

final component

Figure 1.4.1
Bending plies process diagram. Diagram
of bending plies. Drawing by author. A
version of this diagram originally appeared
in *Manufacturing Architecture* (Laurence
King, 2018).

 DOI: 10.4324/9781003299196-6

Bending plies is the manufacturing process that presses thin layers of malleable wood or cellulose-based materials and adhesive into the desired shape until the adhesive sets. For architectural applications, this process uses wood veneers or thin dimensional lumber and can be used to make bent plywood surfaces, laminated veneer components, or large curved glue-laminated structural members. Bending plies can best form single-directional curves but can be used to produce shallow, double curves if needed. Tooling for this process can be made of a range of different materials and capital costs can be kept low, if needed. The amount of curve achievable with this process depends on the type of wood, ply thicknesses, adhesive, overall thickness of the layup, and the press.[1] Generally, bending plywood is time intensive as the adhesive needs to harden before demolding, and cycle times can range from 24 hr to 2 min depending on the layup materials, the allowable spring back, and if the adhesive is air drying or being actively cured.[2]

In bending plies, alternating layers of wood sheet and adhesive are stacked with the wood sheets placed at the top and bottom of the layup. Before the adhesive starts to harden, the layup is placed into the mold. Even pressure is applied to the layup. (Figure 1.4.1 illustrates a closed, matching mold to provide the pressure but clamps or a vacuum bag are also often used.) Once the adhesive has cured, the pressure is released, and the blank is demolded. The edges of the blank are trimmed as the layers are not fully aligned when placing the layup onto the mold. Any further post-production processes, such as drilling, cutting, or sanding, are optional.

Molds for bending plies may be open, open with a partial plug for deep recesses or tight curves, or closed-matched molds. Even pressures need to be applied to the layup during curing and this can be done by a vacuum bag or a press. A vacuum bag can be used with open molds, plugs, or closed molds while presses are used for closed molds. If molding in a vacuum bag, then pressures are low, and molds can be made of foam, corrugated cardboard, or plywood. If molding in a press, then molds are made of plywood (often stacked as in Figure 1.4.1) or metal. If heat induction or radio waves are used to cure the adhesive, then a plywood mold will be covered by an aluminum sheet for conduction. Instead of molds, jigs can also be used for bending plies. Jigs are best suited to manufacturing long linear single curves rather than broad surface curves or double curves. Jigs can be set up either manually or by CNC; however, access to CNC jigs is limited and many manufacturers will not have this capability.

Bending plies may also be known as *bending plywood*, *curving plywood*, *pressing plywood*, or *molding plywood*. The term *molding plywood* is specifically used for creating a complex curved surface. Creating a complex curve in wood is difficult as the material cannot stretch or shrink; therefore, shape-making is more limited than other sheet-forming process for plastic or metal sheet-forming processes (e.g. thermoforming [Chapter 1.2], stamping [Chapter 1.5]).

Wood + Wet Adhesive

Bent plies is a composite material in which the sheets work together with the binding adhesive. Typical woods veneers used are Douglas fir, maple, or birch, but any wood species can be used. In addition, wood products such as thin plywood that is 1/8 in (3 mm) or less in thickness, bendy plywood (e.g. Wacky Wood), MDF, or even cellulose-based materials such as chipboard can be bent. The adhesive is wet applied and can be cured through heat, chemical reaction, radio waves, or water evaporation.[3] The specific adhesive used will depend on the required strength, tightness of the curves, exposure to moisture, and allowable cycle times.

Wellington International Airport South Terminal Expansion in Wellington, New Zealand
By Warren and Mahoney (WaM)

The Wellington International Airport is located less than 5 mi (8 km) southeast of downtown Wellington and is on New Zealand North Island's rocky shoreline. The airport's runway runs north-south with more

Figure 1.4.2
Photograph looking down the South-West Pier Expansion.

Figure 1.4.3
Photograph looking through the custom glue-laminated structure to the tarmac.

than a third of the runway's length bordering Lyall Bay that opens into the Pacific Ocean. The weather of the south coast of New Zealand's North Island is known to be extreme and the airport terminals overlook the tarmac and the bay. According to Warren and Mahoney (WaM) Project Principal Katherine

Skipper, Wellingtonians will visit the airport on Sunday afternoons just to watch the planes against the landscape and sky.[4] WaM's Wellington International Airport South Terminal Expansion includes three parts: (1) the Main Terminal Building Expansion that added 115 ft (35 m) of length to the airport's original

Figure 1.4.4
Close-up photograph of the glulam members.

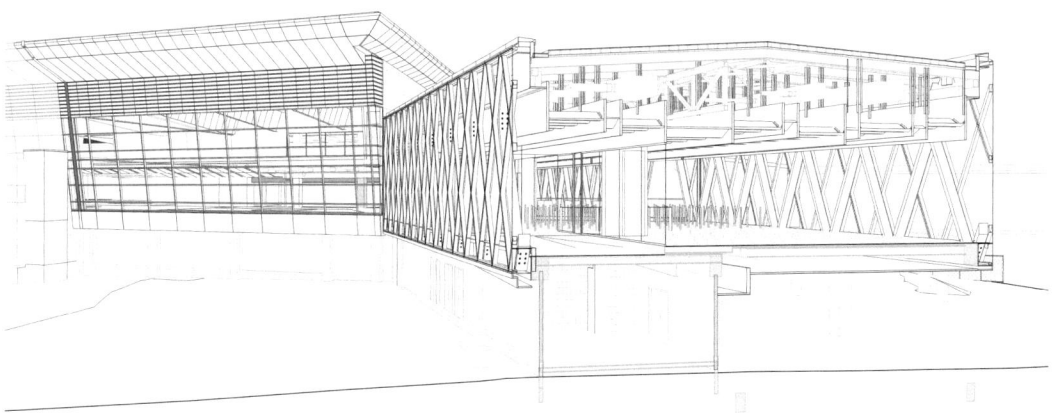

Figure 1.4.5
Cut-away perspective of South-West Pier, looking toward the Main Terminal Expansion.

terminal, following the language of the original architecture, (2) the South Pier Expansion that provides gates for small aircraft, and (3) the South-West Pier Expansion that provides gates with jet bridges for large aircraft.

The South Terminal Expansion design process was highly collaborative between WaM and Beca, the project's structural engineer. The design team knew that they wanted the expansion's structure to be exposed, but Skipper could not recall if the idea of using a glue-laminated structure came from the structural engineers or the architects. Although local architectural writers have referenced hinaki—a type of eel net locally known in New Zealand—when discussing the expansion's design,[5] the X-shape of the members came during the early design process,

with the design teams sketching a series of possible shapes while thinking of the larger project goals of extending and connecting. The X shapes are made by two identical custom curved glulam members that measure 17 ¾ in by 6 in (45 cm by 15 cm) in cross section and 22 ft (6.7 m) tall. The members are separated and yet connected by a custom-fabricated, seismic fuse. The fuse is made with steel plates and pins that are through bolted to each member. Structurally, the X-shaped components support the roof, provide lateral structural resistance, and support the curtain wall; programmatically, their width provides spaces in which passengers can step out of the pedestrian traffic flow to enjoy the views of the tarmac, airplanes, bay, and weather.

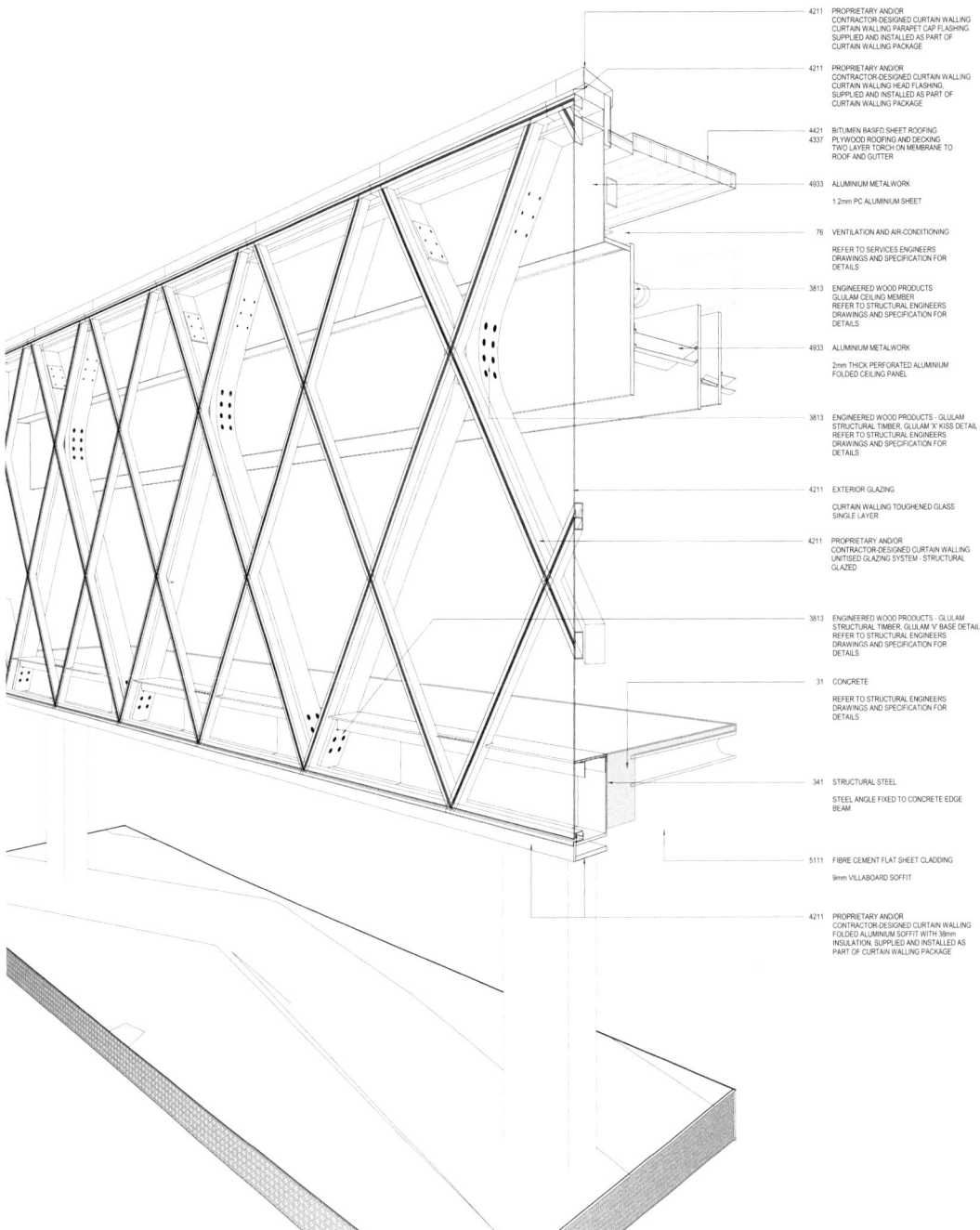

4211 PROPRIETARY AND/OR
CONTRACTOR-DESIGNED CURTAIN WALLING
CURTAIN WALLING PARAPET CAP FLASHING
SUPPLIED AND INSTALLED AS PART OF
CURTAIN WALLING PACKAGE

4211 PROPRIETARY AND/OR
CONTRACTOR-DESIGNED CURTAIN WALLING
CURTAIN WALLING HEAD FLASHING,
SUPPLIED AND INSTALLED AS PART OF
CURTAIN WALLING PACKAGE

4421 BITUMEN BASED SHEET ROOFING
4337 PLYWOOD ROOFING AND DECKING
TWO LAYER TORCH ON MEMBRANE TO
ROOF AND GUTTER

4933 ALUMINIUM METALWORK

1.2mm PC ALUMINIUM SHEET

76 VENTILATION AND AIR-CONDITIONING

REFER TO SERVICES ENGINEERS
DRAWINGS AND SPECIFICATION FOR
DETAILS

3813 ENGINEERED WOOD PRODUCTS
GLULAM CEILING MEMBER
REFER TO STRUCTURAL ENGINEERS
DRAWINGS AND SPECIFICATION FOR
DETAILS

4933 ALUMINIUM METALWORK

2mm THICK PERFORATED ALUMINIUM
FOLDED CEILING PANEL

3813 ENGINEERED WOOD PRODUCTS - GLULAM
STRUCTURAL TIMBER, GLULAM 'X' KISS DETAIL
REFER TO STRUCTURAL ENGINEERS
DRAWINGS AND SPECIFICATION FOR
DETAILS

4211 EXTERIOR GLAZING

CURTAIN WALLING TOUGHENED GLASS
SINGLE LAYER

4211 PROPRIETARY AND/OR
CONTRACTOR-DESIGNED CURTAIN WALLING
UNITISED GLAZING SYSTEM - STRUCTURAL
GLAZED

3813 ENGINEERED WOOD PRODUCTS - GLULAM
STRUCTURAL TIMBER, GLULAM 'V' BASE DETAIL
REFER TO STRUCTURAL ENGINEERS
DRAWINGS AND SPECIFICATION FOR
DETAILS

31 CONCRETE

REFER TO STRUCTURAL ENGINEERS
DRAWINGS AND SPECIFICATION FOR
DETAILS

341 STRUCTURAL STEEL

STEEL ANGLE FIXED TO CONCRETE EDGE
BEAM

5111 FIBRE CEMENT FLAT SHEET CLADDING

9mm VILLABOARD SOFFIT

4211 PROPRIETARY AND/OR
CONTRACTOR-DESIGNED CURTAIN WALLING
FOLDED ALUMINIUM SOFFIT WITH 38mm
INSULATION, SUPPLIED AND INSTALLED AS
PART OF CURTAIN WALLING PACKAGE

Figure 1.4.6
Perspective detail drawing of a typical bay.

Figure 1.4.7
Photograph of prototype, showing custom fabricated
fuse that connects the two members together.

Techlam, located in Levin, New Zealand, just 62 mi (100 km) from Wellington Airport, manufactured the glulam members and bolted them together prior to shipping them to the construction site. Techlam was not engaged with the project until after the construction contract was awarded. During the design phases, Skipper contacted different glulam manufacturers to determine the possible radius that could be formed through this process and determined that 4 ft-7 in (1.4 m) was a reasonable limit and it became the fundamental driver for the structure's geometry. WaM and Beca specified the durability and strength requirements in the technical specifications for Techlam to follow. Techlam had to manufacture the glulam components to tight tolerances so that they could sit properly into the steel-fabricated joint and could support the curtain wall brackets. To support the curtain wall, slightly larger-than-required slots were cut into the glulam and

Figure 1.4.8
Photograph of the South-West Pier under construction. Note the slots in the face
of the glulams that will be used to support the curtainwall.

Figure 1.4.9
Photograph of glulam members being formed against a steel jig. A wood caul is shown
on the lower member to help provide even pressure during clamping.

Figure 1.4.10
Photograph of glulam members after being removed from the jig. Note that the sides
and ends of the layups are unfinished and need to be trimmed.

Figure 1.4.11
Photograph the building under construction. The glulams were connected as a pair, prior to erection.

the curtain wall brackets were pocketed in and then recessed bolts were installed through the glulam members and brackets.

The glulam members are laminated from smaller 3/8 in × 1 ½ in (10 mm × 40 mm) pieces of pine and Techlam formed them by bending the members against a steel jig. The pine used is Monterey pine, or pinus radiata, and is typical in New Zealand construction. This wood is relatively soft, susceptible to mold and insect decay, and must be treated for architectural application.[6] Since the structure would be supporting the building parapet, the design team determined that the wood must be treated to H3.2 level. According to Skipper, in New Zealand, the H3.2 treated wood is stained green at the treatment plant for identification. Techlam sorted the wood pieces so that the greener pieces would be used for the glulam's interior and the less green pieces would be used for the glulam's exterior plies. The glulams were finished with a penetrating sealer that could be reapplied onsite for maintenance.

Notes

1. A curve radius of 2 in (50 mm) can be achieved, but only using a hydraulic press.
2. Active curing can be done with radio frequencies or heat induction.
3. There are post-formable plywoods, like UPM's Grada, that have thin layers of thermoplastic between layers of wood veneer. The post-formable plywood is heated to soften the thermoplastic, molded into shape, and then cooled to room temperature to be demolded.
4. Skipper, Katherine. *Personal Interview.* 6 July 2022.
5. Marriage, Guy. "Wellington International Airport Terminal South Extension : Wellington Airport's Latest Addition, by Warren and Mahoney, Is the Main Entry Point for Wellington's Frequent Domestic Travellers, Featuring a 'Net' of Angled Crossframes That Stretch between Existing Buildings." *Architecture New Zealand*, no. 3, May/June 2017, pp. 66–76.
6. Skipper, Katherine. This appears to be regardless of the wood being used inside, in a controlled environment, and not in contact with the ground.

References

Marriage, Guy. "Wellington International Airport Terminal South Extension: Wellington Airport's Latest Addition, by Warren and Mahoney, Is the Main Entry Point for Wellington's Frequent Domestic Travellers, Featuring a 'Net' of Angled Crossframes That Stretch between Existing Buildings." *Architecture New Zealand*, no. 3, May/June 2017, pp. 66–76.

Skipper, Katherine. *Personal Interview.* 6 July 2022.

Stamping

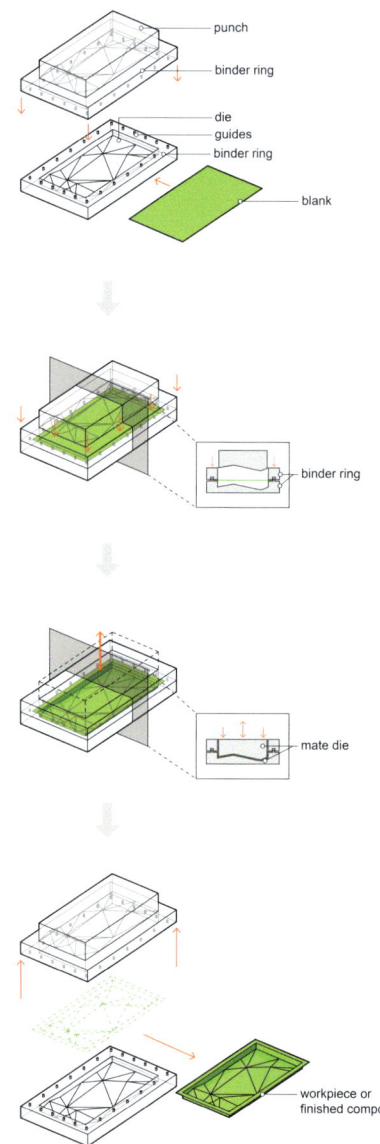

Figure 1.5.1
Stamping process diagram. Drawing by author.
A version of this diagram originally appeared in
Manufacturing Architecture (Laurence King, 2018).

 DOI: 10.4324/9781003299196-7

Stamping is the manufacturing process that uses force in combination with tooling to deform sheet metal. Stamping is a generic term that includes blanking, punching, shearing, and bending. For architectural application, this book focuses on stamping that uses a mate die to deform the sheet metal surface. Stamping can produce a range of shapes and forms, including deep draws, complex curves, ribs, and detailed surface textures, and is comparable to hydroforming (Chapter 1.6). Stamping requires large presses and two metal dies to strike a sheet with great force to deform it. With two dies, capital costs are high and large production runs are often needed to offset those costs. The ability to stamp a particular form on a metal sheet will depend on the metal, the temperature of the metal, sheet thickness, surface area, draw depth, and required detail. It is not uncommon for the press to strike the metal sheet multiple times, in quick succession, to get the deformation needed. Generally, cycle times in stamping are very short, even if the press makes multiple strikes to the workpiece.

In stamping, a lubricated sheet of metal is placed in the press. The press's binder ring closes, holding the edge of the sheet in place. The top die half, the punch, quickly moves down, striking the metal sheet and deforming it to the shape of the die. Once the sheet is fully deformed, the die opens, and the binder ring releases the workpiece. The workpiece is removed and undergoes any required post-production processes such as drilling, trimming, sanding, or finishing. Typically, this is a cold-working process and is done at room temperature, but warm and hot metal stamping are possible.

Stamping tools include two halves of a mated die and the binder ring. Mated dies are designed to stamp precise gages of metal, as the surfaces of the dies are offset based on the blank's thickness. Dies for high-volume productions are typically made from hardened tool steel and may be plated with chromium for increased wear resistance. Dies for mid-volume productions (under 5000 units) may be made from cast Kirksite.

Metal

Ferrous and non-ferrous metals and metal alloys can be stamped, with aluminum, steel, and stainless steel being the most common. The metal should be ductile enough to undergo plastic deformation, and the deeper the draw, the more ductility required. Generally, the metal sheets will have some spring back after the stamping process; to counteract the spring back, manufacturers may overbend the workpiece so that it springs back to its designed form. Some metals have an uneven grain structure, called anisotropy, that may cause the metal to tear or bend differently when bent in one direction when compared to the other. Different metals have different amounts of anisotropy and the type of metal affects what shapes can be stamped. Manufacturers can run computer-aided analysis to determine what troubles may arise before stamping.

Figure 1.5.2
Photograph of the Gnome Parking Garage. During the day, the parking garage
appears as a solid monolith. Photo © Jeroen Musch.

Figure 1.5.3
Photograph of the Gnome Parking Garage entrance. Photo © Ossip van Duivenbode.

Gnome Parking Garage in Almere, the Netherlands
By Mei Architects and Planners

The Gnome Parking Garage is a six-level, open-air parking garage with its four facades visible from the street, an interior block drive, or the neighboring railroad tracks. The garage has 413 parking spaces and 156,077 ft^2 (14,500 m^2) of gross floor area. The garage was commissioned by the government Almere Buiten—one of six districts of the relatively new city of Almere, located in the center of the Netherlands and across Lake IJssel from Amsterdam. Almere is built on land reclaimed from Lake IJssel and, according to Mei Architects

and Planners Partner Robert Platje, Almere has a lot of greenery embedded in its urban fabric.[1] Near the same time that Mei received the commission for the parking garage, Mei participated in a larger group dedicated to design innovation in the Netherlands, and as part of this group, representatives from Mei toured Voestalpine Polynorm BV, a Dutch metal stamping company that stamps parts (e.g. seat bases and side panels) for automobiles. Voestalpine participated in the tour because they were interested in expanding into architecture. Platje believed that they would be good collaborators and liked the idea of using an automotive manufacturer to produce the panels for a parking garage.

Figure 1.5.4
Close-up photograph of the panels and planter boxes. Photo © Jeroen Musch.

Figure 1.5.5
Photograph of garage at dusk, as the interior is lit, and the panels appear
translucent. Photo © Ossip van Duivenbode.

Figure 1.5.6
Photograph from garage interior, looking out. Panels appear nearly transparent. Photo © Jeroen Musch.

Figure 1.5.7
Photograph of the die being CNC milled from tool steel. Photo © Mei architects and planners.

The panels are designed with relief images of Almere and its province, Flevoland, and include birds, a windmill, trees, reed beds, a garden gnome, and a bird house. The metal used for stamping is 40% perforated with small round holes so that during the day, from outside the garage, it has a monolithic look, but inside the garage, it is filled with light, and one can easily see outside. During the evening and at night, the garage's internal lights light the facade and spill onto the sidewalk like a lantern. The panels are stainless steel; are .05 in (1.2 mm) thick, 4 ft (1.24 m) wide, and 9 ft 10 in (3 m) tall; and span from one level of the garage to another. A total of 1164 panels were stamped for this project, including the partial panels at the top level of the garage.[2]

The matched, stamping die was fabricated from CNC-milled tool steel (see Figure 1.5.7). The surface of the die is relatively rough, with the CNC tool marks still visible (see Figure 1.5.8). This is unusual for stamping dies, as a rough surface can leave scratches on the workpiece. Dies are often polished smooth and may be plated with chrome as this allows the blank to flow more easily over the die surface. Platje thought that Voestalpine made a cost-saving decision not to polish the surface as the panel perforations would hide any scratches.

During the design process, Voestalpine digitally analyzed Mei's proposed relief pattern and determined that there were some areas of concern for the panel's stiffness. Based on their original embossing pattern, additional stiffeners would need to be added to the back of the panel so that it could span from floor to floor. The design team did not want the added stiffeners, as they wanted the cost of the stiffeners to be put toward into the cost of the panels themselves, and presumably, the stiffeners would reduce the facade's transparency. Instead, Mei modified the embossing pattern by adding and modifying the relief elements so that the panels would be stiff enough to be self-supporting.

Figure 1.5.8
Photograph of the die surface. Photo © Mei architects and planners.

Figure 1.5.9
Photograph of the die in front of the press used to form the garage panels.
Photo © Mei architects and planners.

Figure 1.5.10
Photograph of the panels being installed during construction. Note that the
panels span from floor to floor without additional substructural supports.
Photo © Mei architects and planners.

Despite the garage being a publicly funded project, Mei did not use a traditional design-bid-build process. Instead, to find the building contractor through tendering, the design team fully engineered the facade, providing drawings, specifications, and a fixed budget that the contractor was required to meet to fulfill the contract. In a process that Platje called *demarcation*, any suppliers for the facade elements (e.g. stamped metal panels and rolled steel substructure) returned highlighted construction drawings and specifications that illustrated what they delivered to the construction site and the general contractor would be responsible to provide the balance. According to Platje, this is a technique that Mei developed, and they use it for architectural projects that they believe the contractors may reject for being too difficult or expensive to build.

Notes

1. Platje, Robert. *Personal Interview*. 13 July 2022.
2. All the partial panels at the garage top are half panels, in which only the bottom half of the panel was used, requiring a full stamping cycle for each partial panel.

Reference

Platje, Robert. *Personal Interview*. 13 July 2022.

CHAPTER 1.6 | Hydroforming

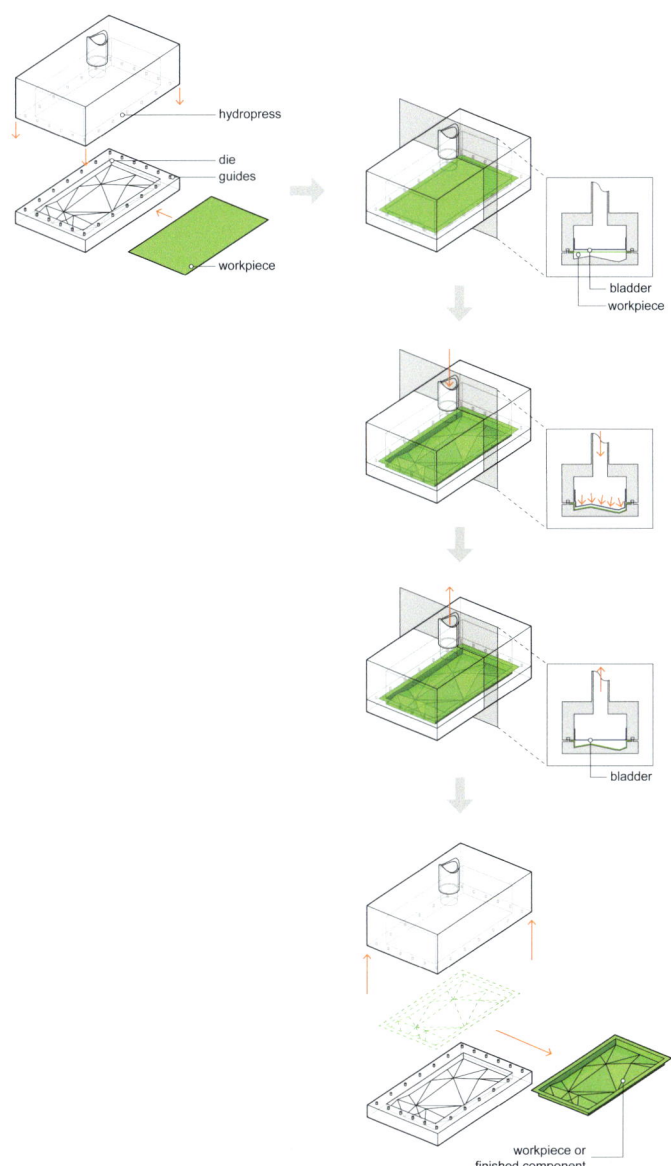

Figure 1.6.1
Hydroforming process diagram.
Drawing by author. A version of
this diagram originally appeared
in *Manufacturing Architecture*
(Laurence King, 2018).

Hydroforming is the manufacturing process that uses hydraulic pressure in combination with tooling to deform metal. Hydroforming is a generic term for any metal deformation that uses hydraulic pressure and can include tube hydroforming. For architectural applications, this book focuses on *sheet hydroforming* (SHD) that deforms sheet metal against an open die. Hydroforming can produce a range of shapes and forms, including deep draws, complex curves, ribs, and detailed surface textures, and is comparable to metal stamping (Chapter 1.5). Hydroforming requires a large press that uses a computer to control the hydraulic pressure exerted on the sheet during deformation. The ability to hydroform a particular form will depend on the type of metal, sheet thickness, surface area, draw depth, and required detail. Generally, tooling costs are approximately half of those of metal stamping, but cycle times are significantly longer than stamping.

In hydroforming, lubricant is used between the blank and the die face and an optional slip-sheet of flexible material is used between the blank and the bladder to protect the bladder, extending its life. The blank is placed in the press and an optional binder ring holds it in place. The press closes and hydraulic pressure fills the bladder, pressing the blank against the die face. During the cycle, the amount of hydraulic pressure varies, helping the metal flow into its intended shape with minimum spring back. Once fully deformed, the pressure is released, the press opens, and the binder ring is removed. The workpiece is removed from the press and undergoes any required post-production processes such as drilling, trimming, sanding, or finishing.

Hydroforming tools include an open steel die and may include a rubber bladder, blank holder, and inserts if needed. The die is made from hardened tool steel and can be plated with chromium for increased wear resistance. The bladder is sized for the component and is often purchased for the component's production as part of the tooling package. Inserts help place pressure at the bottom of narrow, deep crevices that a bladder would not necessarily be able to reach. Inserts also can be used to make small modifications to the die between cycles, allowing for some customization with little increased costs. Since hydroforming uses an open die, compared with the closed mated die required for stamping, the tooling costs are significantly lower in hydroforming than in metal stamping.

Other variants of SHD include *hydraulic stretch forming* (HSF), *hydromechanical deep drawing* (HDD), and *dual-sheet hydroforming*. HSF pushes the mold into the blank, rather than the blank onto the mold, and can produce components in similar forms as SHD. HDD is used primarily for deep draws and dual-sheet hydroforming is often used to produce hollow components such as tanks or storage containers.

Metal

Ferrous and non-ferrous metals and metal alloys can be hydroformed. Generally, all metals that can be stamped are suitable for hydroforming; in addition, hydroforming may be better than stamping for lightweight metals such as aluminum. Typically, with most manufacturing processes in Manipulating Sheet (Part 1), the blank face that is in contact with the tool is the finished face, as dimensions are held by the tool. Unfortunately, with stamping and hydro-forming, the dies are often made from hardened tool steel and therefore are hard enough to possibly scratch the surface of the workpiece. The value of hydroforming is that one side of the workpiece is not in contact with the tool, allowing the inside face to be safe from damage and possibly reducing required post-production finishing.

Figure 1.6.2
Aerial photograph of Science and Engineering Complex site. Photo by Brad Feinkopf.

Figure 1.6.3
Photograph of SEC, taken from the south. Photo by Brad Feinkopf.

Figure 1.6.4
Photograph of SEC laboratory screen walls. The facade includes a triple-glazed glass curtain wall with operable windows where appropriate and the custom, hydroformed sunshade panels mounted on tension rods. Photo by Brad Feinkopf.

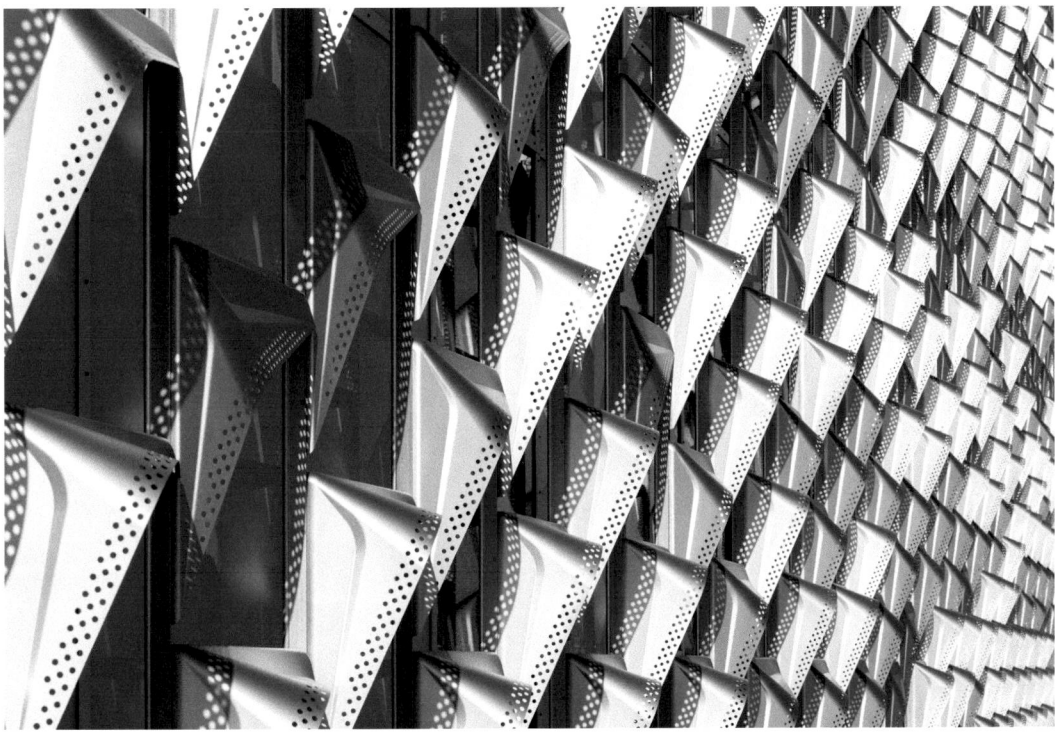

Figure 1.6.5
Close-up photograph of stainless-steel sunshade panels with custom
perforations at edges. Photo by Brad Feinkopf.

Harvard University Science and Engineering Complex (SEC) in Boston, Massachusetts
By Behnisch Architekten

The Science and Engineering Complex (SEC) at Harvard University houses the John A. Paulson School of Engineering and Applied Science (SEAS) in 544,000 ft^2 (50,530 m^2) of gross floor area and eight levels of building with almost 500 ft (150 m) of street frontage. The building was designed to reduce greenhouse gas admissions by 50% and to reduce cooling loads by 65% when compared to similar laboratory buildings. The SEC is certified Platinum by Leadership in Energy and Environmental Design

(LEED) and has a Living Building Challenge (LBC) petal certification in Materials, Equity and Beauty. According to Behnisch Architekten Boston Partner Robert Matthew Noblett, since laboratories are building types that are generally energy intensive and cooling-load driven, the firm was particularly mindful of reducing heat gain during warm days.[1] Behnisch designed hydroformed sunshade panels to shield the interior laboratory spaces from summer solar heat gain while admitting early morning sun during the winter for passive preheating of rooms. Because the sunshade panels are reflective, they also reflect indirect light deep into the building's interior, reducing the building's artificial lighting needs.

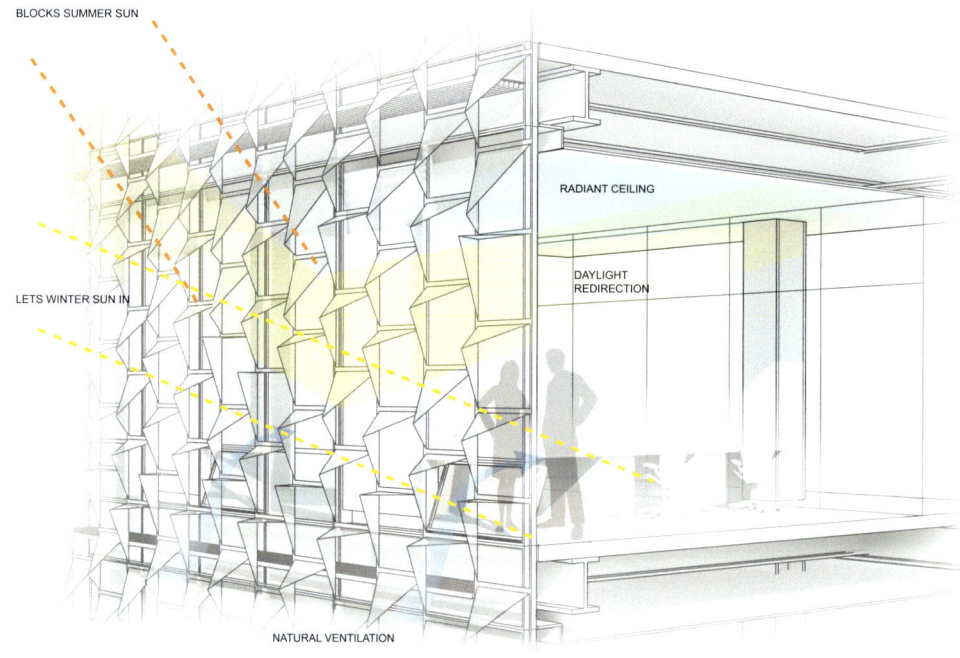

Figure 1.6.6
Diagram of facade performance. Drawing by Behnisch Architekten.

Behnisch wanted a sun-shading system that would be flexible in reducing direct sunlight and dynamic in its aesthetics. During initial design, Behnisch identified two potential options: (1) use one panel shape for the entire building with motors that would mechanically move the shape as the sun moved, or (2) individually fabricate unique panels from welded aluminum plate for each sunlight condition and the panels would be static. Behnisch dismissed the motorized system as it increased the electrical load on the building, and they find that most American clients have concerns about the lifespan and maintenance of exterior motorized components. The second option was dismissed as the design team did not think that the design and fabrication of the individual shades would be successful. When reflecting on their options, Noblett admitted, "In retrospect, it was all very clunky and heavy." Instead, Behnisch pursued a third option that would take the technology out of the shade system and move it to the design phase. The design team used parametric modeling and digital analysis tools to solve the geometric issues of sun shading and shade placement. They did this by developing a system of similarly shaped panels that could be custom trimmed based on their specific light conditions.

Figure 1.6.7
Photograph of building interior, looking through the window at the screen panels. Photo by Brad Feinkopf.

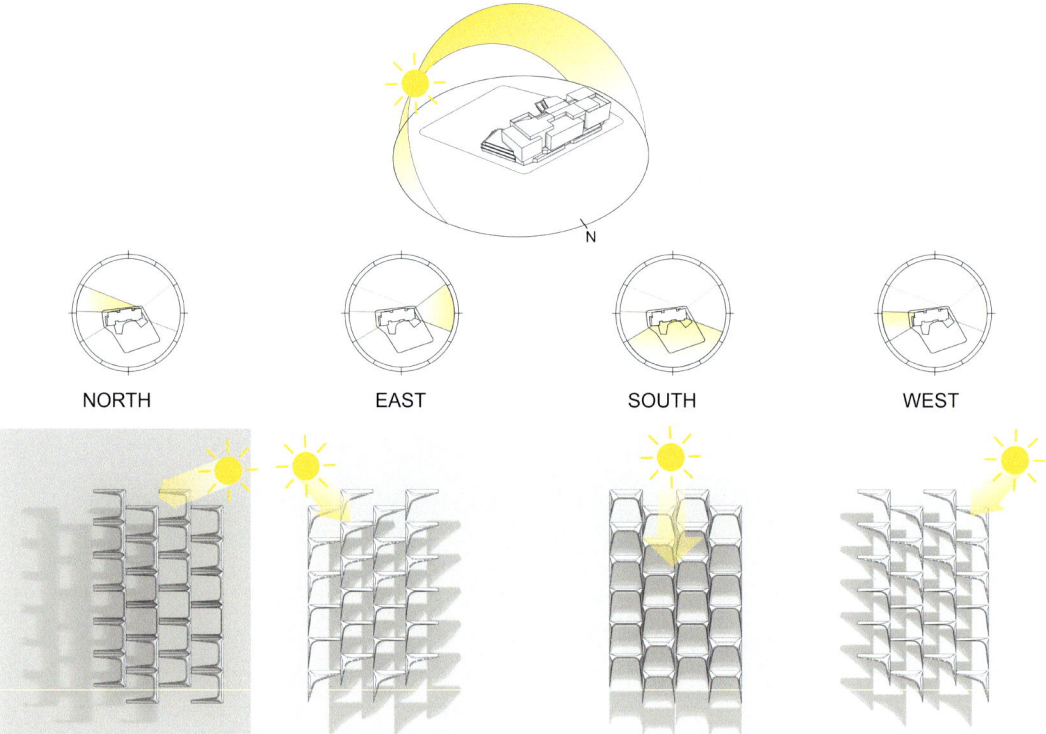

Figure 1.6.8
Drawing of basic sunshade panel shapes, based on the facade orientation
and sun angles. Drawing by Behnisch Architekten.

During the design phase, the building's facade contractor, Josef Gartner GmbH, was the leading front-runner to secure the construction contract, and they were the ones to put Behnisch in contact with the hydroformer Edelstahl-Mechanik GmbH. Edelstahl manufactures components for a wide range of industry sectors, including medical, automotive, agricultural, and environmental engineering. In addition to hydroforming, Edelstahl does computer numeric controlled (CNC) and robotic machining, breaking, welding, and material finishing and had the capacity to do all the manufacturing, fabricating, and finishing required for the panels. Once Edelstahl was engaged in the project, Behnisch communicated with them directly to develop the design. Noblett stated that Edelstahl are the masters of their processes, and the manufacturers gave feedback on what would and would not work. As the design developed, Behnisch shared the CATIA files with Edelstahl to run digital simulations to analyze potential problems with sheet deformation from hydroforming.

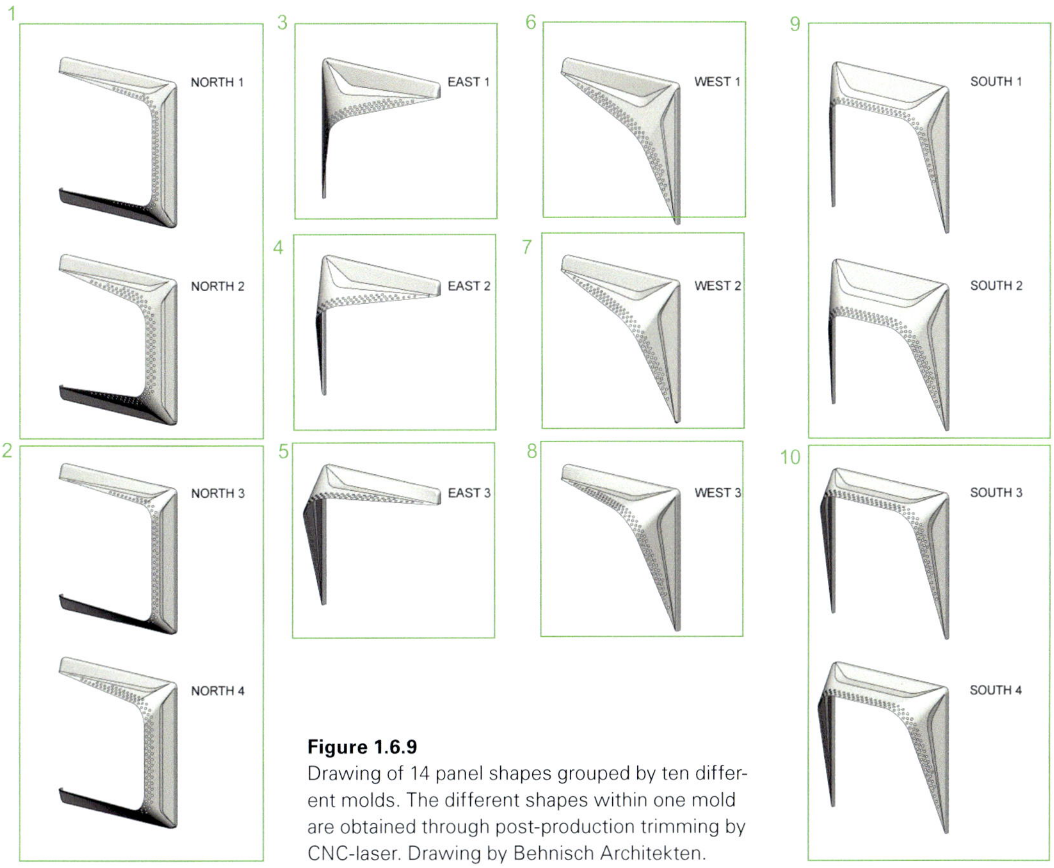

1

NORTH 1

NORTH 2

2

NORTH 3

NORTH 4

3

EAST 1

4

EAST 2

5

EAST 3

6

WEST 1

7

WEST 2

8

WEST 3

9

SOUTH 1

SOUTH 2

10

SOUTH 3

SOUTH 4

Figure 1.6.9
Drawing of 14 panel shapes grouped by ten different molds. The different shapes within one mold
are obtained through post-production trimming by
CNC-laser. Drawing by Behnisch Architekten.

Figure 1.6.10
Quality control at Edelstahl-Mechanik GmbH. Photo by Behnisch Architekten.

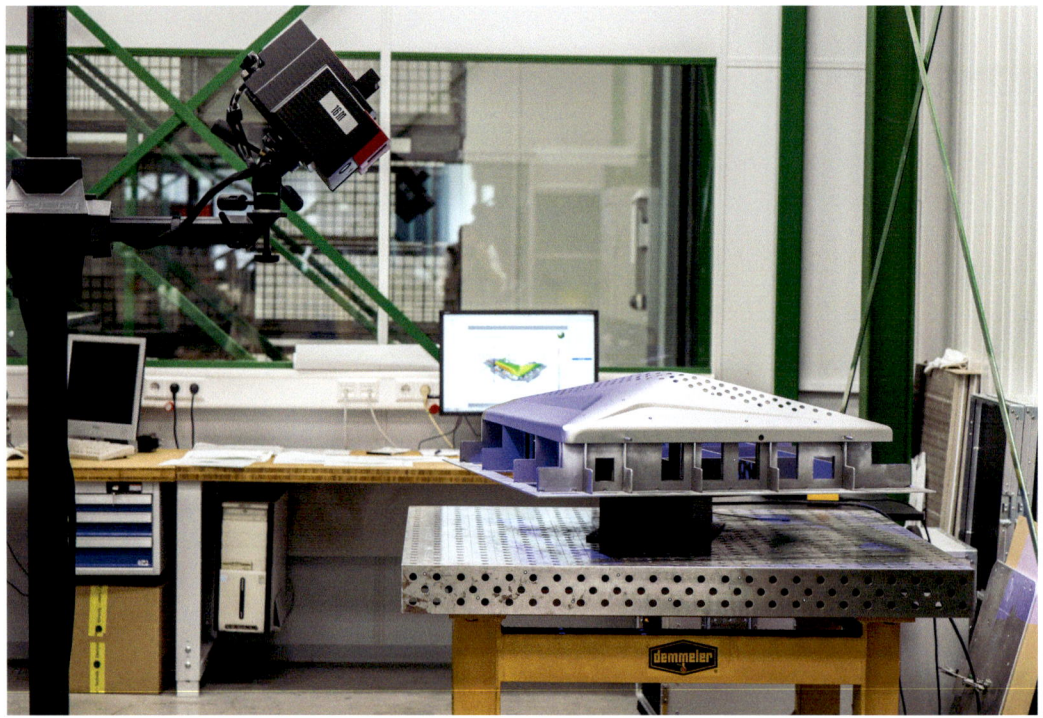

Figure 1.6.11
3D scanning of trimmed panels for quality control. Photo by Behnisch Architekten.

There are over 12,000 sunshade panels on the building facade in 14 different panel shapes that are grouped by their facade's orientation—north south, east, or west. Variants within each orientation group offer visual variety and allow views to the outside from the building interior or to block views from the outside, where appropriate. Ten different hydroforming molds were used to produce the 14 different shade panels with 4 of the variants being achieved through post-production trimming (see Figure 1.6.9). The sunshades are made from 1.5-mm stainless-steel sheets with bends added in the hydroforming process to stiffen and add strength to the thin sheets.[2] Each panel is approximately 3 ft by 3 ft (.9 m × .9 m) and weighs approximately 10 lbs (4.5 kg). The panels course vertically so that there are six panels per building floor and their horizontal coursing aligns with the curtain wall mullions and interior wall partitions where appropriate.

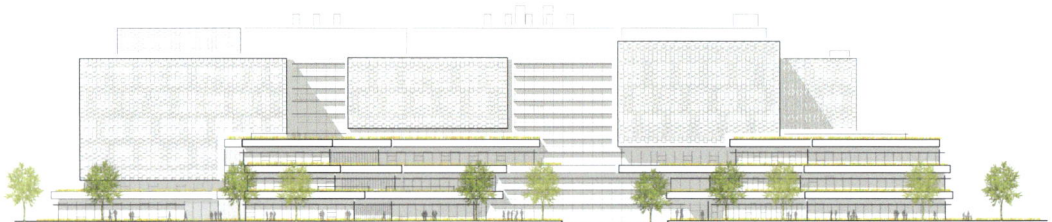

Figure 1.6.12
South elevation. Drawing by Behnisch Architekten.

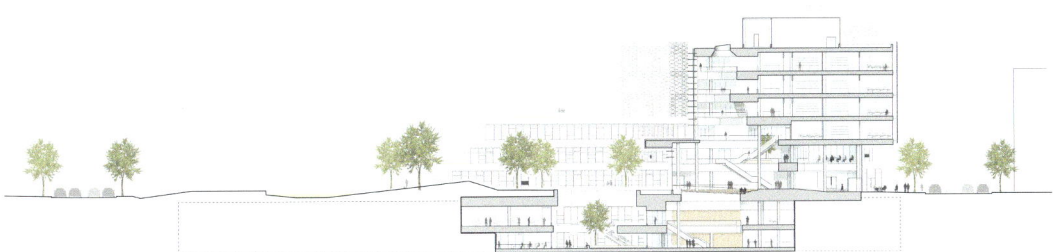

Figure 1.6.13
North-south section. Drawing by Behnisch Architekten.

STAINLESS STEEL
TENSION ROD

STRUCTURAL FOLD

PERFORATED EDGE

INTEGRATED HOLES
FOR ATTACHMENTS

WEST SHADING PANEL
.75 m x .75 m x .25 m
HYDROFORMED SHEET OF
1.5 mm STAINLESS STEEL

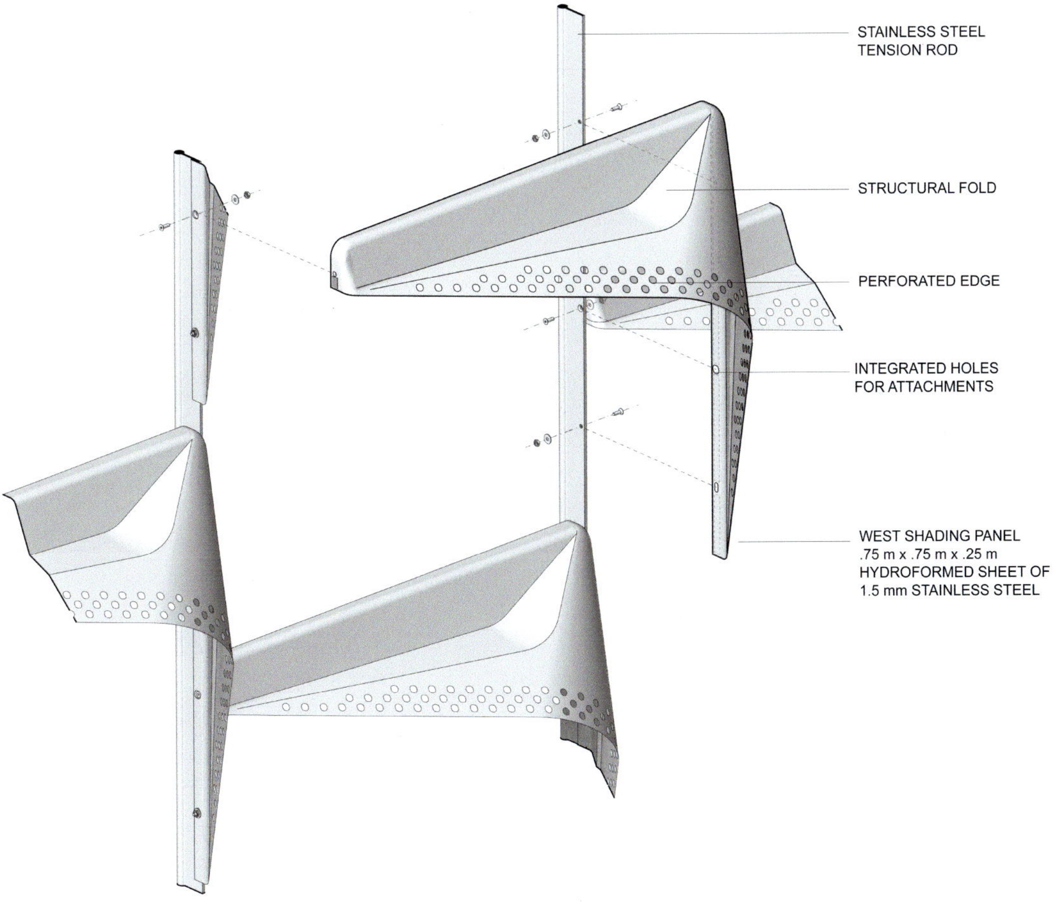

Figure 1.6.14
Details of sunshade panels mounting. The sunshades are supported by spring-tensioned
stainless-steel rods that have pre-drilled tabs on the front for the mechanical connections.
Oblong-shaped holes were used to allow some adjustability during mounting. Diagram by
Behnisch Architekten.

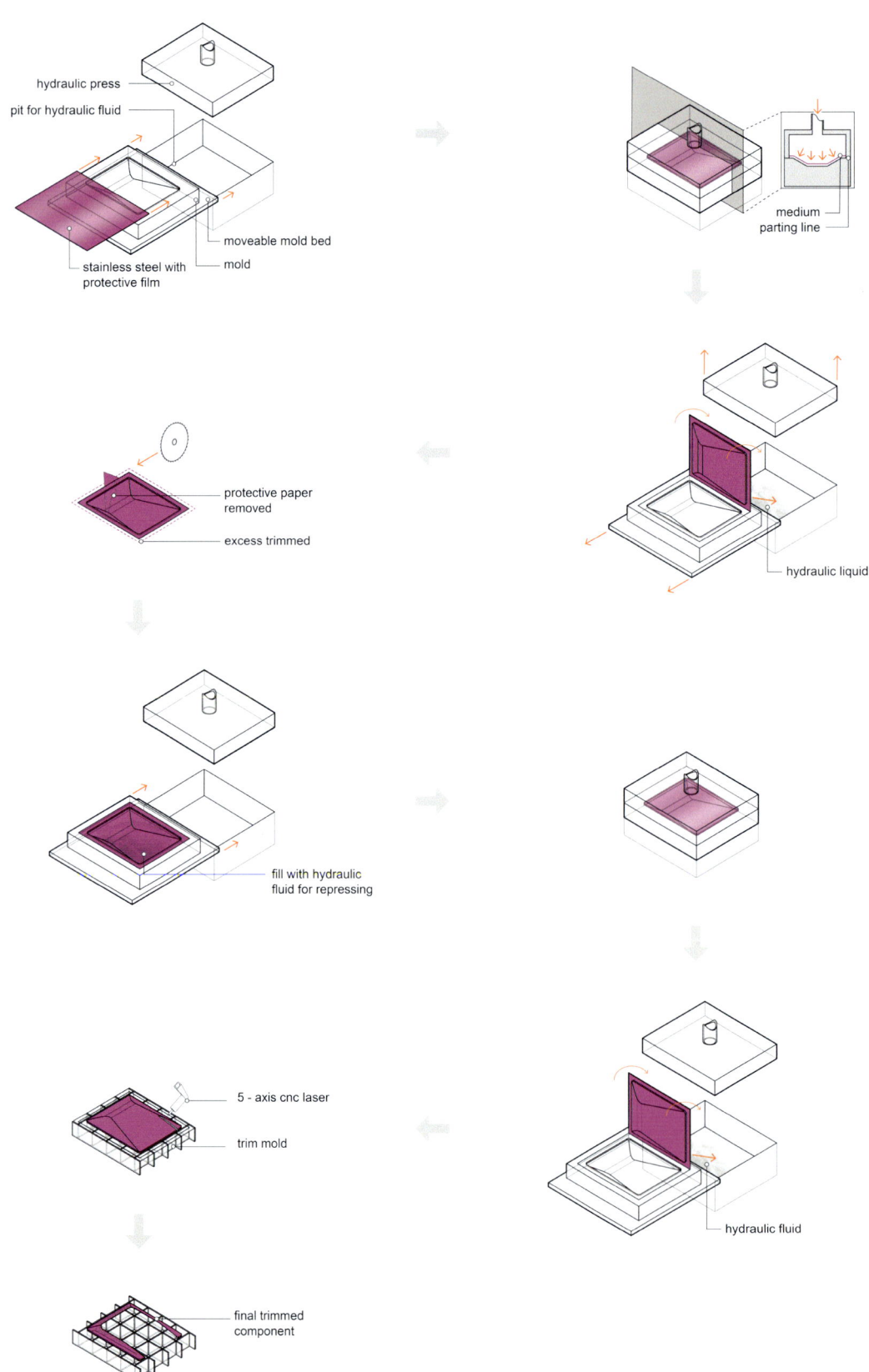

Figure 1.6.15
Diagram of SEC panel manufacturing.

Edelstahl does not use a bladder to contain the hydraulic fluid; instead, the fluid is in direct contact with the workpiece[3] (see Figure 1.6.16). When each workpiece was removed from mold after hydroforming, any the fluid trapped in the components' cavity would be poured into the hydroformer's trough and reused. Then workers transferred the workpiece to a trim die to properly position the workpiece to be trimmed by a five-axis laser cutter. The laser cutter cut the panels into their final shape and added perforated holes at the leading edge for glare control and mounting hardware holes on the panel sides. After post-production trimming, Edelstahl bead blasted the panels to produce a satin-like finish. For quality control, Edelstahl 3D scanned the panels to ensure that they met the project specifications. Noblett estimated that it took less than six months to produce the sunshade panels, including all post-production trimming and finishing.

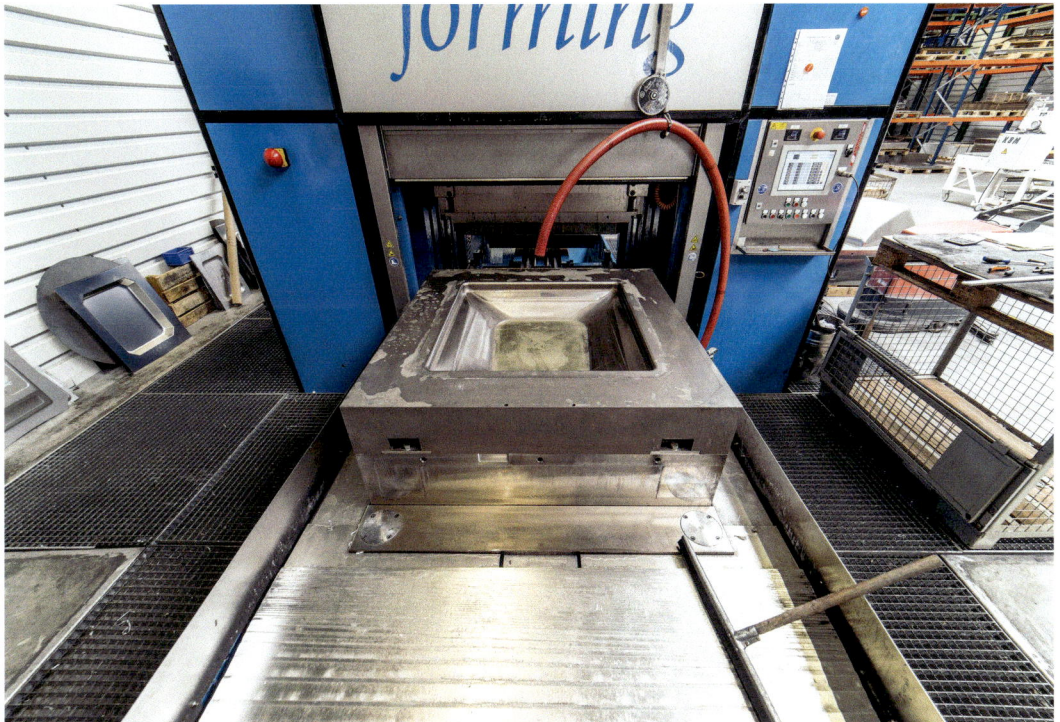

Figure 1.6.16
Photograph of panel being reformed after first post-forming trim. Note the hydraulic fluid in the panel cavity. Photo by Behnisch Architekten.

Figure 1.6.17

Photograph of component and mold after initial forming. Photo by Behnisch Architekten.

Figure 1.6.19

Close-up image of panels at Edelstahl. Edelstahl did the post-production trimming and finishing of the panels. Photo by knippershelbig.

Figure 1.6.18

Photograph of different molds used to hydroform the different panels. Photo by knippershelbig.

Notes

1. Noblett, R. Matthew. *Personal Interview*. 28 April 2022.
2. Behnisch and Edelstahl had researched the possibility of forming the sunshades from 1-mm thick sheets, but it was analyzed that the material would tear often enough during the hydroforming process that they decided to use 1.5-mm sheet thickness.
3. As most bladders are purchased as part of the tooling package for a particular job, not having a bladder reduced tooling costs.

Reference

Noblett, R. Matthew. *Personal Interview*. 28 April 2022.

Shaping Continuous

2

This part includes those manufacturing processes that form a continuous cross section along a component's length. Manufacturing processes include Chapter 2.1, Extrusion and Chapter 2.2, Pultrusion. Materials in this part are varied and include metal, plastic, clay, stiff mud, and fiber-reinforced plastic. Although the cross sections are consistent, the lengths of the components can easily be customized. Theoretically, components can be formed with infinite lengths, but in reality, lengths will be restrained by material strength capacities, shipping restrictions, or erection practices.

DOI: 10.4324/9781003299196-9

CHAPTER
2.1

Extrusion

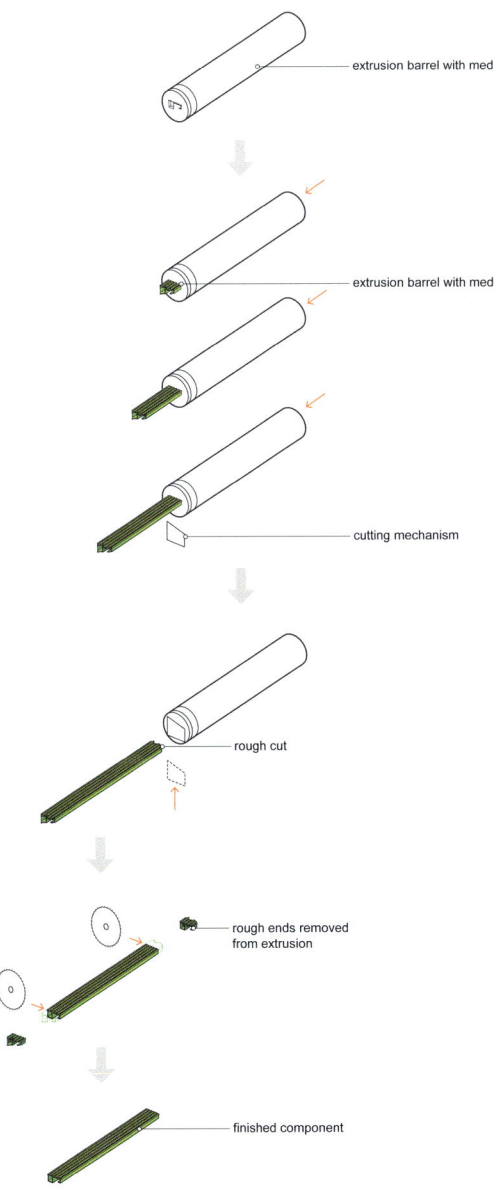

extrusion barrel with medium

extrusion barrel with medium

cutting mechanism

rough cut

rough ends removed
from extrusion

finished component

Figure 2.1.1
Extrusion process diagram. Drawing by author.
A version of this diagram originally appeared in
Manufacturing Architecture (Laurence King, 2018).

DOI: 10.4324/9781003299196-10

Extrusion is the manufacturing process that pushes a medium through a die to create a continuous cross-sectional shape. For architectural applications, extrusion uses aluminum, clay, concrete, stiff mud, and thermoplastic as its medium. Extrusion produces components that are solid, semi-hollow, or hollow with simple or complex cross-sections. Equipment and tooling used for extruding clay can be small and simple, but equipment and tooling used in extruding stiff mud, aluminum, or plastic are larger and more complex, often requiring a large production run for those materials. Extrusion is a continuous process and new material can be added during the process, theoretically producing components of an infinite length; however, lengths are limited by the manufacturing facility or transportation restrictions. Extruded components' lengths can easily be customized, as needed, with little-to-no increase in costs. Generally, the production speed for this process is fast with the specific rate of extrusion depending on the medium, the complexity of the cross section, and the allowable tolerances.

In extruding, the extrusion medium is placed into an optionally heated barrel and is pushed through a die opening. As the extrudate exits the die, it is rough cut and may go through secondary processes such as rolling, sizing, or stretching, as needed. The components are then trimmed to their final lengths during post-production. The die face of the extruded component is typically considered the finished face, whereas the cut surfaces are considered unfinished and are hidden.

The tooling required for extrusion includes the die and any addition in-line processes such as rollers for rolling surface textures.[1] For low-volume productions of clay extrusions, pressures are low, and dies can be made from flat plywood, acrylic sheets, or sheet metal attached on the end of a pug mill. For high-volume productions of extruded clay, or for extruding stiff mud, plastic, or metal, dies will be made of tool steel and may be reinforced with carbide or be plated with chrome. The steel dies are often designed to channel the extruding medium to the die opening to maintain a consistent flow speed when the extrudate exits the die. For extruding plastic, the tool package is expensive as it includes a long die to manage an even flow of plastic and a sizer to maintain the extrudate shape as it cools after exiting the die.

Aluminum

Many metals and their alloys, such as aluminum, copper, tin, and titanium, can be extruded, but aluminum and its alloys are most often used for architectural applications. Metals may be extruded cold (at room temperature), warm, or hot extruded. A manufacturer may use a variant process, called indirect or reverse extrusion, in which the die is pushed against a stationary billet. Aluminum tends to flow unevenly while being extruded, causing a curve to the extrudate. Manufacturers stretch the rough-cut extrudate to straighten it before final trims or other post-production processes.

Clay

Clay is used for extruding terracotta components. Clay is free of stone, debris, and organic materials and may contain small particles of grog as fillers. Clay has enough water content that it is moist to touch and can easily be formed with little-to-no pressure, but not so much water that it will collapse under its own weight. Generally, clay is more plastic than stiff mud and therefore requires less extruding pressure, will often exit the die without surface tears, and can be used to form more complex profiles with thinner walls than stiff mud.

Stiff Mud

Stiff mud is the material used for extruding brick. Stiff mud includes clay, with larger particles (e.g. sand and feldspar) and less water than clay alone. The particles in stiff mud do not necessarily adhere to one another, require pressure to stick to one another, and when hand squeezed do not leave a wet residue behind. Stiff mud must be extruded at higher pressures than clay, often requiring thicker walls and resulting in a rougher surface texture. To control the surface of the stiff-mud extrudate, the extrusion speed can be slowed, or in-line rollers can be added.

Figure 2.1.2
Photograph of the Angle Lake Transit Station in SeaTac, Washington.
Photo by Ben Benschneider/Courtesy of Brooks + Scarpa.

Figure 2.1.3
Photograph of the Angle Lake Transit Station parking garage clad in
anodized iridescent blue custom aluminum extrusions.
Photo by Ben Benschneider/Courtesy of Brooks + Scarpa.

Figure 2.1.4
Detail photograph of the ruled line surface. Photo by Ben Benschneider/Courtesy of Brooks + Scarpa.

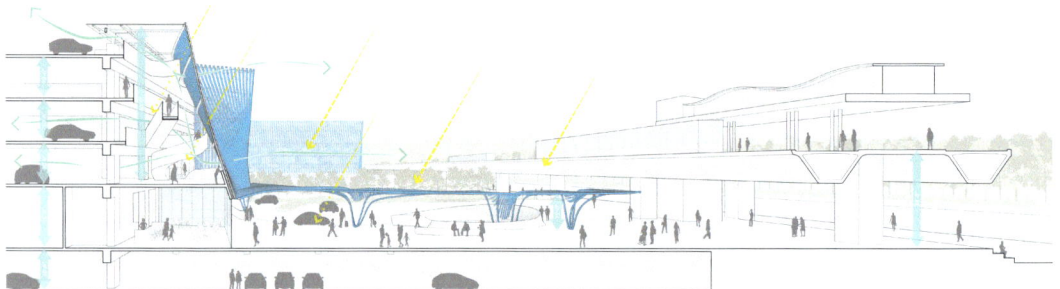

Figure 2.1.5
Transit Station section perspective through plaza. Drawing by Brooks + Scarpa.

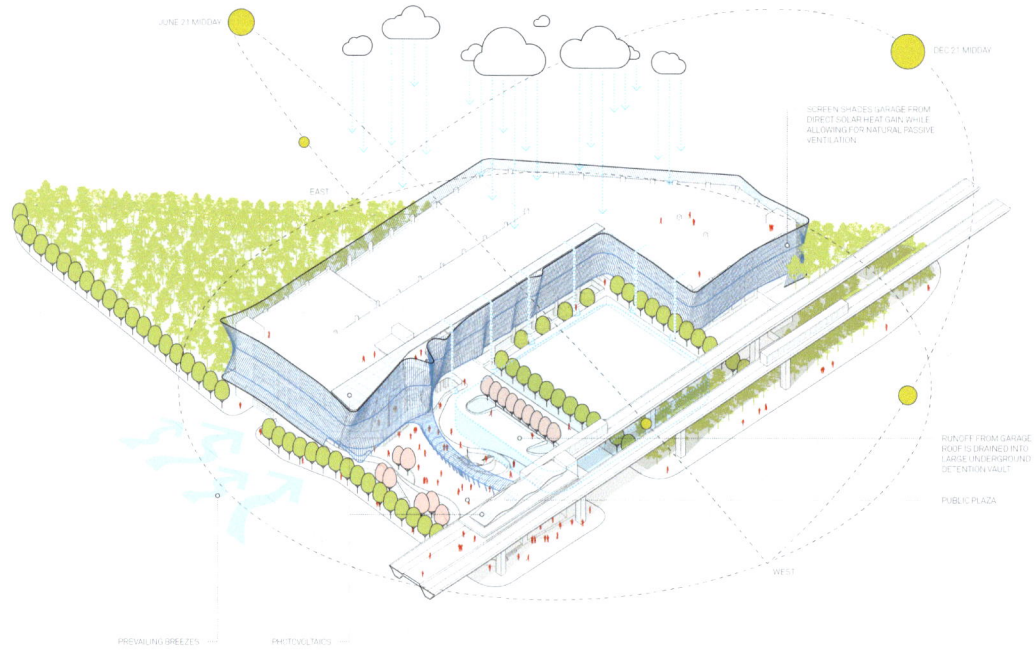

Figure 2.1.6
Transit Station plan oblique. There are several sustainable strategies used in the project, including water collection and native plants. Note that the ruled lines screen the parking garage and provide a shade canopy for pedestrians walking to the raised light rail platform. Drawing by Brooks + Scarpa.

Angle Lake Transit Station in SeaTac, Washington, United States
By Brooks + Scarpa

Angle Lake Transit Station is a light rail station that connects SeaTac, Washington—just south of the Seattle-Tacoma Airport—to downtown Seattle. This project was part of an international design-build competition, in which architect design proposals were evaluated along with a contractor-guaranteed construction budgets. The station includes a parking garage with 1050 spaces, surface parking with 70 spaces, drop-off, retail space, a plaza, and secure bike storage all within a one-block walk to the

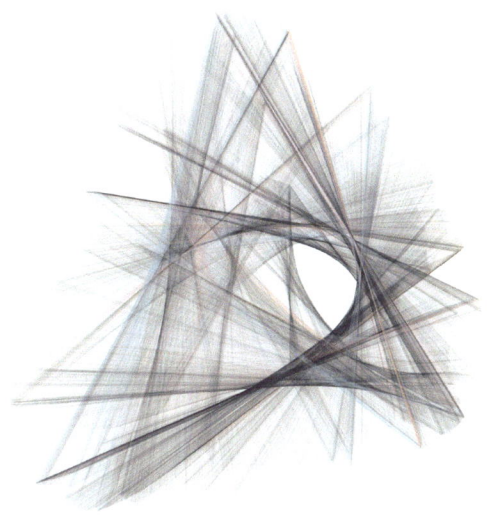

Figure 2.1.7
Early explorations of ruled lines creating complex geometries. Drawing by Brooks + Scarpa.

RapidRide bus line. Although the project program is complex, the project is dominated by the parking garage. Prior to the competition, Lawrence Scarpa had been exploring using ruled lines to create complex geometries and thought that they could be applied to this project. "'I was interested in repetition,' he says, 'And how you can transform repetition from a boring thing to something that is dynamic, interesting, interactive, and exciting.'"[2]

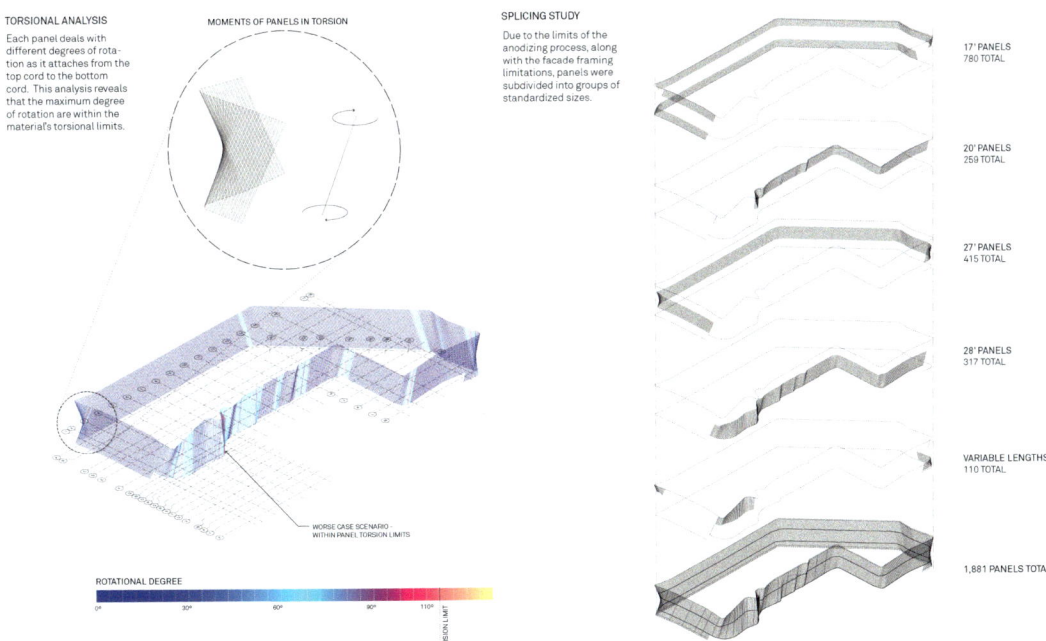

Figure 2.1.8
Torsion analysis and placement of extrusion lengths across the facade. Apel made 1881 extrusions in total for this project. Diagram by Brooks + Scarpa.

Figure 2.1.9
Extrusion mounted on steel supports, located at ever other level of the garage. Photo by Brooks + Scarpa.

In the original competition submission, the ruled surface was going to be made from a stock, 8-in (203-mm) wide, corrugated aluminum profile; however, as the team started developing the design, they realized that the stock profile would not meet the performance requirements for the design and project budget. First, to reduce material and labor costs, the design team maximized the span of the aluminum extrusions so that they would be only supported at

every other level of the garage. Second, although the ruled lines of the aluminum extrusions are straight, the horizontal steel supports located at every other floor are curved. This meant that the aluminum profiles needed a certain amount of torsion flexibility so that each extrusion could be mounted tangential to its curved steel support. Third, Scarpa wanted an extruded profile that would have depth and shadow.[3] The component's rounded corners combined with its blue iridescent anodized color provide a shimmering effect as one travels past the building. Fourth, the profile shape can hide a male connector sleeve made of a standard aluminum profile, to which each custom profile is mechanically attached (see Figure 2.1.10). This results in a virtually seamless connection from one extrusion to another. The financial benefits of the custom profile outweighed the cost of the die, and the contractor supported the design having a custom profile.

Apel Extrusions produced the custom extrusions. According to Scarpa, the collaboration between the design team and the extruder was straightforward. Apel outlined what they can extrude, what is easiest to extrude, and general costs, and the design teamed worked within those parameters. Apel validated any profiles that the design team sent for extrudability and had few-to-no concerns about what was sent. Since the design maximized the extrusion lengths—17, 20, 27, and 28 ft (5.18, 6.1, 8.23, and 8.53 m, respectively)—and Apel did not have the capacity to anodize extrusions of that length, they were anodized by another company in Redding, California.

Installation of the aluminum extrusions was labor and cost efficient. It took four crane operators on boom lifts, or "cherry pickers," less than two weeks

Figure 2.1.10
Extrusions being mechanically fastened onto the steel supports. Note the round, standard male
sleeves used to connect one extrusion length to another. Photo by Brooks + Scarpa.

to install the ruled lines. All the curved horizontal steel members were attached with supporting arms onto the parking garage's perimeter beams. Next, each aluminum extrusion was attached only at its top and bottom, either to a horizontal support or to another extrusion using the hidden sleeve connector. Finally, the extrusions were then pushed or pulled to be tangential to the intermediate curved supports and attached with self-tapping screws and neoprene washers.

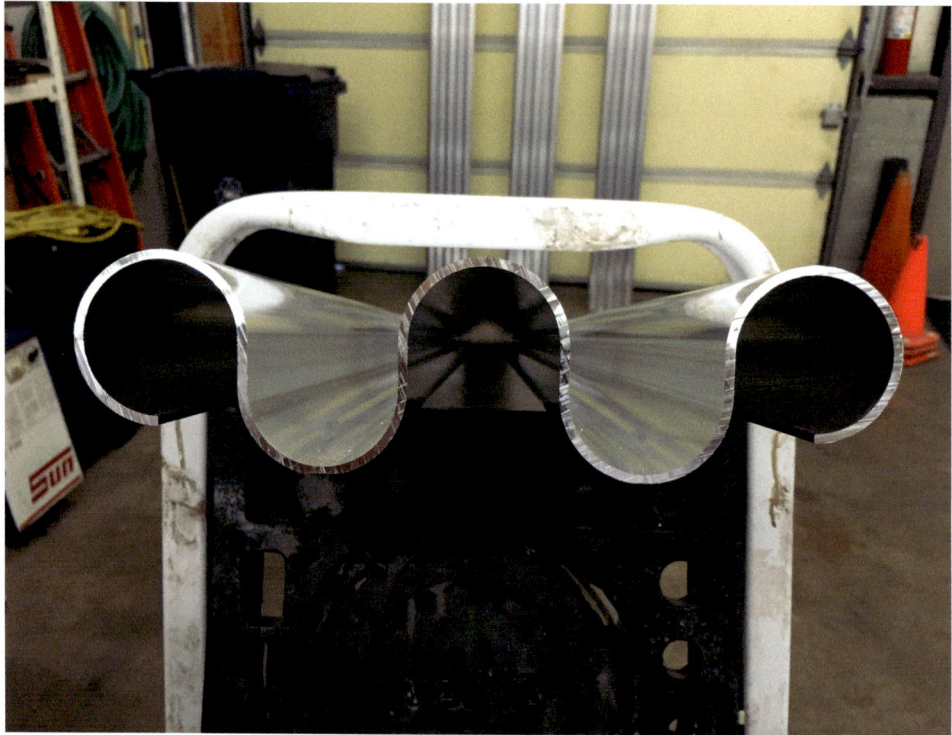

Figure 2.1.11
Close-up photograph of the extrusion at Apel, post final trim. Photo by Brooks + Scarpa.

Figure 2.1.12
Extrusions packaged for shipping to Redding, CA to be anodized. Photo by Brooks + Scarpa.

Figure 2.1.13
Clamping extrusions on rack for horizontal anodizing. Photo by Brooks + Scarpa.

Figure 2.1.14
Anodizing bath. Photo by Brooks + Scarpa.

London Wall Place in London, England
By Make Architects

London Wall Place (LWP) is in central London, just north of St. Paul's Cathedral. The site's south boundary is London Wall Street that follows the urban grain created by the ancient Roman wall that surrounded London. The project site is a 2-acre campus that includes two office buildings, with over 500,000 ft² (46,452 m²) of workspace; 35,000 ft²

(3252 m²) of roof terrace; and retail. 1 London Wall Place is the new headquarters for global asset manager Schroders; it is 13 stories and is located on the east end of the site on the wedge-shaped parcel between the angled London Wall street and Fore Street to the north. 2 London Wall Place has multiple, multi-floor tenants; it is 17 stories and is located on the west end of the site next to the Alban Gate building that spans London Wall. Approximately half the site is dedicated to public gardens and raised pedestrian walkways that link the Barbican Center

Figure 2.1.15
Aerial Photograph of London Wall Place looking South, with St. Paul's Cathedral and River Thames seen beyond. 1 LWP is on the left, 2 LWP is on the right, and the gardens connect the two buildings together. Photo © Make Architects.

Figure 2.1.16
London Wall Place as seen looking down London Wall toward 2 LWP. The remnants of St. Alphage Church
can be seen in the public gardens between the buildings. Photo © Make Architects.

Figure 2.1.17
1 LWP stepped facade, overhanging London Wall. Photo © Make Architects.

and neighborhood to the north with the center of London to the south. Tiered, the LWP gardens connect to the block's existing public St. Alphage gardens and the private Salters' Gardens. The LWP gardens provide a backdrop and public access to the previously hidden remnants of the Roman wall and St. Alphage Church, a medieval church built against the wall. Make Architects consider the office buildings as state of the art, but it is the combination of the buildings and the public realm that has attracted the leasers to the two office buildings.[4]

Make designed the massing of 1 LWP and 2 LWP in response to the local site conditions and history. The site and surrounding area were heavily destroyed during World War II bombing, with only parts of St. Alphage and the Roman wall remaining. According to Make Architects Chris Jones, after World War II, the cleared site became an opportunity for a development masterplan designed by Sir John Burnet Tait & Partners that included commercial office buildings along London Wall and housing for the area that is now the Barbican.[5]

Make's commission was part of a design competition held by the original site developers, and Make used the city wall remnant to organize the site for their design. 1 LWP has a bigger footprint than the previous post-war office tower that occupied the site. However, 1 LWP is stepped in its massing to respect the sunlight requirements of the existing residential units to the north so that only about a fourth of its floor area rises to its full 13-story height, thus matching the previous post-war tower's cast shadow. In addition, 1 LWP's massing was driven by London View Management Framework that protects views from key locations of historic and important buildings and monuments. For this site, the protected views were of St. Paul's Cathedral, approximately half a mile (.8 km) from the site, and 1 LWP steps back accordingly.[6] In contrast, 2 LWP is on the western edge of the site and follows more closely the post-war masterplan. 2 LWP is similar in height to its neighboring, office towers,[7] while its east face is stepped back to allow light into the Salters' Garden to its north.

Figure 2.1.18
1 LWP overhanging London Wall sidewalk. The building materials allude to flint
found in medieval sections of the Roman wall. Photo © Make Architects.

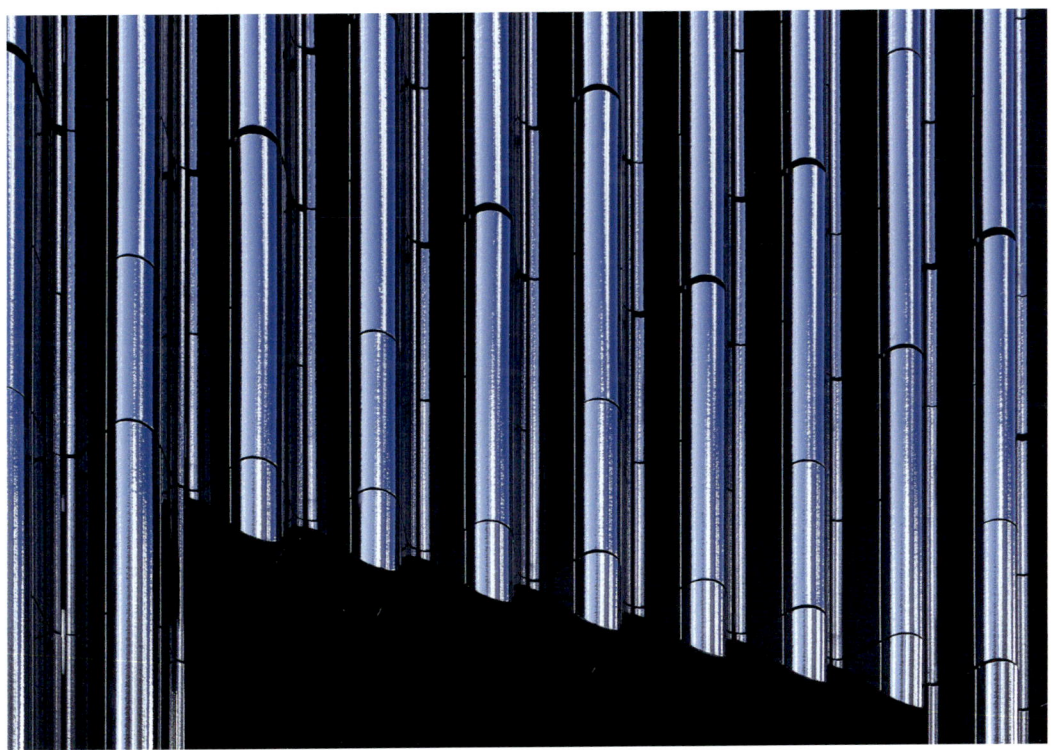

Figure 2.1.19
The dark glazed terracotta can appear blue on sunny days. Photo © Make Architects.

LWP exterior building materials include white glass fiber reinforced concrete (GFRC) panels, dark iridescent glazed terracotta custom profiles, and glass. Jones stated that the exterior materials reference flint found in the medieval-dated sections of the site's Roman wall. Flint is a form of mineral quartz. Often the outside has a chalky white layer that is associated with the GFRC panels; whereas the inside is a crystal, and can be a dark, purply iridescent color and is associated with the glazed terracotta. Make further contrasted the materials by the GFRC panels designed with clean edges, sharp corners, and aligned joints, while the terracotta elements were designed with curves, rounded

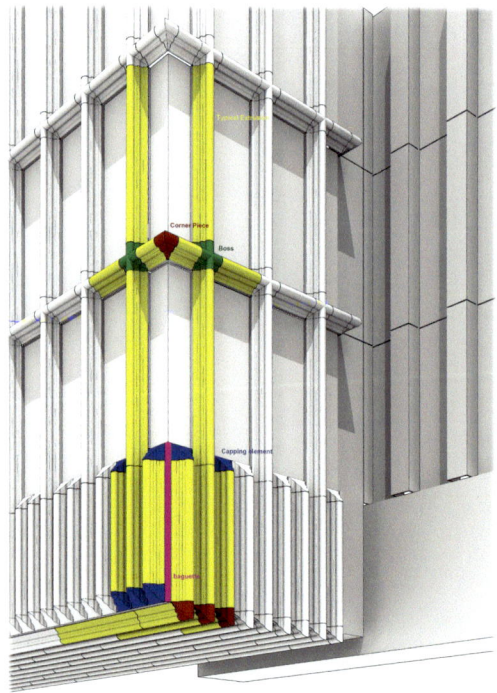

Figure 2.1.20
Diagram by Make the depicts the special conditions: the four-way intersections in green, the 90-degree corners in red, the fused capping element in blue, and the corner baguette in pink.[8] Photo © Make Architects.

Figure 2.1.21
Foam full-scale study of the profile and window surround. Photo © Make Architects.

Figure 2.1.22
Image of the terracotta elements made from two different profiles. Photo © Make Architects.

or eased corners, and offset joints. According to Jones, the additional benefit of the curved terracotta profiles is that its geometry with the high-gloss finish of the glaze refracts light despite its dark color.[9] On cloudy days, the terracotta appears almost black, but on sunny days the profiles reflect the sky and appear to be a deep blue color.[10]

Make knew the basic profiles that they wanted for the terracotta and studied the material, manufacturers' capabilities, and different glazes and colors. They worked with several manufacturers to see how the profiles might be formed, where the joints between elements would go, how the profiles would be shaped so that the glaze would not flow during firing and leave thin spots, the necessary internal structure to support the profile's self-weight before

firing, and how the components could be replaced after construction for maintenance purposes. In addition, Make had to resolve the different profile conditions that included a four-way intersection, a 90-degree corner, a capping profile at the lower windows and soffit, and the baguettes used at the building corners (see Figure 2.1.20). Make discussed options with several manufacturers, including Palagio Engineering, James + Taylor, NBK, Ceramica Cumella, and Boston Valley Terra Cotta. Some of the manufacturers proposed making the elements from only extruding, while others suggested a combination of slip casting, molding, and extrusion. Make evaluated the uniformity between components and procurement risks, and they selected NBK as the manufacturer.

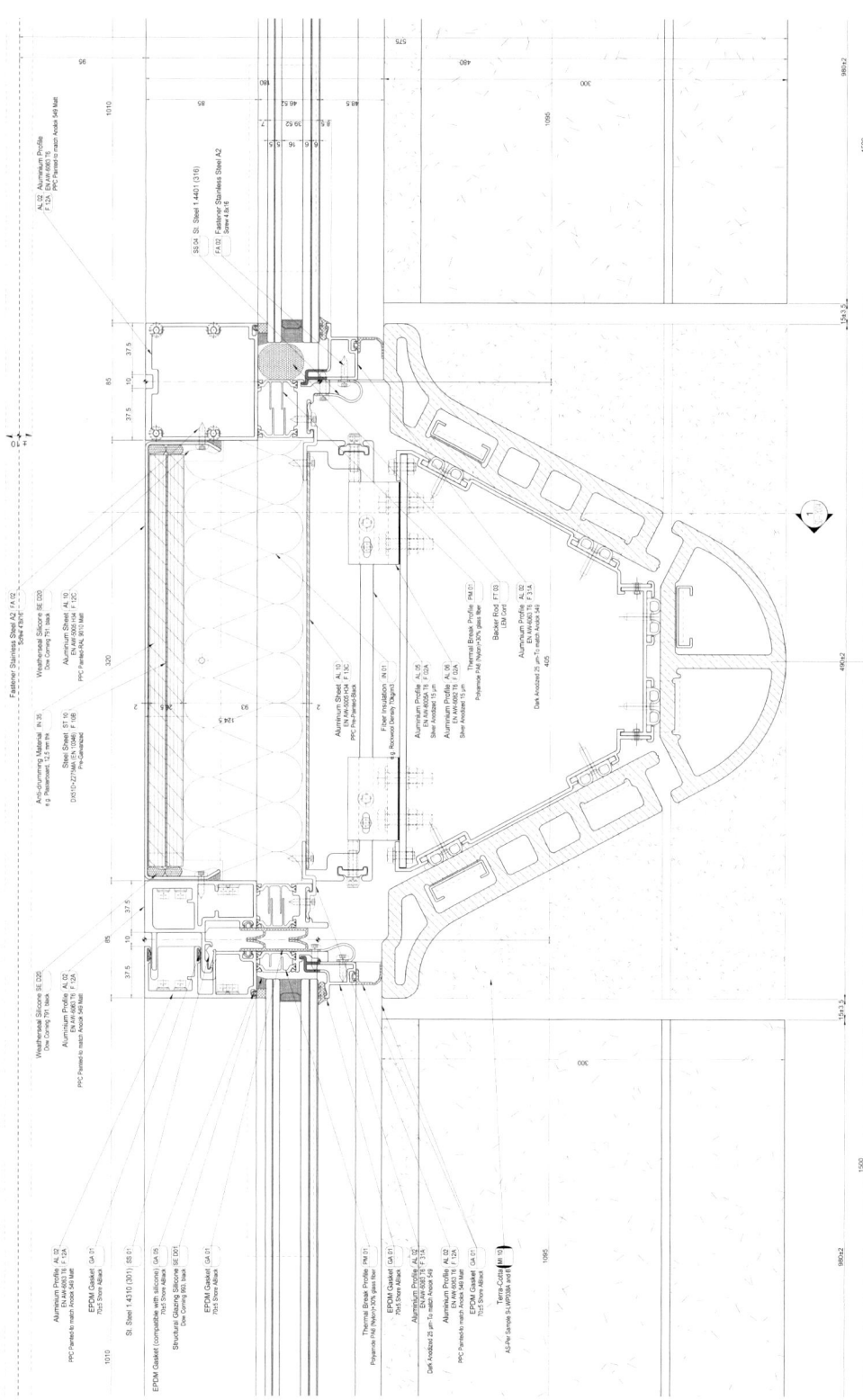

Figure 2.1.23

Construction document plan detail of terracotta window surround, drawn at full scale. © Make Architects.

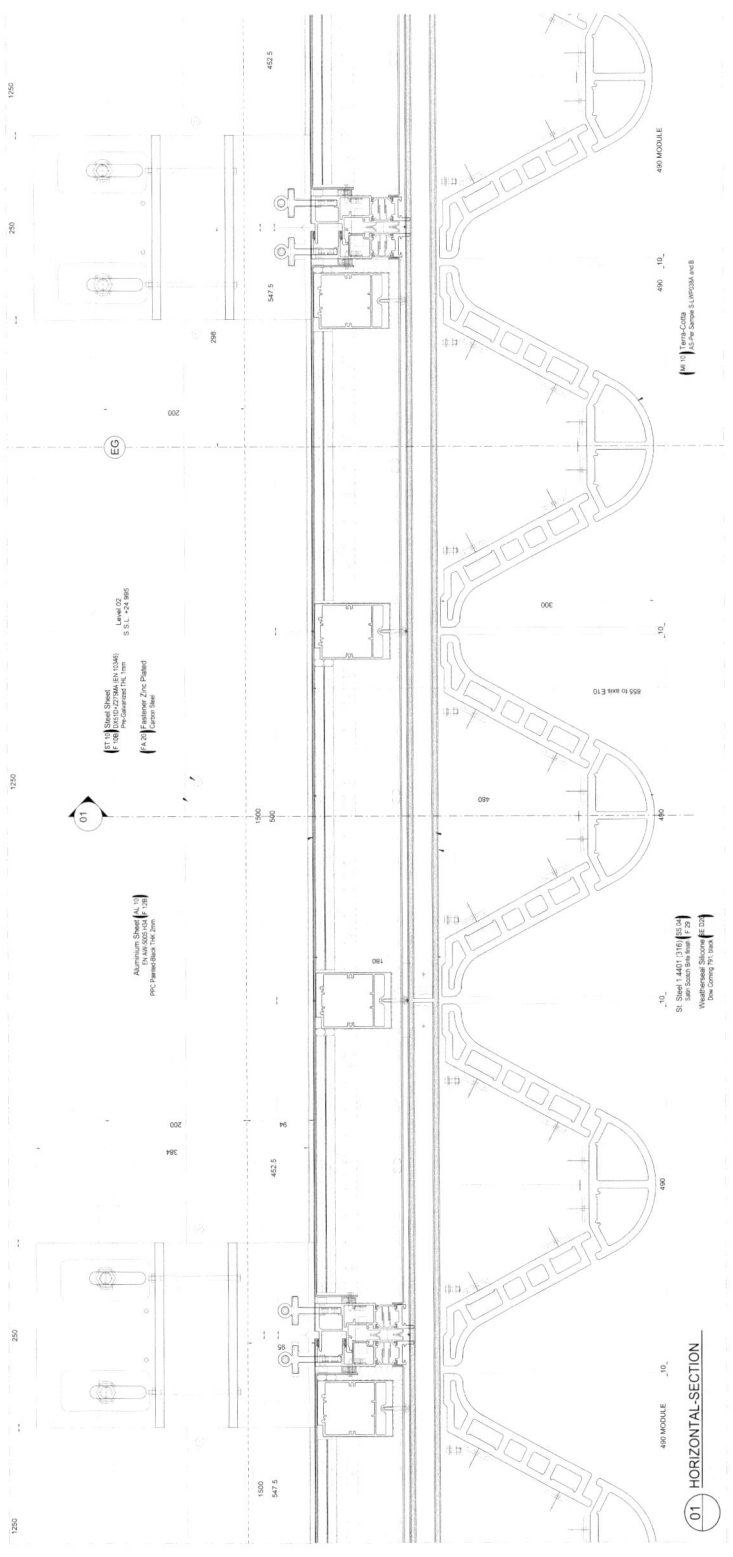

Figure 2.1.24

Construction document plan detail of lower level windowsills, drawn at half scale. © Make Architects.

Figure 2.1.25
Photograph of the four-way corners bonded and mechanically fastened together. Photo © Make Architects.

Figure 2.1.26
Photograph of a preassembled element with overlapping vertical joints and offset horizontal joints. Photo © Make Architects.

The custom curved terracotta elements are made of three parts in two different profiles. The cap is one profile, and the sides share the same profile, flipped. The elements form the window surrounds, lower level spandrel panels, and the ground-level soffits. The element is 19 ¼ in (490 mm) wide, 11 ¾ in (300 mm) deep, and project 13 ¾ in (348.5 mm) from the outside face of the glass, providing some solar shading.[11] NBK produced the 36,000 terracotta elements almost entirely through extrusion. NBK fabricated the four-way intersections and 90-degree corners by mitering the ends of the profiles, then those were bonded and mechanically fastened with an aluminum substructure off-site (see Figure 2.1.25). For the lower level windowsills and ground-floor soffits, NBK capped the

profiles by adhering flat clay slabs onto the extrusions with clay slip while the clay was still moist. The linear profiles were preassembled off-site and mechanically fastened to an extruded aluminum substructure that facade contractor, Permasteelisa Group, used to attach as preassembled units onto the facade. The longitudinal joints between the cap and side profiles overlap and remain hidden. The transverse joints between the cap or side lengths are staggered in the preassembled unit to give them a monolithic appearance (see Figure 2.1.26). The joints within the preassembled units are narrower than the 9/16 in (15 mm) joints between the preassembled units. The 9/16 in dimension allows for movement between the preassembled units.

Figure 2.1.27
Detail image of the capped lower level windowsills and ground-floor soffits. © Make Architects.

Figure 2.1.28
Detail image of the larger joints between the terracotta elements at the four-way intersection and 90-degree corner and compared to the smaller, overlapping and offset joints within the preassembled elements. Photo © Make Architects.

Lantern House in Manhattan, New York, United States
By Heatherwick Studio

Lantern House is a multi-family apartment building in the Chelsea neighborhood of Manhattan on West Eighteenth Street between Tenth and Eleventh Avenues. Lantern House has 181 residential units and consists of two towers, separated by the High Line.[12] The Lantern House east tower is on 10th Avenue and is ten stories tall. Its west tower, in the middle of the block, is 22 stories tall and has views of Chelsea Piers and the Hudson River. In addition to the apartments, the building amenities include a fitness center and pool, spa, shared entertainment room, residential lounges, and parking for 100 vehicles. The building lobby is under the High Line, with its roof suspended between the two towers and the riveted steel columns of the High Line landing in the middle of its space. Lantern House shares the block

with Gehry Partners' InterActive Corp (IAC) headquarters (2007) and Selldorf Architects' 520 West Chelsea (2009) and is steps away from Atelier Jean Nouvel's 100 11th Avenue (2010).

The building's exterior is brick and has two-story high, lantern-shaped bay windows that curve toward the outside within the thickness of the wall (see Figures 2.1.31 and 2.1.32). Heatherwick Studio incorporated multiple aspects when designing the facade that included referencing the historic building materials in the Chelsea neighborhood,[13] a desire to create a sense of permanency,[14] and real estate market studies of neighboring apartment buildings and their exterior building materials.[15] Since the project is in Manhattan, floor areas are at a premium, and Lantern House maximized its building footprint while working within the mandated city setbacks. According to Heatherwick Studio Senior Associate Carlos Parraga-Botero, the design team quickly concluded that any design solutions for the facade

Figure 2.1.29
Lantern House as seen from the High Line, looking down 10th Avenue. Photo by Kevin Scott.

Figure 2.1.30
Two towers of Lantern House, separated by High Line. Photo by Kevin Scott.

Figure 2.1.31
Lantern House at ground level, with radial corners and 270-degree bay windows. Photo by Kevin Scott.

Figure 2.1.32
Detail photograph of two-story lantern windows and building corner. Photo by Kevin Scott.

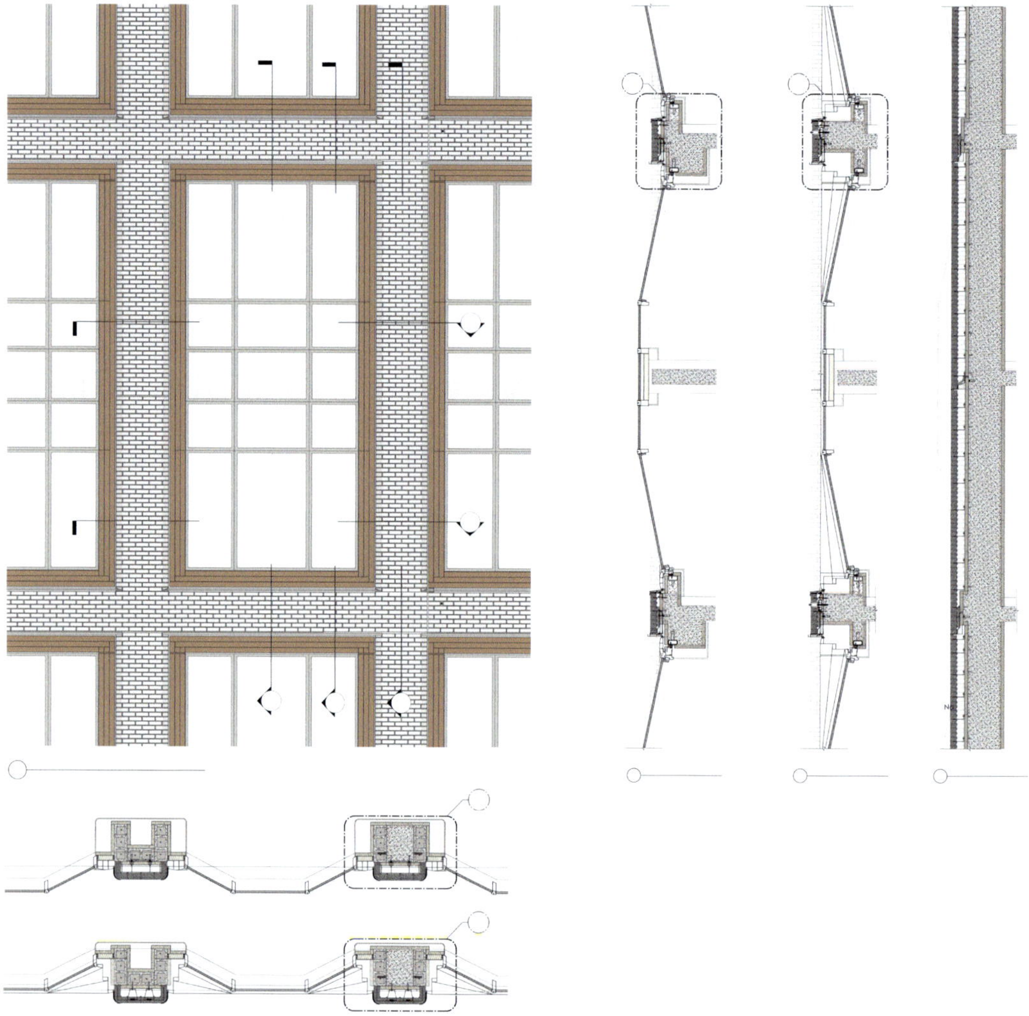

Figure 2.1.33
Bay elevation, sections, and plan details.

could impact only 1-ft facade depth; because anything more would negatively affect the floor areas,
and thus the real estate price, of the apartments.
The use of the lantern-like bay windows came out
of the constraint of designing within the 1 ft while
providing a notable architectural feature. The design
team did investigate other facade materials, such
as precast concrete and terracotta, but the advantage of the bricks over the other materials was that
bricks' small, repeatable dimensions ensured that
they could accommodate the project's different bay
sizes, pier dimensions, floor to floor heights, and
radial building corners.

Figure 2.1.34
Detail photograph of brick on ground floor during construction. Taylor made both the stretcher and custom
bullnose through extrusion.

Figure 2.1.35
Custom bullnose bricks stacked at Taylor Clay Bricks.

Figure 2.1.36
Worker at Taylor Clay hand carving the third brick shape.
This brick is made from a double-height modular brick
and is used at the upper corners of the ground-level
openings.

Figure 2.1.37
Lantern House under construction, view from street level. The special, hand-carved bricks are at the upper
corners of the ground-floor openings and provide a transition from the bullnose jamb to the bullnose soffit.

Figure 2.1.38
Standard stretcher die with bullnose profile attached after the die opening and the modified core hole replaced a standard rectangular core hole. By modifying existing dies, Taylor could keep costs low.

The Lantern House client and developer was Related Companies. Related had worked previously with Belden Bricks, and so Belden was a consultant during initial design discussions. Belden put Heatherwick in contact with Taylor Clay Bricks, a brick manufacturer in North Carolina that focuses its manufacturing on custom and specialty bricks. Lantern House has three different brick shapes. Two of the brick shapes—a standard dimension modular brick and a custom-shaped bullnose brick—are extruded, and the third brick shape is hand carved.[16] The modular bricks are the standard

Figure 2.1.39
Slurry and mineral coatings added in-line after the clay was extruded. This was done for both the stretcher and bullnose bricks.

stretcher in a running bond and form the towers' approximately 33-ft (10-m) diameter radial corners. The bullnose bricks form the soffit headers for the ground-level openings and all the window jambs. Taylor Clay Brick National Territory Manager Anita Edwards said that Taylor made some changes to an existing die to make the special bullnose profiles.[17] To create the bullnose, Taylor attached a steel plate on top of the standard modular brick die and changed one of the core hole shapes (see Figure 2.1.37). This allowed them to extrude the custom profile without the added cost of a full die. Edwards estimates that there were 200,000 custom bullnose bricks used for Lantern House.

In addition to the custom bullnose and hand-carved shapes, Taylor and Heatherwick Studio collaborated to customize the clay mix and brick finish. Parraga-Botero stated that most of the studio's effort in the brick design was to refine the brick blend, "Even though the brick is the basic element, we were very intrigued by the artisanal part of making the brick." In addition to the three different shapes, Heatherwick designed the brick as a blend of three different brick textures, including die finished, wire cut, and blade cut bricks. In the assembly line, just after the bricks exited the extruder, Taylor applied a slurry finish to the brick surface and tossed random ceramic minerals onto the bricks (see Figure 2.1.38). After each type of stretcher brick was fired, they were then blended by hand, because according to Edwards, hand blending leads to a better result in the final building than blending mechanically or by robot.

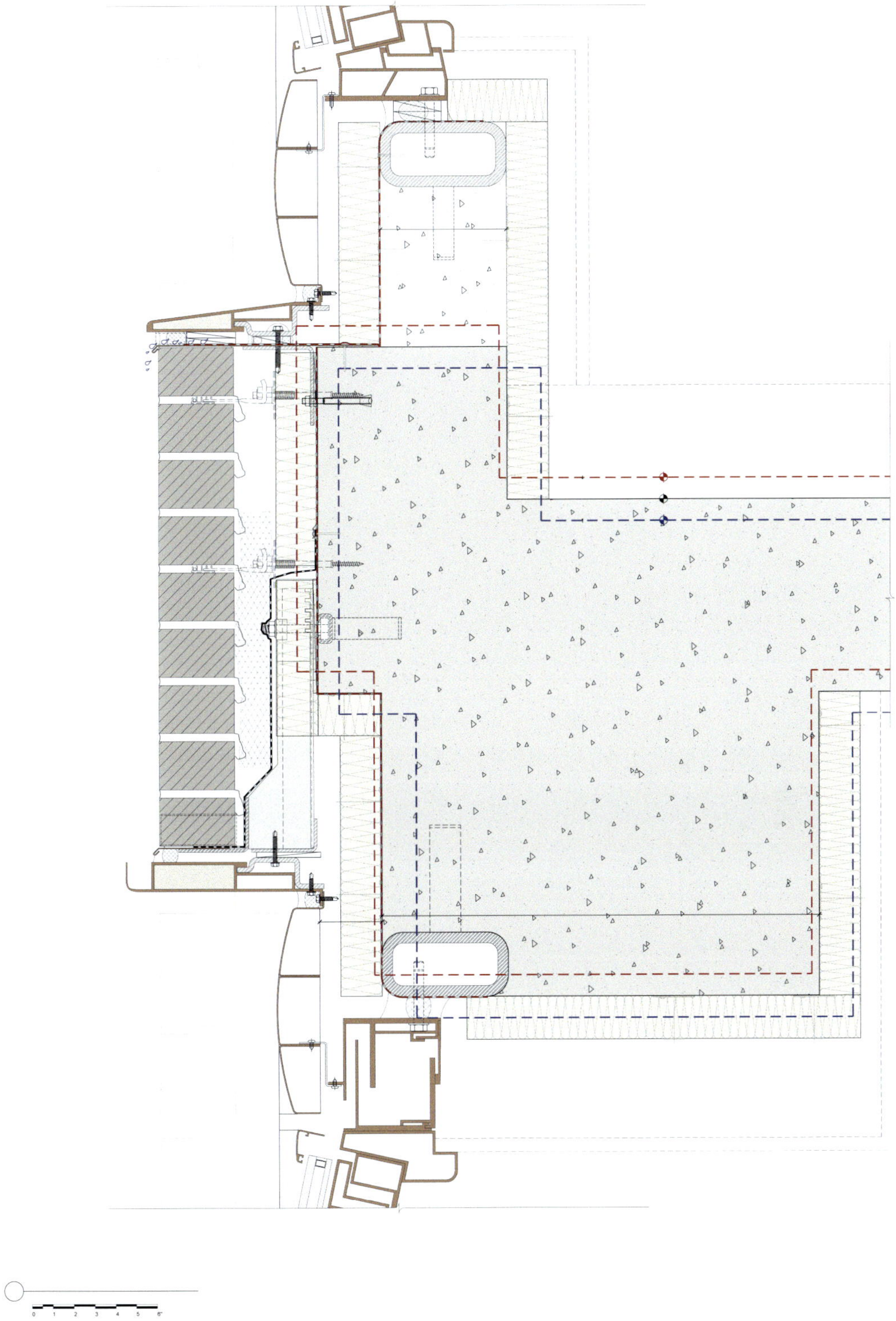

Figure 2.1.40
Detail for windowsill and header at bay center. Red and blue dashed lines note
the tolerances of the site-cast concrete structure.

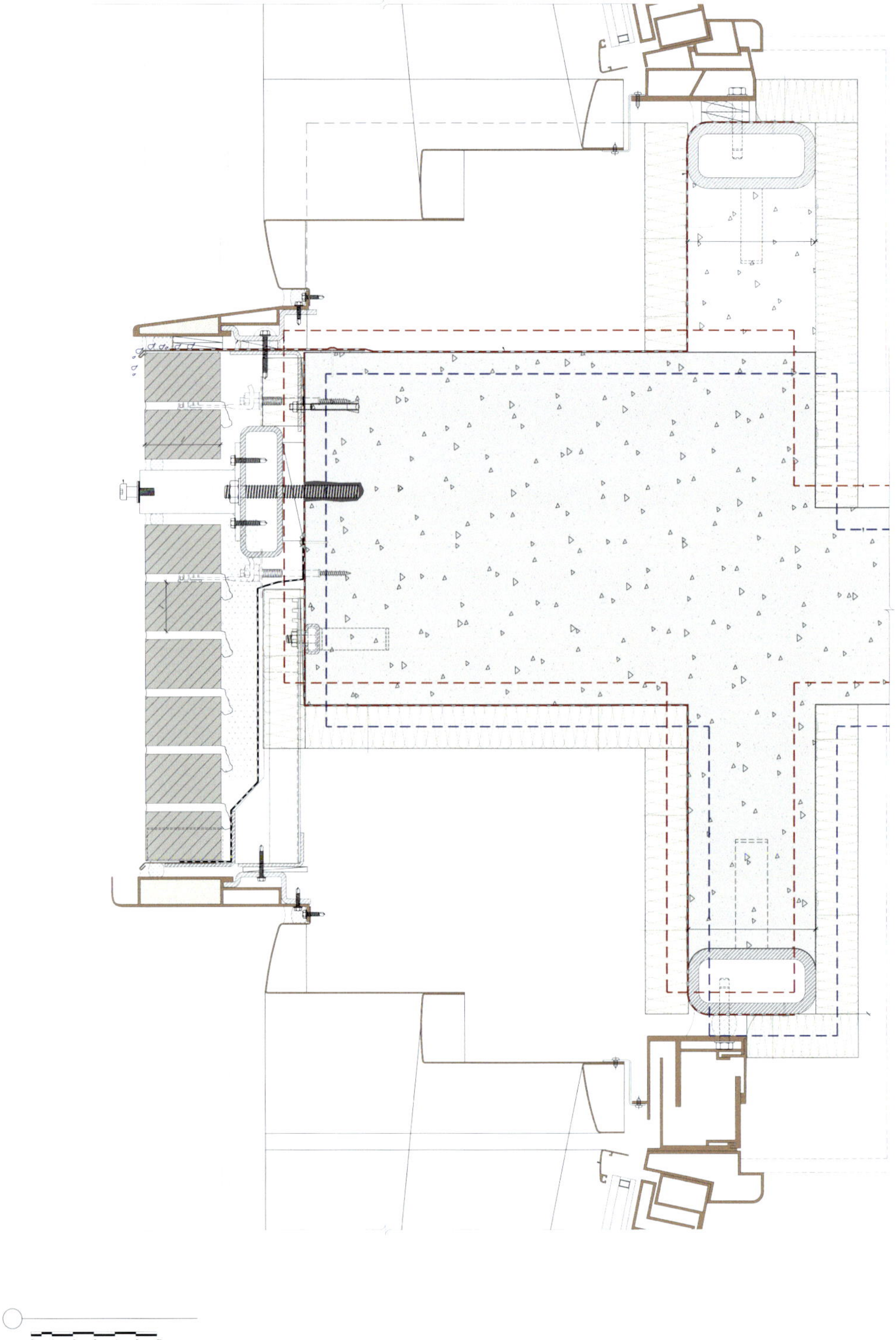

Figure 2.1.41
Detail for windowsill and header at bay edge. Note that at the upper levels, the aluminum window surround resolves the curved geometry of the bay with the orthogonal planes of the brick. Red and blue dashed lines note the tolerances of the site-cast concrete structure.

Notes

1. For thermoplastic extruding, sizers will be required as part of the tooling package. In thermoplastic extrusion, the extrudate leaves the extruder as a liquid. Sizers maintain the dimensions of the cross-sectional shape as the plastic cools enough to maintain its shape.
2. Rothman, Tibby. *Ordinary and Extraordinary Brooks + Scarpa.* Los Angles, Gulf Pacific Press, 2018. 135.
3. Scarpa, Lawrence. *Personal Interview.* 23 May 2022.
4. Griffiths, Alyn. "Make Inserts Public Gardens Between Office Towers at London Wall." *Dezeen*, 29 August 2018. https://www.dezeen.com/2018/08/29/make-office-towers-london-wall-place-bridges-gardens-architecture/. Accessed 24 October 2022.
5. Jones, Chris. *Personal Interview.* 7 November 2022. Lee House, Sir John Burnet Tait's project at 125 London Wall, was demolished in 1990 and replaced by Alban Gate.
6. 1 LWP's neighboring buildings include Fore Street by HKR Architects, which steps down to protect the view corridors, and Moor House by Foster + Partners, which is curved in its massing to hide behind St. Paul's.
7. This includes Alban Gate by Sir Terry Farrell, 5 Aldermanbury Square by Eric Parry, London Museum by Powell, Moya &Partners, and 88 Wood Street by Richard Rogers.
8. The drawing illustrates an earlier profile iteration.
9. Make Architects worked with lighting consultant Studio Fractal to best determine how to light the ground-floor soffits for the public realm.
10. NBK Keramik GmbH. "London Wall Place." https://nbkterracotta.com/project/london-wall-place/. Accessed 24 October 2022.
11. Griffiths.
12. The High Line is an elevated greenway built on disused sections of the raised New York Central Railroad West Side Line that served the city's Meatpacking District.
13. Parraga-Botero, Carlos. *Personal Interview.* 15 December 2022.
14. Heatherwick Studio. *Lantern House.* https://www.heatherwick.com/project/lantern-house/. Accessed 13 December 2022.
15. Roux, Caroline. "The Latest Starchitect Project to Come to NYC's High Line Is a Geometric Wonder." *Departures*, 5 May 2020. https://www.departures.com/lifestyle/architecture/thomas-heatherwick-lantern-house-nyc. Accessed 13 December 2020.
16. Taylor made the hand-carved bricks from a double-height modular brick that are at the upper corners of the ground-level openings. The tooled groin vault-like shape transitioned from the vertical bullnose of the jamb to the horizontal bullnose of the soffit.
17. Edwards, Anita. *Personal Interview.* 15 December 2022.

References

Edwards, Anita. *Personal Interview.* 15 December 2022.

Griffiths, Alyn. "Make Inserts Public Gardens Between Office Towers at London Wall." *Dezeen*, 29 August 2018. https://www.dezeen.com/2018/08/29/make-office-towers-london-wall-place-bridges-gardens-architecture/. Accessed 24 October 2022.

Heatherwick Studio. *Lantern House.* https://www.heatherwick.com/project/lantern-house/. Accessed 13 December 2022.

Jones, Chris. *Personal Interview.* 7 November 2022.

NBK Keramik GmbH. "London Wall Place." https://nbkterracotta.com/project/london-wall-place/. Accessed 24 October 2022.

Parraga-Botero, Carlos. *Personal Interview.* 15 December 2022.

Rothman, Tibby. *Ordinary and Extraordinary Brooks + Scarpa.* Los Angles, Gulf Pacific Press, 2018. 135.

Roux, Caroline. "The Latest Starchitect Project to Come to NYC's High Line Is a Geometric Wonder." *Departures*, 5 May 2020. https://www.departures.com/lifestyle/architecture/thomas-heatherwick-lantern-house-nyc. Accessed 13 December 2020.

Scarpa, Lawrence. *Personal Interview.* 23 May 2022.

CHAPTER

2.2

Pultrusion

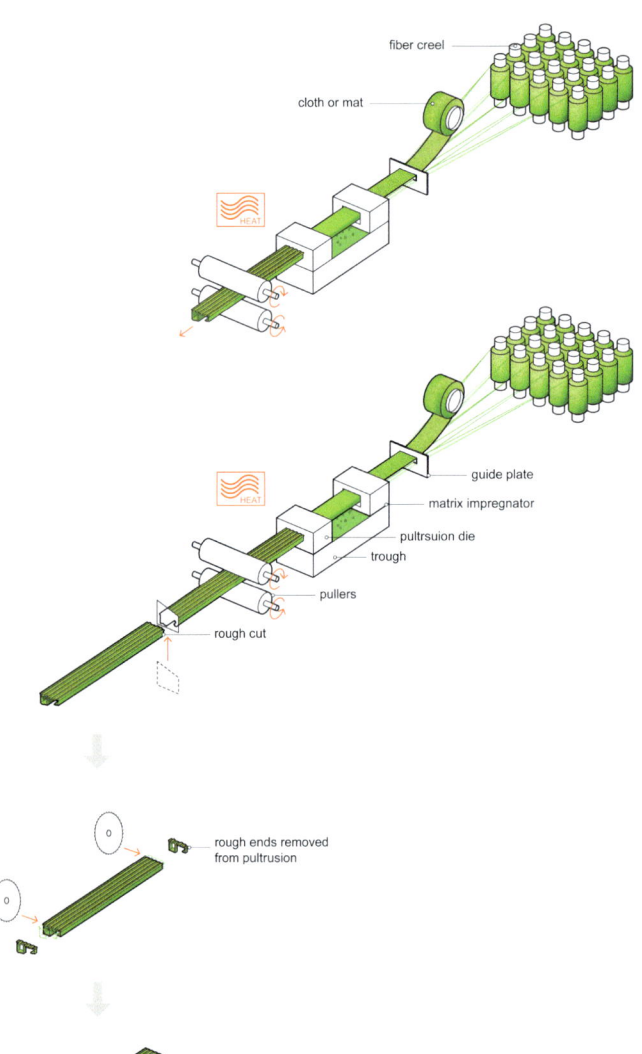

fiber creel

cloth or mat

HEAT

guide plate

matrix impregnator

pultrsuion die

trough

pullers

rough cut

rough ends removed
from pultrusion

workpiece or
final component

Figure 2.2.1
Pultrusion process diagram. Drawing
by author. A version of this diagram
originally appeared in *Manufacturing
Architecture* (Laurence King, 2018).

DOI: 10.4324/9781003299196-11

The word *pultrusion* comes from combining the words *pull* and *extrusion*. Pultrusion is the manufacturing process that pulls a composite medium through a die to create a continuous cross-sectional shape. For architectural applications, pultrusion uses fiber-reinforced plastic (FRP) or fiber-reinforced concrete (FRC) with fibers being made from glass, carbon, hemp, or cellulose, but the process can use other fibers and matrix-like materials such as plaster. Pultrusion produces components that are solid, semi-hollow, or hollow with simple or complex cross sections. Pultrusion is a continuous process as the fibers and matrix can be continually added during manufacturing, theoretically producing components of an infinite length; however, pultrusion lengths are limited by the manufacturing facility or transportation restrictions. Pultruded component lengths can easily be customized, as needed, with little to no increase in costs. Generally, production speeds are slower with pultrusion than extrusion (Chapter 2.1). Set-up times can be long as the fibers need to be organized prior to pulling.

In pultruding, reinforcing fibers and mats or fabrics are organized on a creel and fed through a guide plate or a series of guide plates. The plates keep the fibers in place and pre-shape mats or fabrics before they enter the die. The reinforcement fibers are impregnated with the matrix and pulled through the die to form the cross-sectional shape. The composite medium is pulled slowly through the die so that it cures or is stable enough to maintain its shape when exiting the die. The pultrusion then passes by a set of rollers that places a continuous tension on the composite, essentially pulling the medium throughout the entire manufacturing sequence. The composite is then rough cut after the rollers with final cuts made during post-production. Typically, the die face of the pultruded component is considered the finished face, whereas the cut surfaces are considered unfinished and are hidden.

The tooling required for pultrusion includes a two-part die and one or more guide plates. Dies are made in two parts so that the halves can be closed around the fibers rather than the fibers being fed through the die, thus reducing set-up times. The manufacturer can add shims between the die halves to pultrude variations of a profile without making another die. Dies are typically made of hardened tool steel with optional chromium or ceramic plating. During the pultrusion process, most of the die wear is on the entry side of the die. If the component's profile is symmetrical, then the die can be flipped during production so that the composite enters the other end of the die, essentially doubling the life of the die. The guide plates may be made of plywood, plastic, or steel.

Glass Fiber-Reinforced Plastic (GFRP)

The most common materials used in pultruding are thermoset plastic as the support matrix and glass fibers as the reinforcement. When pultruding thermoset plastic, the resin needs to harden in the die before exiting; therefore, the linear feet pultruded each minute is lower than extruding.

The specific type of thermoplastic used for pultruding will depend on the required performance, manufacturing speed, and profile details. Additives and fillers can be added to the plastic to reduce costs, improve pultruding, increase fire resistance, reduce shrinkage, and stabilize against ultra-violet light degradation.

Sea Scouts Base Galveston in Galveston, Texas
By Shipley Architects

The Sea Scouts Base Galveston (SSBG) is a non-profit marine and maritime education center that was made possible by a private

Figure 2.2.2
Sea Scouts Base Galveston from Offatts Bayou, on Galveston Island. Photo by Arlen Kennedy.

Figure 2.2.3
Photograph of the upper levels of the dormitory and seminar training facility. Photo by Arlen Kennedy.

benefactor—a former captain, who wanted to educate young people about boats and the water. The Sea Scouts is a program of the Boy Scouts of America and similarly promotes leadership, skill-building, and preparedness for young men and women between the ages of 14 and 20. The 70,030-ft^2 (6506 m^2) building includes dormitory facilities with common areas; indoor, outdoor, and floating classrooms; a banquet hall;

and seminar training facility that can accommodate 250 people for multifunctional events. The SSBG is on Galveston Island in the Gulf of Mexico on a 4.5-acre (1.8-hectare) site with 750 ft (228 m) of waterfront exposure. With its location, the building was designed to withstand a Category 4 hurricane, its ground level was designed to allow flooding, and its exterior materials were selected to withstand saltwater corrosion.

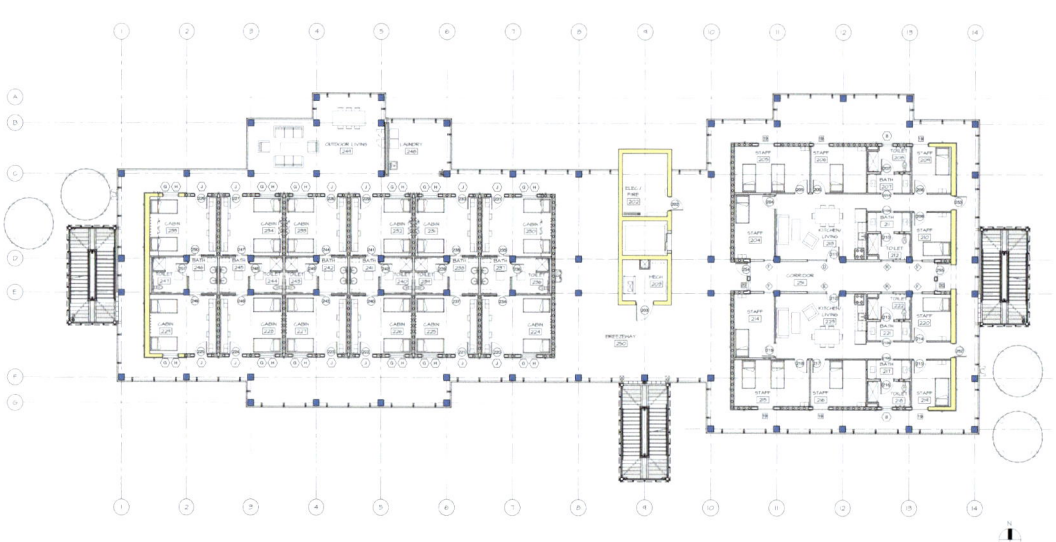

Figure 2.2.4
Plan of dormitory cabins and exterior common spaces. Drawing by Shipley Architects.

Figure 2.2.5
Photograph during construction of exterior dormitory corridors with FRP louvers.
Photo by Shipley Architects.

According to Shipley Architects Principal Dan Shipley, the concept for the SSBG was that the dormitory, which occupies the building's middle three floors, would be similar to housing facilities on a boat.[1] Although the dorm rooms, or cabins as they are called, are conditioned, the hallways that wrap the building's exterior and the dorm's common spaces are open to the weather. Shipley stated that the design team had considered using pipe rails as the hallway guard rails but decided to use evenly spaced louvers for the three dormitory floors both to unify the facade and to keep within the vernacular tradition of the South.

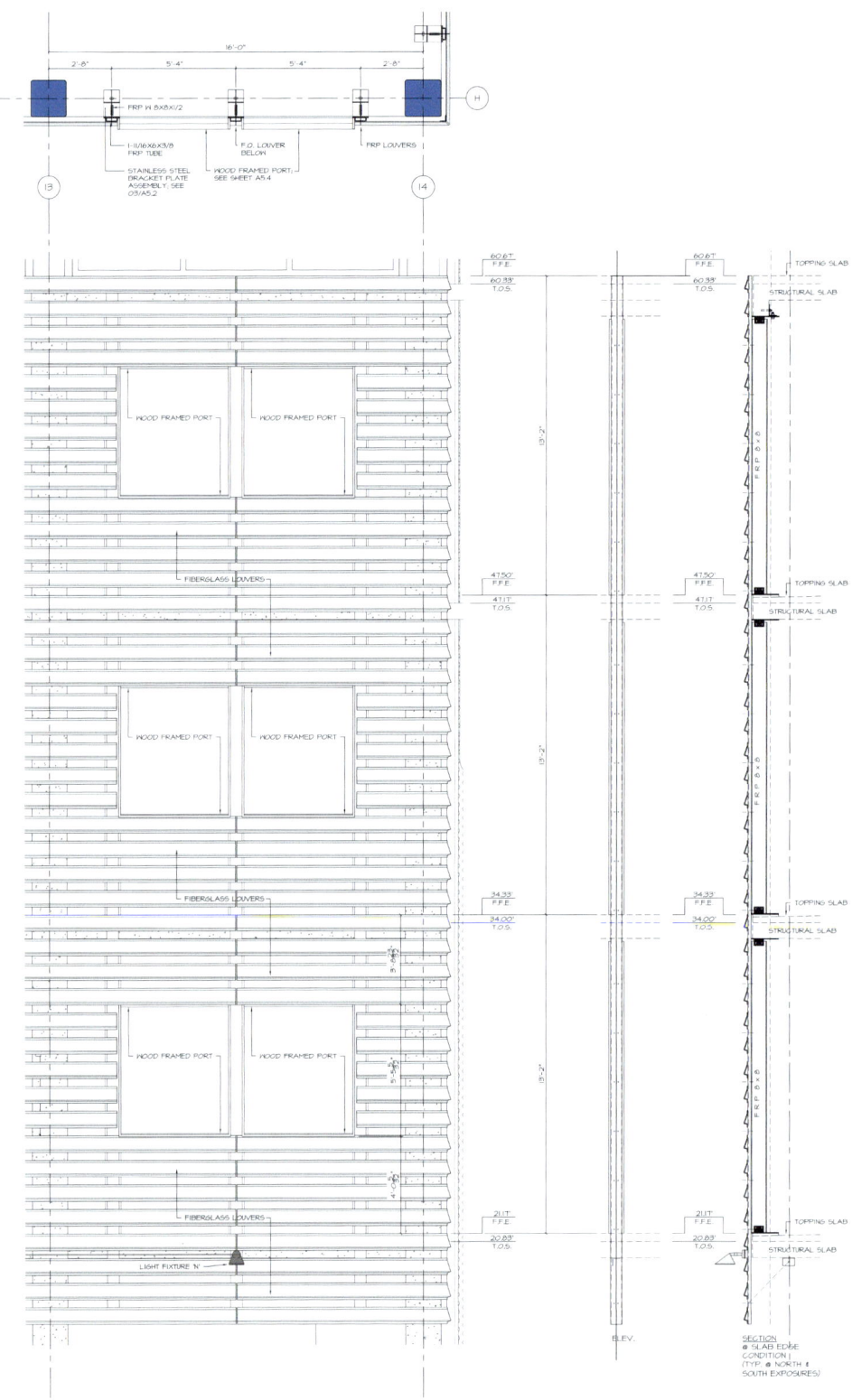

Figure 2.2.6
Detail elevation of louvers at exterior building corner. Custom louvers are face screwed with stainless-steel fasteners to pultruded, standard-cross-section components. Drawing by Shipley Architects.

Figure 2.2.7
Water view of building. Fire stairs clad with hot-dipped-galvanized screen, while dormitory clad in FRP louvers. Photo by Arlen Kennedy.

During early design, the design team considered using a FRP as a screen for the building's fire stairs and had reached out to Strongwell and other FRP manufacturers about available products. However, as the design developed, the team was concerned with fire spread issues typically associated with FRP, and they moved to a hot-dipped galvanized metal screen for the fire stairs. See Figure 2.2.7. During the same time, the team had investigated using stainless steel for the evenly spaced dorm louvers but rejected the metal because of potentially sharp edges and that it could become too hot to touch when exposed to the Galveston sun. In the end, the team decided to use an FRP louver for the dormitory.

Originally, the design team investigated using standard pultruded profiles for the louvers, but the stock profiles did not meet their structural needs. The profiles needed to span over 5 ft (152 cm) and structurally resist the hurricane-force winds. Shipley recalled that Strongwell raised the possibility of designing a custom pultruded profile that could meet the necessary requirements. Since the building had 22,000 ft (6706 m) of louvers in the project, the custom profile was cost effective. Strongwell sent to Shipley Architects prototypes of the profile, and the design team was happy with the shape, size, and particularly the rounded edges that offered a tactile softness that the stainless-steel louvers would not have. In the end, Shipley was so taken with the material that he would have used more on the project.

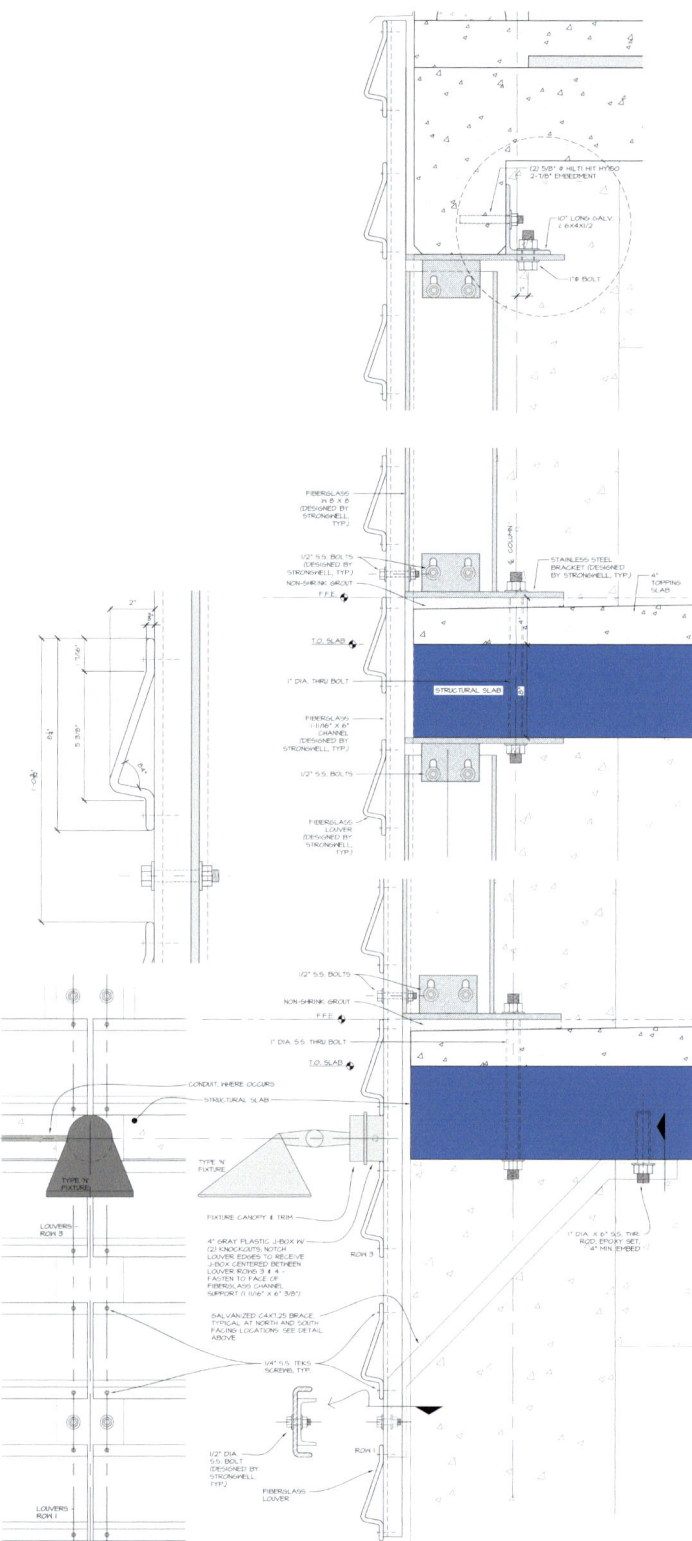

Figure 2.2.8
Architectural detail of louvers. The louvers measure 8¼ in. (210 mm) high and 2 in. (51 mm) deep with their lengths varying as needed. Drawing by Shipley Architects.

Figure 2.2.9
Strongwell sent multiple samples to Shipley
Architects for review and color selection.
Photo by Shipley Architects.

Figure 2.2.10
Detail photograph during construction. Exterior and
interior corners were mitered with a small gap to allow
for thermal movement. Photo by Shipley Architects.

Note

1. Shipley, Dan. *Personal Interview*. 1 June 2022.

Reference

Shipley, Dan. *Personal Interview*. 1 June 2022.

Making Thin or Hollow

3

This part includes those manufacturing processes that form thin-walled components in either open or closed molds. Manufacturing processes include Chapter 3.1, Contact Molding; Chapter 3.2, Vacuum Infusion Process (VIP); Chapter 3.3, Filament Winding; Chapter 3.4, Rotational Molding; Chapter 3.5, Spin Casting; and Chapter 3.6, Slip Casting. Materials in this part are varied and include plastic; fiber-reinforced plastic, gypsum, and concrete; traditional concrete, and clay. The surface of the component in contact with the mold face during forming is significantly better than the surface not in contact with the mold face. This back surface is often hidden from view.

DOI: 10.4324/9781003299196-12

CHAPTER
3.1

Contact Molding

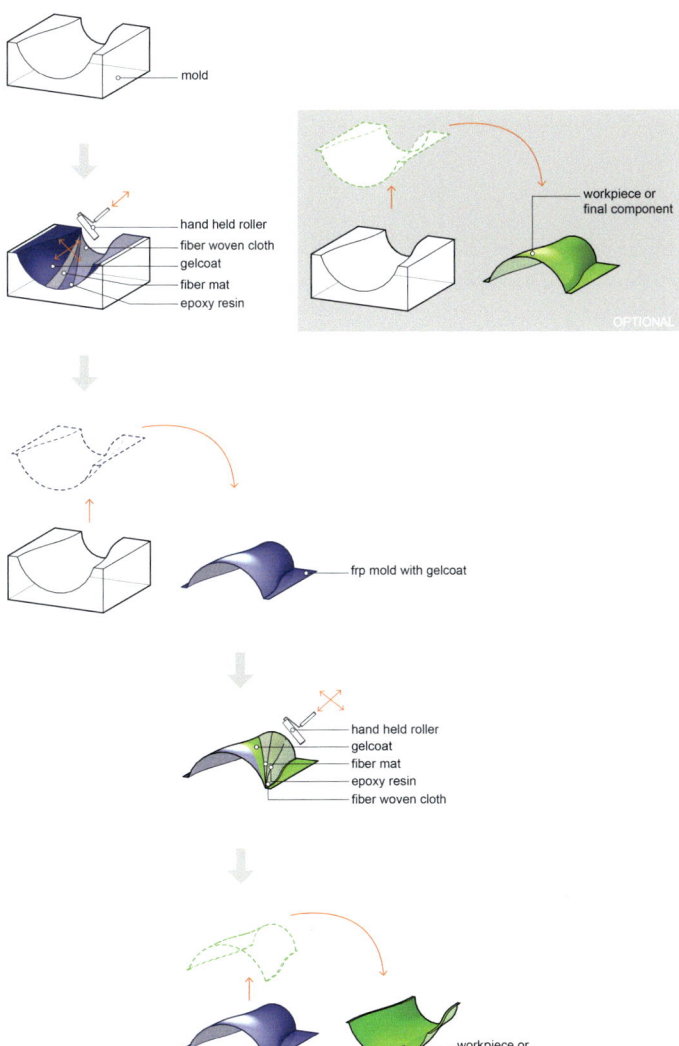

Figure 3.1.1
Contact molding process diagram.
Drawing by author. A version of this
diagram originally appeared in
Manufacturing Architecture (Laurence
King, 2018).

DOI: 10.4324/9781003299196-13

Contact molding is the manufacturing process that lays a relatively thin cross section of a composite medium in contact with a mold. For architectural applications, contact molding uses fiber-reinforced plastic (FRP), fiber-reinforced concrete (FRC), and fiber-reinforced gypsum (FRG) plaster; however, for non-architectural applications, FRP is the dominant material used for contact molding. Contact molding can form large components and is good for complex curves and some surface details. The face of the component in contact with the mold is the component's finished face and the face not in contact with the mold is unfinished and is often hidden from final view. In this process, little pressure is placed on the mold; therefore, tooling can be made of a wide range of inexpensive materials and capital costs are low. Generally, contact molding is a time intensive process and the binder must cure before the components can be demolded. Oftentimes multiple molds are used for parallel productions to reduce the overall production schedule.

In contact molding, a top finish layer (called the gelcoat when using FRP) is applied first to the mold. This first layer is thin, is made of the binder material (e.g. plastic, gypsum plaster, or concrete), and forms the component's smooth final finish face. The finish layer is often allowed to cure before subsequent layers are applied to the mold. Placement of the composite medium onto the mold can be done through hand-layup that alternates layers of binder (e.g. plastic) and fiber matts, hand placement of a fiber-binder mixture (e.g. fibers mixed with gypsum plaster), or a spray-on method that sprays a wet, sticky mixture of a binder and chopped fibers (e.g. FRP, FRG, or FRC) onto a mold using a chopper gun. If using the hand-layup methods, then handheld rollers are used to ensure that the layup is in full contact with the mold, that the binder fully saturates the fibers, and that air bubbles are eliminated. Through this process, multiple layers of composite are added to the layup until the desired thickness is achieved. Any mechanical components needed for attachment or for handling can be incorporated into the composite before curing. After the composite cures, it is demolded and can move into post-production.

Generally, molds for contact molding are typically made of a single piece of material and are open with all surfaces being easily accessible to apply the composite medium. For some specialty components, molds may be made of multiple parts and partially closed with tight interior corners or relatively small openings. These molds will slow down the process as applying the composite will take longer, require more skilled labor to ensure full contact with the mold, and increase demolding and mold setup times.[1] Molds can be made of a wide range of materials, including MDF and wood, foams or plastics, fiberglass, or metal. Manufacturers select the mold material based on cost and how many units the mold needs to produce.

Fiber-Reinforced Concrete (FRC)

Typically, FRC uses Portland cement, fine aggregate, water, and alkaline-resistant glass fibers, although white cement and alternative fibers (e.g. cellulose and steel) can be used. If color consistency is important, then white cement should be used with a color additive instead of Portland cement. FRC is applied to the mold either by trowel or sprayer. Similar to wet-cast concrete (Chapter 4.1), FRC can go through post-production finishing such as acid washing or sandblasting. Molds may be made from foam or plywood for low-volume productions or rubber for mid-volume productions or for fine surface details.

Fiber-Reinforced Gypsum (FRG)

Typically, FRG uses a gypsum-based plaster and glass fibers, although alternative fibers (e.g. animal hair and hemp) can be used. FRG is applied to the mold either by trowel or sprayer. FRG is for interior applications in which the material will not be exposed to moisture or dampness. FRG accepts paint easily and can be painted in the factory or onsite.

Fiber-Reinforced Plastic (FRP)

Typically, FRPs for contact molding use a thermoset plastic (e.g. epoxy or polyester) as the binder and glass (GFRP) or carbon fibers (CFRP), although thermoplastic and alternative fibers (e.g. hemp) can be used. Typically, the initial gelcoat layer is opaque and can be custom matched to any color, with the subsequent composite layers using a

clear plastic. Additives may be added to the gelcoat and subsequent layers to enhance the FRP's performance properties, such as resistance to ultraviolet light degradation and flame spread. The specific thermoset plastic will affect material strength, creep, and its curing, including curing time, the heat generated, warping, and wear on the mold. Molds may be made from foam with a gelcoat for a production run of 10–15 units, from FRP for production runs around 50 units, or from metal for high-volume productions.

The Broad in Los Angles, California, United States
By Diller Scofidio + Renfro

The Broad was built to store and exhibit the growing art collection of Eli and Edythe Broad. Located

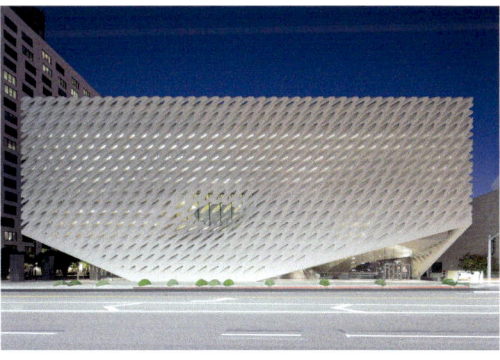

Figure 3.1.2
The Broad, front facade on South Grant Avenue. Photo: Iwan Baan.

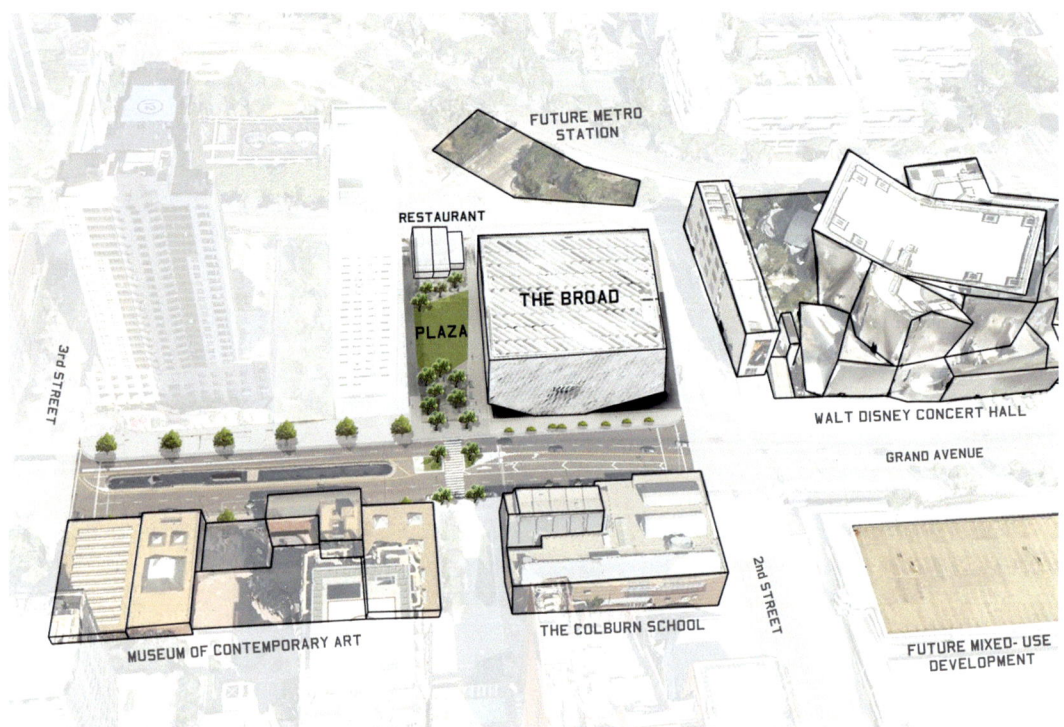

Figure 3.1.3
Site context with notable neighboring buildings. Courtesy of Diller Scofidio + Renfro.

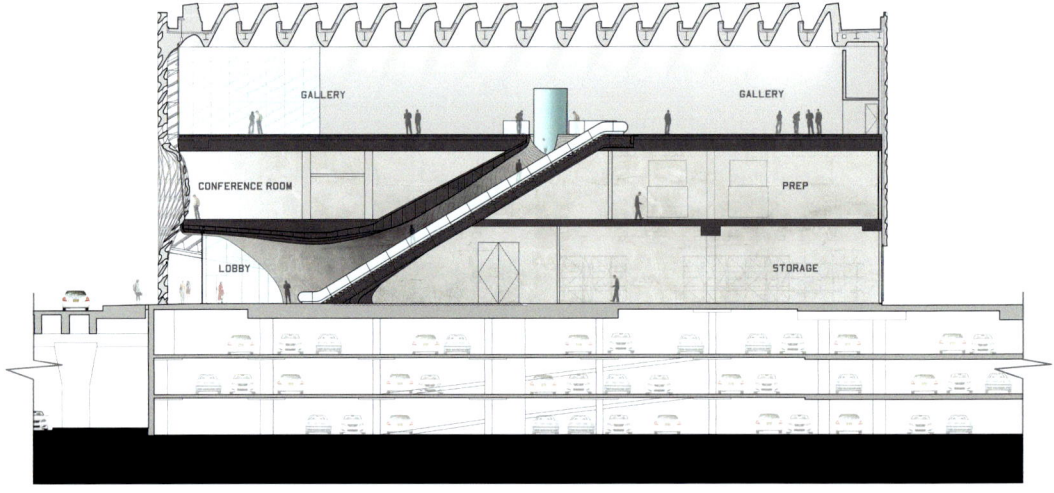

Figure 3.1.4
Transverse section with three levels of below-grade parking. Courtesy of Diller Scofidio + Renfro.

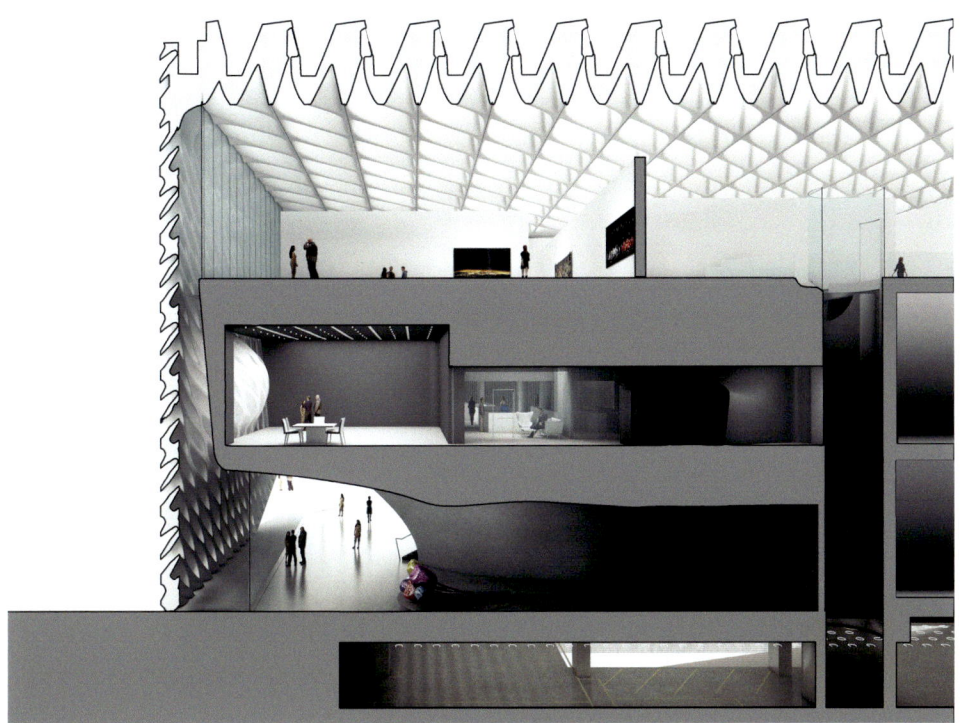

Figure 3.1.5
Building section that shows the relationship between the veil and the vault. Courtesy of Diller Scofidio + Renfro.

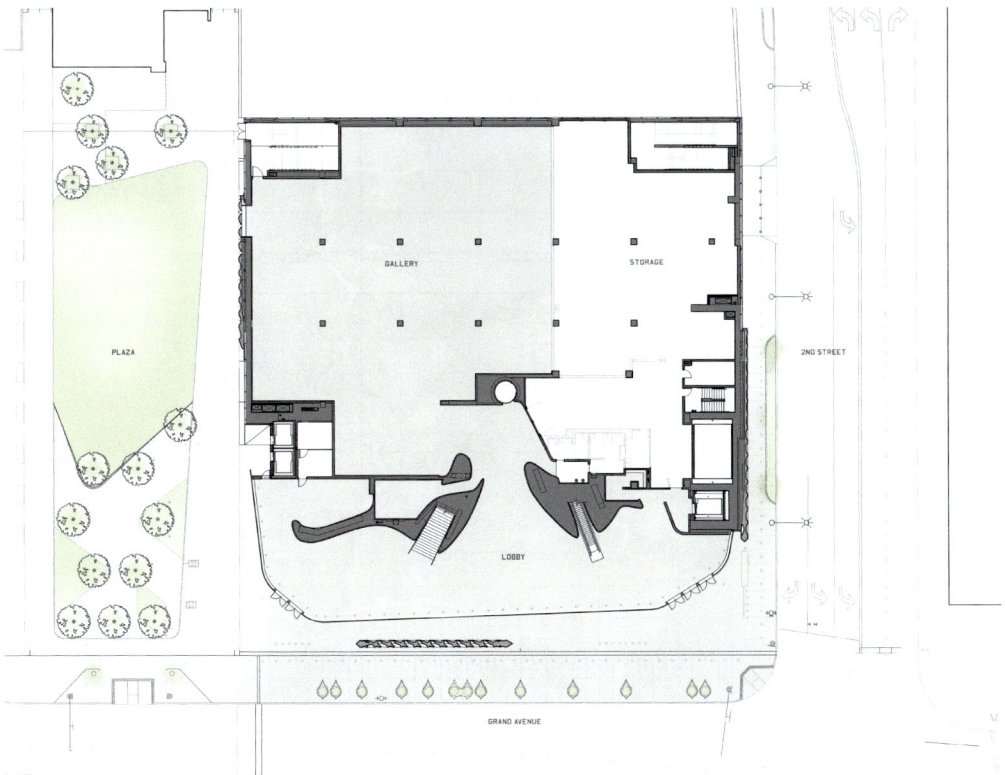

Figure 3.1.6
Entry-level floor plan. Courtesy of Diller Scofidio + Renfro.

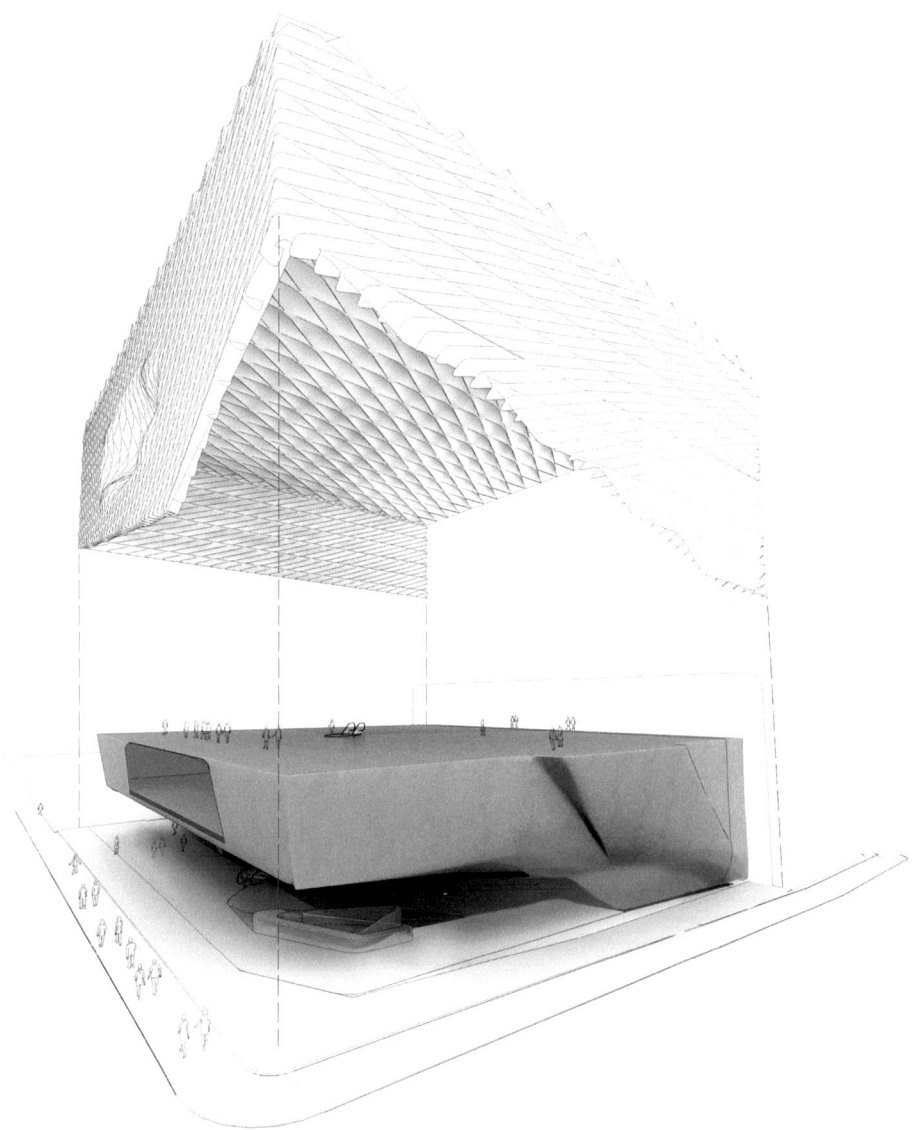

Figure 3.1.7
Building parti diagram with the veil and the vault. Courtesy of Diller Scofidio + Renfro.

on Grand Avenue in Los Angeles, California and part of a development overlay, the Broad is next door to Frank Gehry and Partners' Walt Disney Concert Hall and is a neighbor to Arata Isozaki's Museum of Contemporary Art (see Figure 3.1.3). The Broad includes below-street-level parking, a rear plaza, an inner-block side courtyard, and the main building. The Broad is 120,000 ft^2 (11,148 m^2), three stories tall, and is LEED Gold certified. The building stores the Broad Foundation's art collection, lends parts of its collection to other art institutions, and displays part of the collection and its larger pieces. Diller Scofidio + Renfro (DS+R) developed the building's overall parti into two parts—the veil and the vault. The vault is essentially the program's storage and lending functions, the veil is the outer building wrapper, and the main gallery occupies the space between the two.

Figure 3.1.8
Photograph of how the building corner, parapet, and entry points are resolved with the veil.

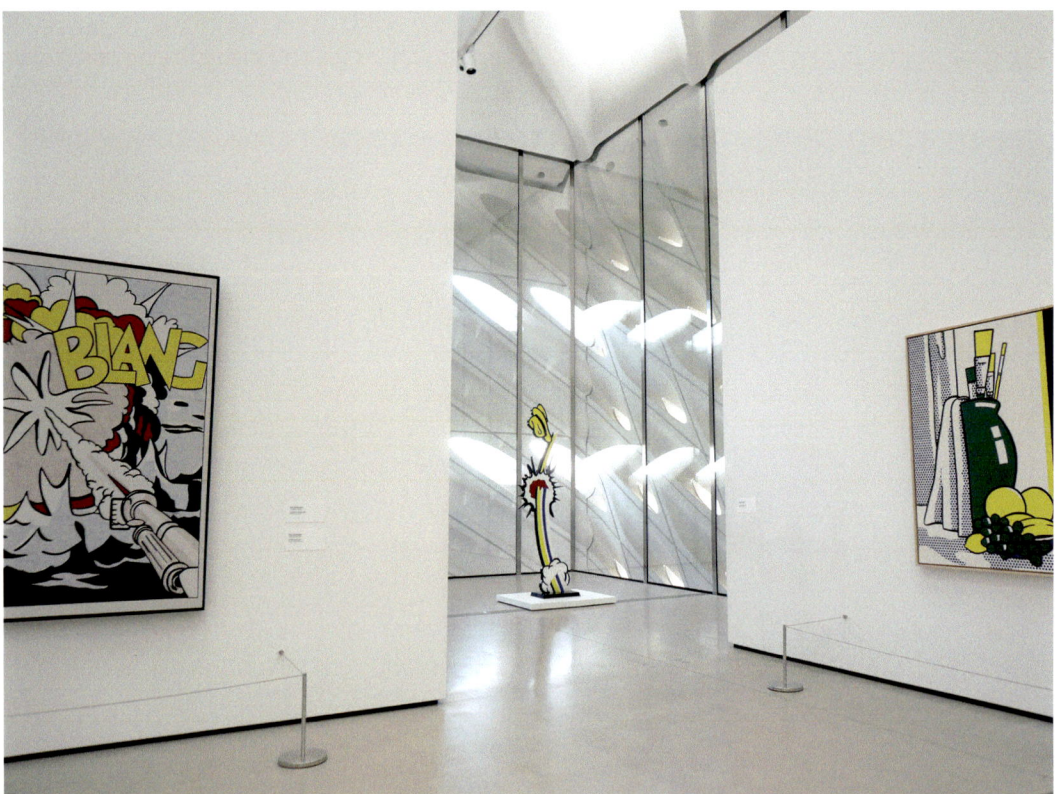

Figure 3.1.9
Interior image, looking toward the veil to Grant Avenue.

The function of the veil is to control light into the building while allowing for glimpses from the museum's interior to the city. According to DS+R Project Director Kevin Rice, the original design of the Broad did not include the oculus, but then the firm decided that without it, the design was too brutalist, too akin to the American Cement Building by Daniel, Mann, Johnson, and Mendenhall, and that without the oculus it could have been built in the 1960s.[2] Some of the units are highly repetitive, while the units around the oculus are unique. By inserting the oculus into the veil, the resulting geometry change would only be possible through contemporary computer-aided digital and computer-aided fabrication (CAD/CAM) technologies as the complex, low, or single-use molds can be fabricated by CNC equipment rather than by hand.

DS+R wanted the veil to be "structurally separate and disconnected from the building as much as possible."[3] Initially, the design team had investigated making the veil of either self-supporting precast concrete elements post-tensioned together or glass fiber–reinforced concrete (GFRC) shells with an internal steel structure. According to Rice, the advantage of the precast concrete was that the veil

Figure 3.1.10
The veil's design was developed to be precast concrete elements, post-tensioned together. This was rejected due to mold costs and was replaced with GFRC. Courtesy of Diller Scofidio + Renfro.

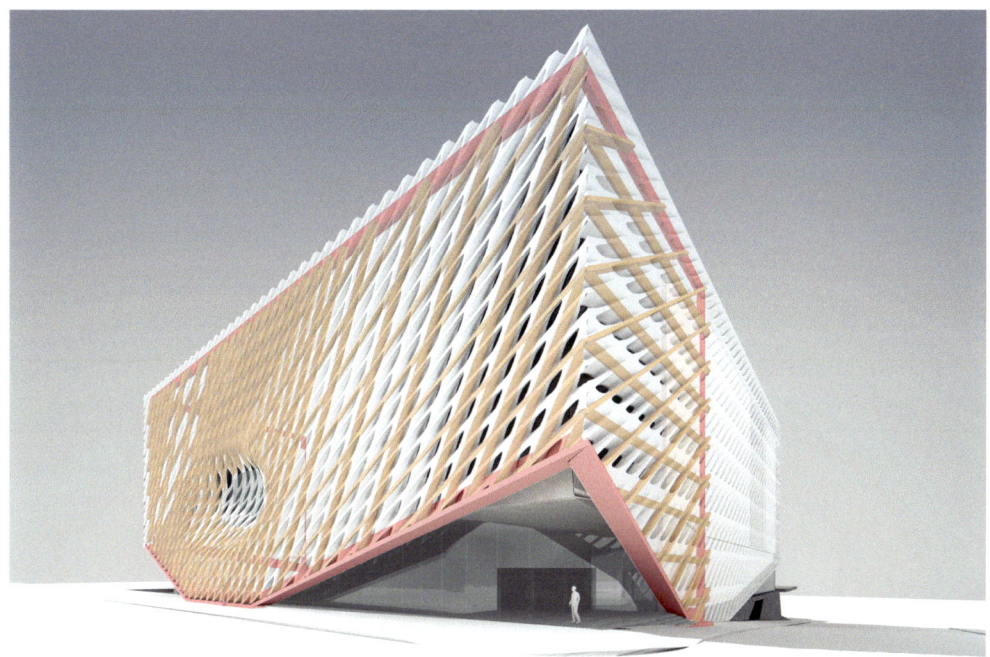

Figure 3.1.11
The veil's design was developed to be precast concrete elements, post-tensioned together. This was rejected due to mold costs and was replaced with GFRC with an internal steel structure. Courtesy of Diller Scofidio + Renfro.

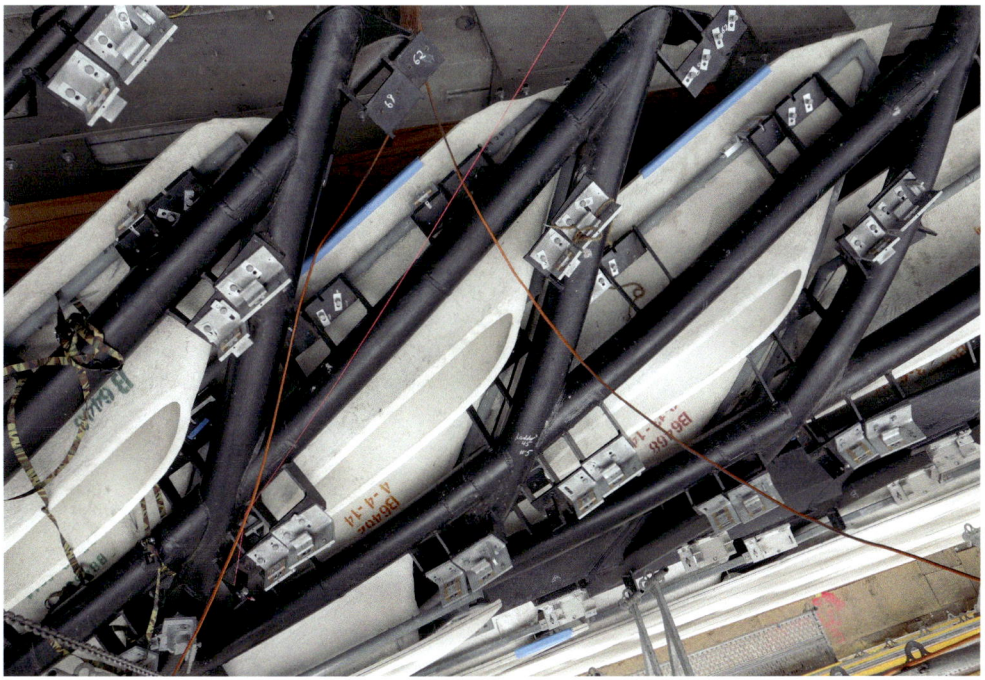

Figure 3.1.12
Interior view of the veil. GFRC panels mechanically mounted on one side of rolled interior steel structure. An outside layer of GFRC panels will be added to the other face, enclosing the steel. Courtesy of Diller Scofidio + Renfro.

Figure 3.1.13
Detail photograph of gasket joint between GFRC panels and their grit-blasted finish.

were designed to have two finished faces—both front and back—and would be cast vertically in a closed mold (see Figure 3.1.10). Generally, GFRC molds are inexpensive and can be made from low-density foam, while the precast molds needed to be fabricated from FRP to withstand the hydrostatic pressure associated with vertical casting. The combination of the more expensive FRP molds and the number of low or single-use molds resulted in the price of the precast molds being ten times more than the original budget. The design team was halfway through construction documents when they realized that they need to change the veil's materials.

The constructed GFRC veil has essentially two 1-in (25 mm) thick GFRC panels, mounted back-to-back, sandwiching an interior, rolled round tube steel structure with a gunnable sealant joint between the front and back panels. Openings between the two GFRC panels align and pass through the interior steel structure to allow for views through the veil. Between the GFRC panels, in the front and back planes of the veil, is a 1¼-in (30 mm) wide, crème-colored rubber gasket that overlaps from one panel to another to keep water out and provide visual continuity between the panels. Rice had originally thought that the joints between the GFRC panels would be sealed with gunnable caulk, but labor costs to caulk all the joints were too high and the seismic resistance of caulk over that of the gaskets would have required wider joints between panels.

DS+R and the project facade consultant, Seele, found Willis Construction. Located in California,

could be thinner than if it was made from GFRC and steel. Evaluating both GFRC and precast, the design team believed that precast concrete was the correct choice, and so their initial selection was to use precast. The precast concrete components

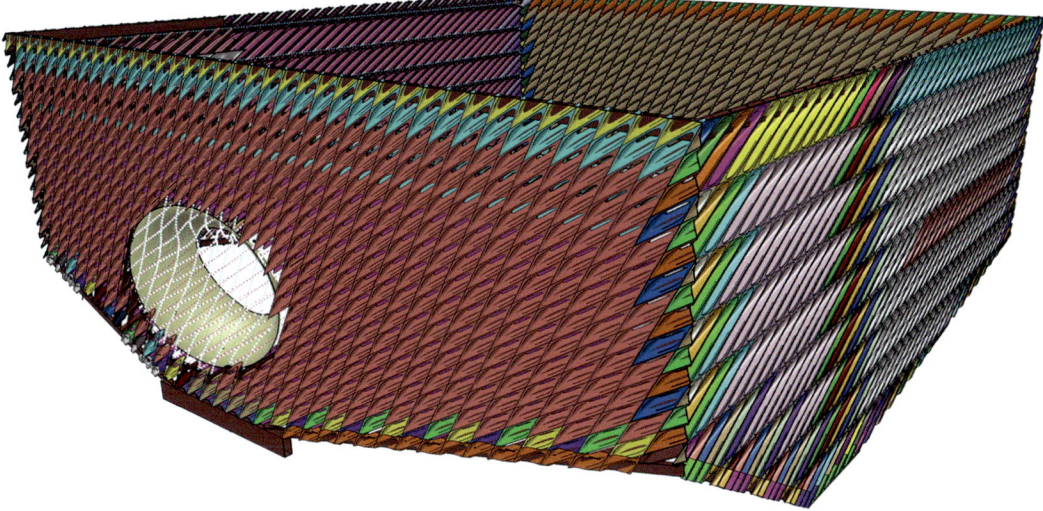

Figure 3.1.14
Digital model with each color representing a different panel type. Note the panels at the oculus are not shown; each of those panels is unique and required unique individual molds. Courtesy of Diller Scofidio + Renfro.

Willis produces loadbearing and non-loadbearing architectural precast (Chapter 4.1), GFRC panels, and structural components. Willis purchased a CNC mill for this project so that they could fabricate the GFRC molds themselves. The project was a close collaboration among the architect, facade consultant, and GFRC manufacturer. DS+R worked on the digital design model to create the design of the veil surface, then Seele would give the GFRC panels a thickness and work out the inner steel structure. Both DS+R and Seele would negotiate the veil hole sizes and locations, and then Willis would check the digital model for manufacturability.

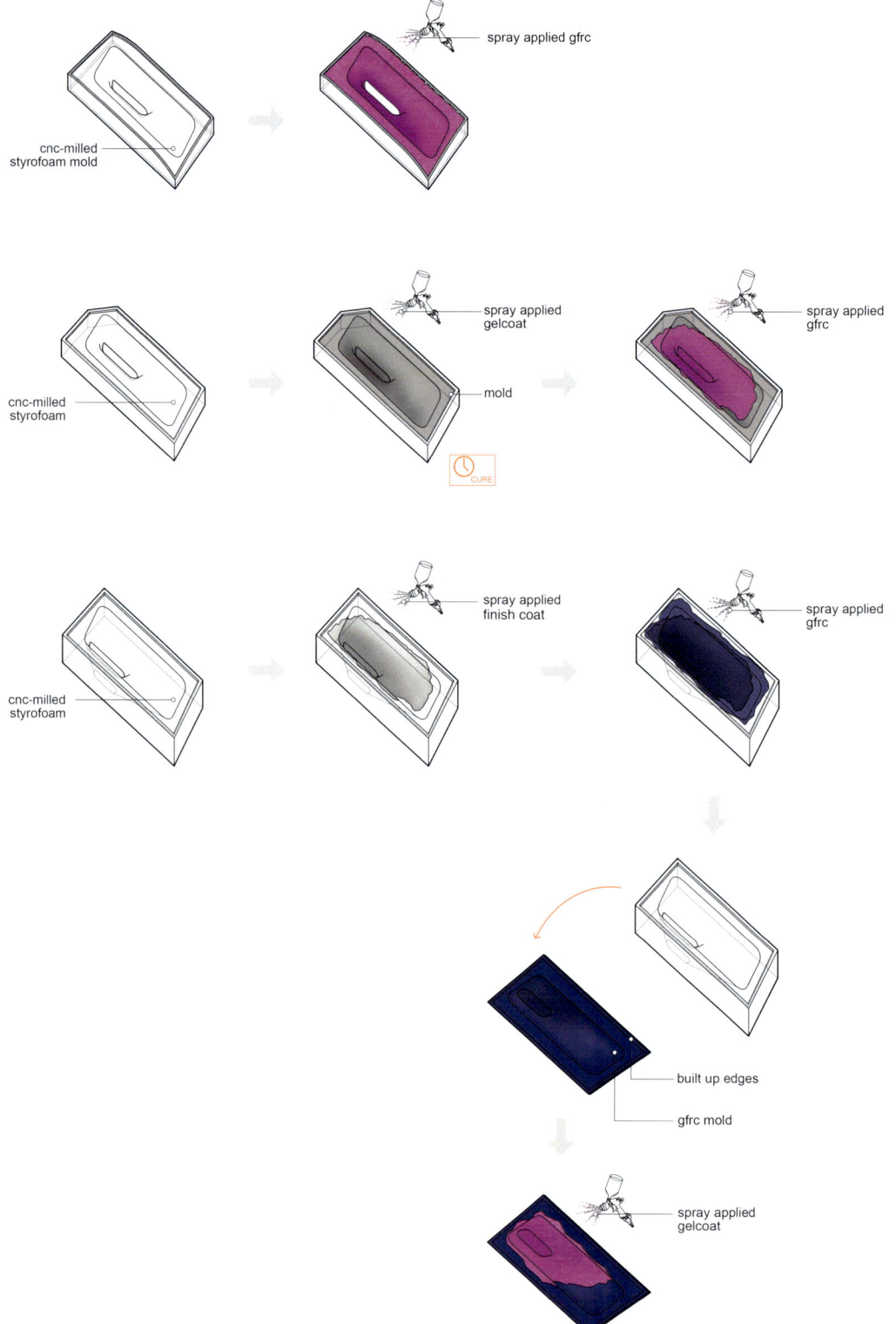

Figure 3.1.15
Diagram of model fabrication. For less than three pulls, Willis would spray GFRC directly onto the foam. For three to ten pulls, Willis would coat the foam with a protective gelcoat and spray the GFRC on the gelcoat surface. For more than ten pulls, Willis would make a GFRC mold from a foam pattern. To shorten the overall production time, Willis could make multiple GFRC molds from a pattern.

Figure 3.1.16
Collection of CNC-milled foam molds. Courtesy of Diller Scofidio + Renfro.

Figure 3.1.17
A GFRC panel being demolded from a GFRC mold. Courtesy of Diller Scofidio + Renfro.

Figure 3.1.18
A GFRC panels mounted to metal frame, stored before shipping. Courtesy of Diller Scofidio + Renfro.

There are 380 different molds used to produce the 2500 GFRC panels with 150 unique molds to form the panels at the oculus and the building corners. If Willis used a mold for less than three panels, then the GFRC would be sprayed directly onto the CNC-milled foam. If Willis used a mold for less than ten panels but more than three, then the foam mold would be coated in a durable coating—such as a gel-coat resin. If Willis used a mold for more than ten panels, then they would make mold out of GFRC. To speed up production for some of the highly repetitive panels, Willis made multiple GFRC molds for parallel productions. After demolding, the panels were grit blasted and grooves cut on the panel sides to receive the friction-fit rubber gasket strips. Then Willis attached each GFRC panel to a metal frame that was then bolted to the veil's structural steel frame.

In addition to the Broad's exterior GFRC panels, the building also incorporates custom interior ceiling coffers made of glass fiber–reinforced gypsum (GFRG) plaster for the third-floor gallery skylights. The coffers are deep, hiding 7-ft (2.133-m) deep girders that span the entire third floor, making the gallery a column-free space. During conceptual design of this project, DS+R was already starting to coordinate with the project's construction manager, MATT Construction. It was during this discussion that the team decided to make the coffers out of GFRG as

Figure 3.1.20
Image during construction of GFRG coffers being assembled and patched at floor level before being installed. Courtesy of MATT Construction.

Figure 3.1.19
Interior image of upper gallery with custom GFRG ceiling coffers to control light from skylights. Photo: Iwan Baan.

it keeps the coffers under the purview of the drywall subcontractor Anning-Johnson. One benefit of using GFRG was the ease of on-site modification, because if any holes needed to be located for utilities (e.g. sprinkler heads) or patching required, then they could be completed by the drywall subcontractor. Moonlight Molds made the molds and manufactured the 11 different GFRG coffer shapes. The coffers were delivered onsite in pieces, they were assembled, the seams patched by the drywaller, and lifted into place.

Tobin Center for the Performing Arts in San Antonio, Texas, United States of America
By LMN Architects

The Tobin Center for the Performing Arts includes the renovation and partial reuse of San Antonio's historic Municipal Auditorium, a 1920's building in the Spanish Colonial style. Approximately, 70% of the original auditorium facade was kept,

Figure 3.1.21
Tobin Center for the Performing Arts is a renovation and partial reuse of San Antonio's historic Municipal Auditorium. The Tobin Center maintains the original building entrance. The new addition can be seen above the 1920s, Spanish Colonial style facade. © Mark Menjivar/image courtesy of LMN Architects.

Figure 3.1.22
The addition connects the building to San Antonio's River Walk with an outdoor plaza that
hosts movies and performances. © Brandon Watts/image courtesy of LMN Architects.

along with the original entrance and entry foyer. The project includes 157,000 ft² (14,586 m²) of new construction with 26,000 ft² (2415 m²) of renovated construction with a new 1762-seat performance hall, a 231-seat studio theater, and an outdoor space that seats 600, connects the building to the city's River Walk, and hosts movies and

performances. The large performance hall is the first in the United States to feature the Gala System in the orchestra area; like the reconfigurable seating for the Dee and Charles Wyly Theater by OMA, the Tobin Center's orchestra area can be remotely reconfigured from a flat open area to tiered fixed seating.

Figure 3.1.23
The lobby sits in between the new auditorium and the original building entrance. On the right are the custom, contact-molded, glass fiber–reinforced gypsum (GFRG) panels cladding the multi-story auditorium wall. © Ed LaCasse/image courtesy of LMN Architects.

Figure 3.1.24
Ground level of the feature wall with openings to an upper level walkway, beyond.
© Mark Menjivar/image courtesy of LMN Architects.

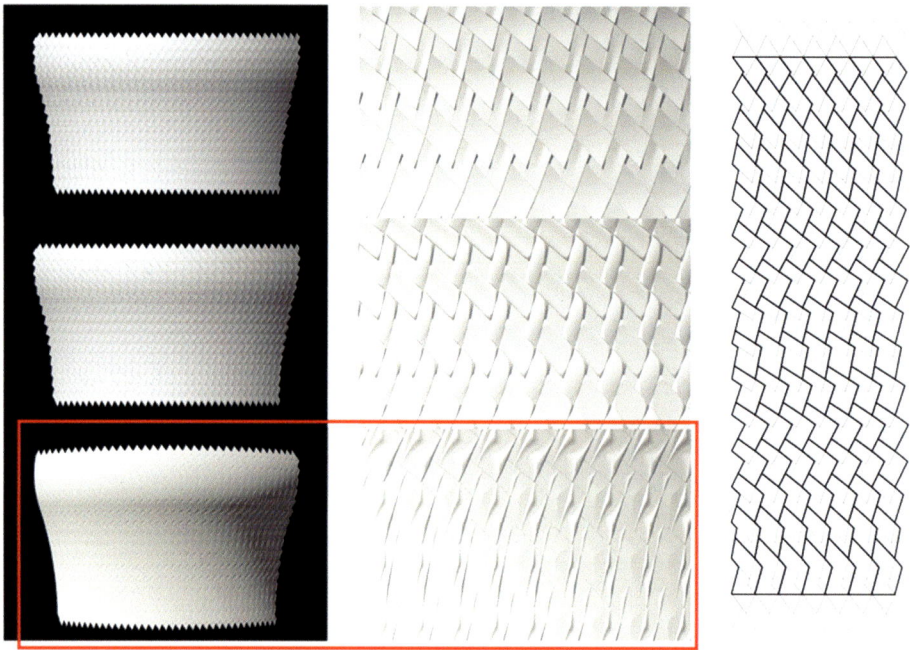

Figure 3.1.25
LMN design studies for the feature wall. Image courtesy of LMN Architects.

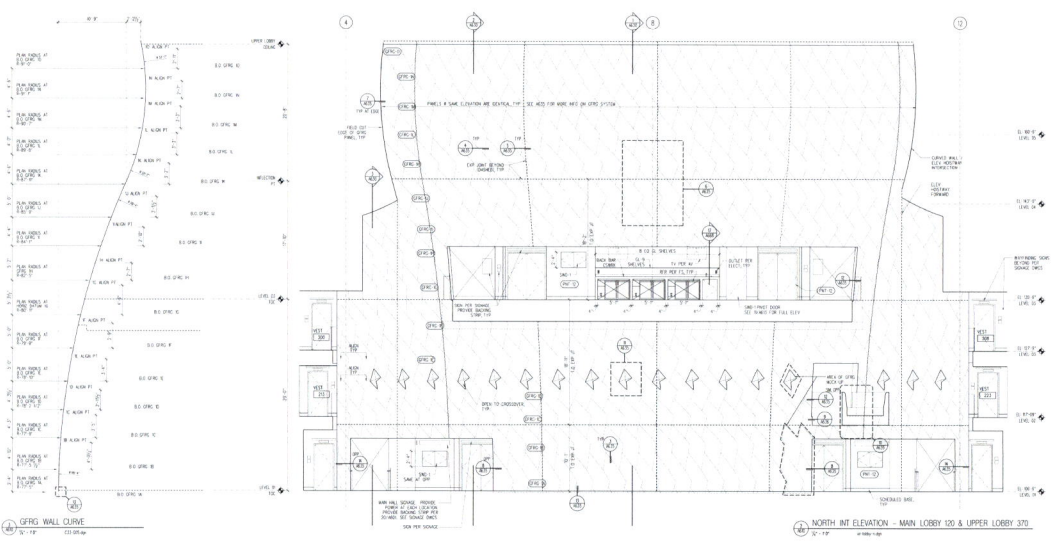

Figure 3.1.26
Elevation and section of feature wall with panel callouts. Underlaying expansion
joints are indicated with dashed lines. Image courtesy of LMN Architects.

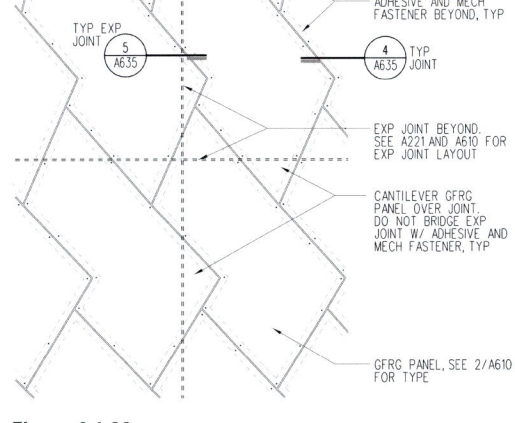

Figure 3.1.27
Photograph of GFRG mold at Stromberg. The mold is made of coated, CNC-milled foam with built-up plywood edges. Image courtesy of LMN Architects.

Figure 3.1.29
Elevation detail of panels near expansion joint. Image courtesy of LMN Architects.

According to LMN Architects Partner Mark Reddington, the building's main lobby is "a kind of in between space" where one side is the renovated historic facade and entry foyer and the other side is the newly built performance hall.[4] In the lobby, the performance hall forms a four-story high feature wall, clad with custom GFRG panels that range in size from 3 ft 4 in to 5 ft tall (101–152 cm). The panels are L-shaped and interlocked with one another. They have a wave-like surface texture that is reminiscent of the San Antonio River that bounds the site to the north, and they create patterns that reference traditional Spanish tile work of the area.[5] The feature wall matches the shape of the performance hall volume and is a gentle complex curve that is C-shaped in plan and S-shaped in section. There are 18 different panel designs—15 are arranged in 15 rows of matching panels, and 3 are specialty-type panels found at the underside of the third-floor balcony level, the openings for the performance hall entrances, or the portal-like openings into a second-floor hallway. The back of the GFRG panels is curved

Figure 3.1.28
Photographs of GFRG panels at Stromberg (a) front of panels, after being stripped from mold (b) back of panels, with the finished side down. Image courtesy of LMN Architects.

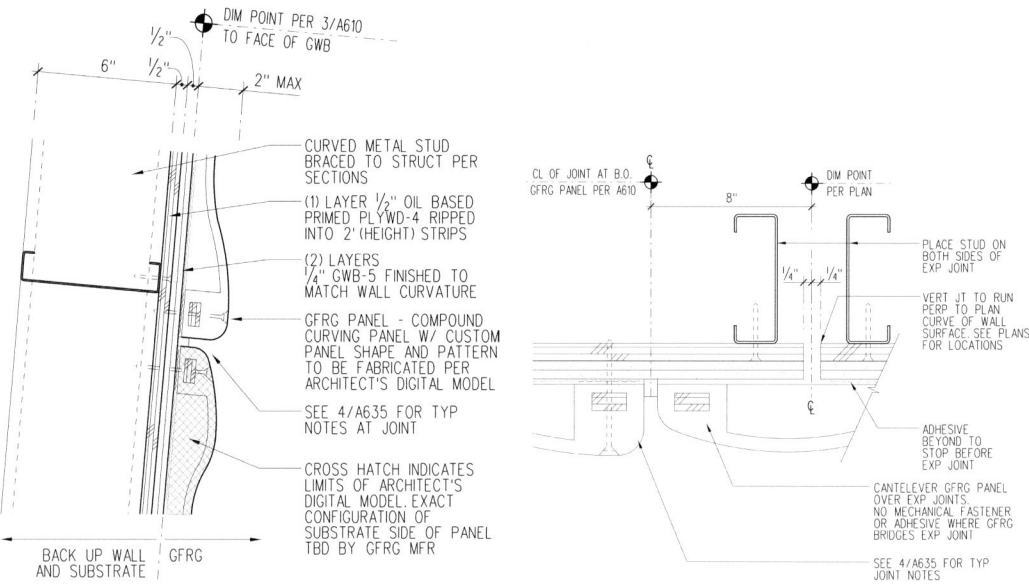

Figure 3.1.30

Construction details of GFRG panels (a) standard joint detail (b) joint detail near expansion joint. Note plywood reinforced edges and mechanical attachments with patching. Image courtesy of LMN Architects.

to meet the shape of the performance hall wall, with the specific curve being dependent on the panels' location.

Stromberg Architectural Products manufactured the GFRG panels by spraying the wet GFRG mixture onto coated foam molds with built-up plywood edges. After applying the first layer of GFRG, Stromberg laid in a plywood frame on the wet layup and then sprayed GFRG over the plywood to embed it fully within the panel. For installation, drywall contractors screwed through the finished face of the GFRG and their embedded plywood frames, to mechanically fasten the panels to the performance hall wall's layers of gypsum board and plywood.[6] They sunk the screw heads into the panel's finished surface, and the holes were patched and sanded as needed. To allow for movement, the performance hall wall has several expansion joints. When the GFRG panels are laid over an expansion joint, the panels are fastened to only one side and essentially cantilever to the other side of the joint, allowing the panel to float over the joint. On site, installers caulked the panel joints, further reducing the likelihood of cracks appearing in the panels' surface. After installation, the feature wall was primed and painted.

This project was delivered as a construction manager at risk with the contractor, Linebeck, providing a guaranteed maximum price through the design process. LMN Principal Rich Johnson did not recall having multiple bidders for the GFRG panels and instead only recalled interacting with Stromberg.[7] LMN was already familiar with GFRG and how it can be formed; therefore, LMN did not coordinate with Stromberg until the construction document (CD) phase. During CDs, LMN did share a three-dimensional digital model (modeled in Rhinoceros 3D) with Stromberg for coordination and technical input, and Stromberg's shop drawings closely matched LMN's provided model.[8]

Stromberg built a mockup of the feature wall at their manufacturing facility. The mockup replicated approximately five rows of different panels and the design team used it to evaluate the lighting effects on the wall surface, the wall curvature, and the portal-like openings on the second floor. Originally, the construction documents called for closed-cell foam tape to be used between the panels; however, the mockup demonstrated that detail was unacceptable as the foam tape compressed, not turning panel corners with a consistent thickness, and leaving some gaps in the joints. In response to

Figure 3.1.31
Photograph of mockup with raking lighting at Stromberg. After evaluating the mockup, the design team changed the joint between the panels from foam tape to gunnable caulk. Image courtesy of LMN Architects.

Figure 3.1.32
With the mockup, the design team realized that the foam tape would compress when turning corners, leaving small gaps. Image courtesy of LMN Architects.

the mockup, the design team changed the panel joints from foam tape to a gunnable caulk. There was a bond breaker at the back of the caulked joint so that the caulk would not bond to the supporting wall surface, allowing for movement, while the joint between panels was fully sealed. In the end, the sealed joints give the wall a smooth finish in which its wave-like surface dominates the eye more than the joints between the panels.

In general, mockups are a key tool for LMN's practice. According to Reddington, ideally LMN would build mockups before bidding every project, and their office has a fabrication shop that they use to build their own mockups. Johnson also values mockups, "We are building a prototype every time. If you are doing something unusual, you can look at it on the computer screen and draw all the details, but you will be surprised out in the field." Reddington noted that in recent years physical mockups have gained importance in their practice's design processes. Digital design tools, direct-to-fabrication processes, and mockups work together for custom building components and having those components realized as envisioned. In the end, Reddington noted, "Mockups always pay for themselves."

Figure 3.1.33
Detail photograph of the feature wall. © Mark Menjivar/image courtesy of LMN Architects.

Figure 3.1.34
Newtown School junior and senior buildings with the ADS-designed facade.

The Newtown School in Kolkata, India
By Abin Design Studio (ADS)

While Newtown School was under construction, the client, Savitri Education Foundation, commissioned Abin Design Studio (ADS) to design new building facades and interiors to give the project a greater architectural identity than its original design. The previous design team had massed the 160,000-ft^2 (14,865 m^2) program into two, six-story high volumes—one for the junior school and the other for the senior school. Each building is organized around a central, covered courtyard, with recreational spaces between the buildings. According to ADS Founder Abin Chaudhuri, ADS designed the new facades as the school's signage, developing panels that included Roman letters, numbers, and symbols—such as ampersands, pi, and omega.[9]

The new screens were to unify the buildings, give the school a distinct character separate from its generic surroundings,[10] and provide some protection from the harsh Kolkata sun.

Figure 3.1.35
Image of the Junior School covered inner courtyard.

Figure 3.1.36
Image of the recreation space between the two buildings.

Figure 3.1.37
Photograph detail of panels.

Initially, ADS investigated using aluminum for the screens, but that was dismissed as being too costly. Since the screen was to be supported by an already-constructed facade and structure, it needed to be lightweight, stiff, and not transmit a heavy wind load. Another client of ADS introduced the design team to FRP and to a manufacturer that makes FRP panels for car bodies. That manufacturer made a sample panel approximately 10 ft × 10 ft (3 m × 3 m) and the design team discovered that the panel would only weigh 121 lb (55 kg). After the initial prototype, ADS approached other FRP manufacturers to potentially produce similar panels for the building but found that those manufacturers required greater draft angles, a larger radius on the corners, and were more expensive than the original manufacturer. In the end, the team returned to the original manufacturer as they were able to produce the panels to the quality that ADS wanted.

Figure 3.1.38
Photograph of the panels being installed on painted hollow square sections.

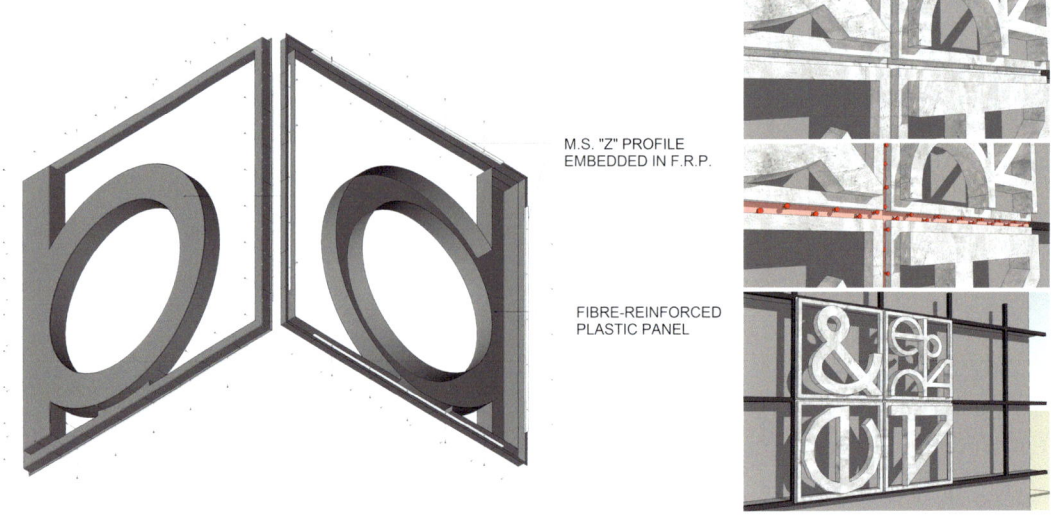

M.S. "Z" PROFILE
EMBEDDED IN F.R.P.

FIBRE-REINFORCED
PLASTIC PANEL

Figure 3.1.39
ADS original design drawings. Note the U-shaped cross section of the panel and the embedded Z clips.

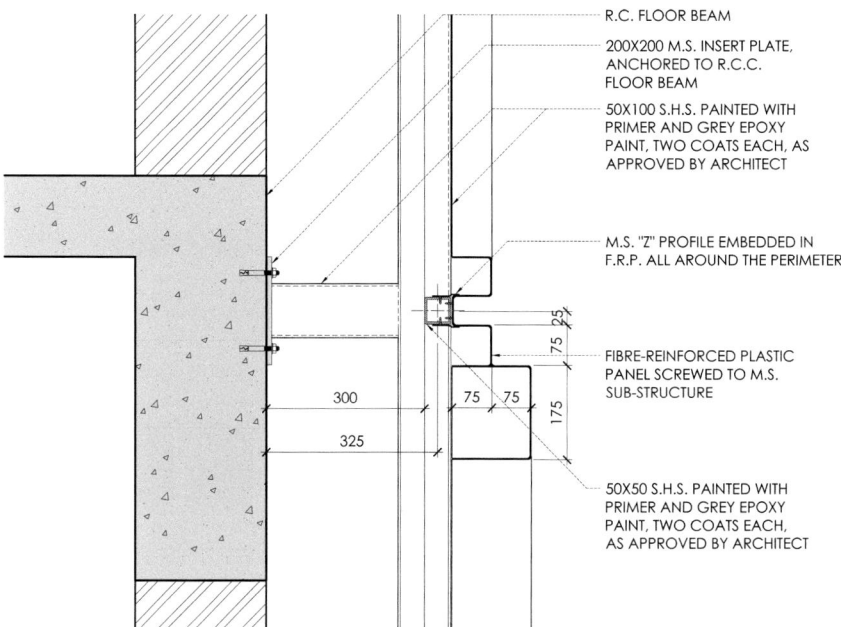

R.C. FLOOR BEAM

200X200 M.S. INSERT PLATE,
ANCHORED TO R.C.C.
FLOOR BEAM

50X100 S.H.S. PAINTED WITH
PRIMER AND GREY EPOXY
PAINT, TWO COATS EACH, AS
APPROVED BY ARCHITECT

M.S. "Z" PROFILE EMBEDDED IN
F.R.P. ALL AROUND THE PERIMETER

FIBRE-REINFORCED PLASTIC
PANEL SCREWED TO M.S.
SUB-STRUCTURE

50X50 S.H.S. PAINTED WITH
PRIMER AND GREY EPOXY
PAINT, TWO COATS EACH,
AS APPROVED BY ARCHITECT

300

325

75   75

25

75

175

Figure 3.1.40
Planned section detail of the screen. The 50-mm gap between the panels was eliminated
by the manufacturer surveying the facades prior to fabricating the molds.

There was little direct collaboration between ADS and the FRP manufacturer beyond coordinating through shop drawings; Chaudhuri did not visit the facility in which the panels were made, and all coordination was done through the client. The FRP manufacturer did visit the site to measure the already-built facades to determine the sizes of the panels. It was these measurements that the manufacturer used to fabricate the open molds. Chaudhuri said that the FRP panels were so precise in their tolerances that the joints remained flawless and there were no uneven gaps between panels. Originally, ADS had designed the screen to allow for 2-in (50-mm) gaps between the FRP panels, but in the end, they were manufactured and erected with "paper-thin joints" between each panel[11] (see Figure 3.1.38).

Newtown School has 13 different panel designs. Each panel is a square and can be rotated 90 degrees, allowing for four different orientations. The FRP panels have a U-shaped cross section with only ¼ in (6 mm) of resin thickness, which keeps the panel light but stiff. Each panel includes the symbol or symbols and a peripheral frame that is bolted with integrated Z-shaped clips to 2-in (50-mm) painted, mild-steel hollow square sections. FRP often warps during manufacturing due to uneven curing of the resin; however, the Z clips on all four sides of the panels ensure that any warping is corrected by the steel frame. The panels were painted in the factory with a nitrocellulose paint to give them a matte white finish, while the backs of the panels are unfinished.

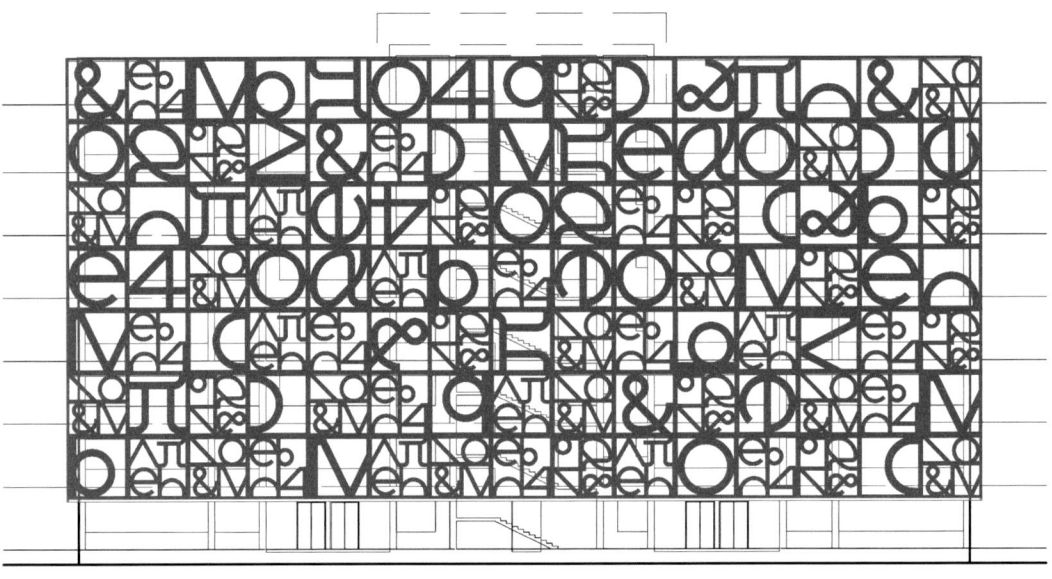

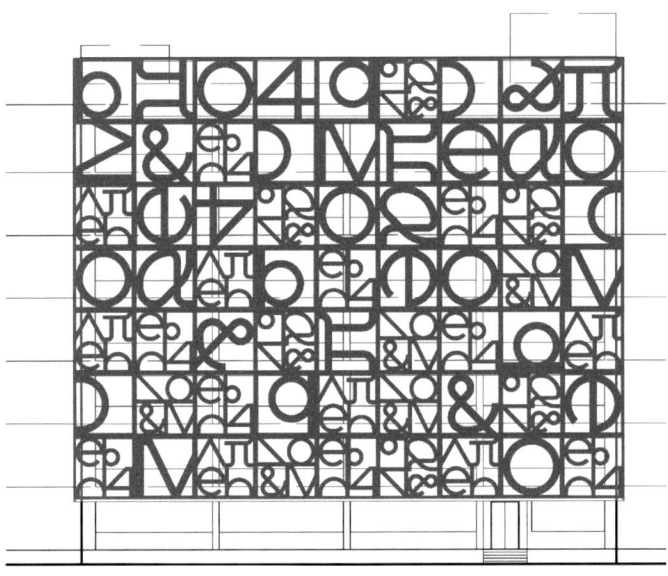

Figure 3.1.41
Typical elevations. The screen panels span from floor to floor with no intermediate support.

Figure 3.1.42
Interior of the classroom, with views out the window, looking at the back of the screen.

Notes

1. A variant of this process, called *bladder inflation molding (BIM)*, uses a closed model that comes in two or more parts. The composite is applied to each mold part and then the mold parts are closed around an internal bladder. The bladder is then inflated to fuse the composite layups together and to ensure the component is in full contact with the closed mold.
2. Rice, Kevin. *Personal Interview*. 24 May 2022. Daniel, Mann, Johnson and Mendenhall. American Cement Building. 1961, Los Angles, California. In addition to Rice's reference of the American Cement Building, other popular press writers have made similar analogies between the buildings, including Sarah Amelar and Edwin Heathcote in separate articles in *Architectural Record*. Amelar, Sarah. "L.A. Screenplay." *Architectural Record*, September 2015, pp. 64-69. Heathcote, Edwin "The Last Thing Grand Avenue Needs in Another Icon." *Architectural Record*, January 2016, pp. 32–41.
3. Rice.
4. Reddington, Mark and Rich Johnson. *Personal Interview*. 20 July 2022.
5. Capps, Kriston. "Tobin Center for the Performing Arts." *Architect*, 5 October 2011. https://www.architectmagazine.com/project-gallery/tobin-center-for-the-performing-arts_o. Accessed 5 August 2022. Spencer, Ingrid. "A New Landmark." *Texas Architect*, Dec., Nov. 2014, pp. 68–72.
6. The plywood was added beneath the gypsum board so that the GFRG panels could be mechanically fastened as needed without concern for the supporting metal stud locations.
7. Reddington and Johnson.
8. *Ibid.*
9. Chaudhuri, Abin. *Personal Interview*. 19 July 2022.
10. Abin Design Studio. *The Newtown School, Kolkata*. http://www.abindesignstudio.com/projects/details/2-the-newtown-school-kolkata. Accessed 12 July 2022.
11. Chaudhuri.

References

Abin Design Studio. *The Newtown School, Kolkata*. http://www.abindesignstudio.com/projects/details/2-the-newtown-school-kolkata. Accessed 12 July 2022.

Amelar, Sarah. "L.A. Screenplay." *Architectural Record*, September 2015, pp. 64–69.

Capps, Kriston. "Tobin Center for the Performing Arts." *Architect*, 5 October 2011. https://www.architectmagazine.com/project-gallery/tobin-center-for-the-performing-arts_o. Accessed 5 August 2022.

Chaudhuri, Abin. *Personal Interview*. 19 July 2022.

Heathcote, Edwin "The Last Thing Grand Avenue Needs in Another Icon." *Architectural Record*, January 2016, pp. 32–41.

Reddington, Mark and Rich Johnson. *Personal Interview*. 20 July 2022.

Rice, Kevin. *Personal Interview*. 24 May 2022.

Spencer, Ingrid. "A New Landmark." *Texas Architect*, Dec., Nov. 2014, pp. 68–72.

CHAPTER
3.2

Vacuum Infusion Process (VIP)

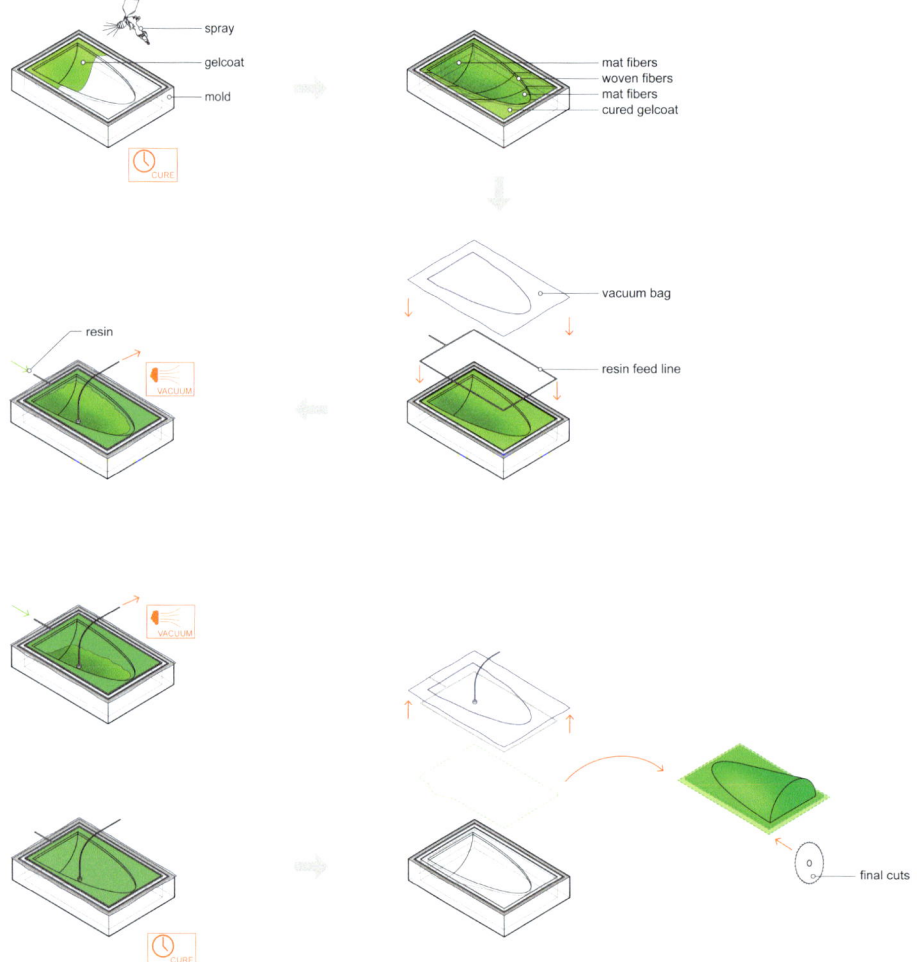

Figure 3.2.1
Vacuum infusion process diagram.

DOI: 10.4324/9781003299196-14

Vacuum infusion process (VIP) uses vacuum pressure to pull a liquid resin through dry fibers already laid in a mold. For architectural applications, the resin is typically a thermoset plastic with additives to better performance, although alternative resins can be used. Like contact molding (Chapter 3.1), VIP can form large components with complex curves; however, with the pressure of the vacuum, VIP is better suited to more intricate designs, with deeper draws, and more surface details than contact molding. Like contact molding, the face of the component in contact with the mold is the component's finished face; however, the back-side face of the VIP component is smoother in texture than that of contact-molded components, because the vacuum bag puts pressure on the back surface. VIP reduces the resin-to-fiber ratio, compacts the fiber-reinforced plastic (FRP), and reduces air bubbles and the potential for trapped moisture, essentially making VIP-formed FRPs stronger than those formed by contact molding. Since more pressure is placed on the mold in VIP than in contact molding, VIP molds are more expensive than those of contact molding.

In VIP, a gelcoat, which is a thin layer of resin only, is applied first to the mold to form a smooth finish face. This cures before dry fiber mats, precut into shape, are applied to the mold. A resin supply line is added, and a polymeric vacuum bag is placed over the mold and layup. The edges of the bag are sealed against the mold. The vacuum is turned on and it pulls the resin through the dry fiber mats while the vacuum bag provides even pressure on the layup. Once the fibers are fully impregnated by the resin, the vacuum is turned off and the component stays in the mold until the resin is cured enough to demold and be moved into optional post-production.

Molds for VIP are typically made from a single piece of material and are full and open, as the vacuum bag needs to be in full contact with the component. Generally, sharp corners and edges on which the bag might tear should be eliminated. For deep recesses and tight interior corners, where the bag may not stretch enough to reach, plugs, inserts, or partial molds can be added so that pressure from the bag is applied to the layup at these locations. These modifications will slow down the process and increase labor costs and mold set-up times. The vacuum does reduce the volatile organic compounds (VOCs) emitted by the resins during curing, making this process safer for the workers than contact molding. Since VIP requires additional equipment, not all contact molders are able to offer this process.

VIP is different than *vacuum bag molding*, which uses a vacuum bag to apply even pressure to a wet hand layup or spray applied FRP. Although vacuum bag molding does have less air bubbles and moisture content compared with traditional contact molding, VIP has the added benefit of a lower resin-to-fiber ratio, making VIP-formed FRPs stronger than comparable FRPs made from vacuum bag molding.

Fiber-Reinforced Plastic (FRP)

Typically, FRPs for VIP use a thermoset plastic (e.g. epoxy or polyester) as the binder and glass (GFRP) or carbon fibers (CFRP), although thermoplastic and alternative fibers (e.g. hemp) can be used. The gelcoat is opaque and can be custom matched to any color, and the injected resin clear or colored to match the gelcoat. Additives may be added to the gelcoat and subsequent resin layers to improve the plastic's resistance to fire and reduce its degradation to UV light. The specific thermoset plastic will affect material strength, creep, curing time, heat generated during curing, the warping of the component, and wear on the mold.

Gas Receiving Station (GOS) in Dinteloord, the Netherlands
By Studio Marco Vermeulen

Dinteloord is a small village, approximately 30 miles (50 km) south of Rotterdam in the

Figure 3.2.2
Gas Receiving Station (GOS), located in an agricultural and food cluster near Dinteloord, the Netherlands. Photo by © Ronald Tilleman.

Figure 3.2.3
Detail image of panels. The letters in the panel depict the chemical composition
of gas in its ratio of hydrogen, carbon, and nitrogen. Photo by © Ronald Tilleman.

Netherlands, and is known for its sugar factory. The Gas Receiving Station, or *Gasontvangststation* (GOS), is part of an agricultural and food cluster—a large 1480-acre (600-ha) greenhouse production center just east of Dinteloord. When it was first built, the GOS managed natural gas, but as part of a Studio Marco Vermeulen (SMV) masterplan, the station transitioned to biogas generated from an onsite bio-digestion system that uses vegetable waste. The GOS is constructed of masonry with a parge coat and a rainscreen of bio-composite panels made by VIP.

Figure 3.2.4
GOS under construction. The bio-composite panels are mounted to a substructure that has been painted at the joints to match the panels. Photo by © Studio Marco Vermeulen.

Figure 3.2.5
Test samples of color and quality. Initially, SMV had thought that the panels would
be translucent, formed without the gelcoat. In the end, the color comes from
gelcoat additives for fire resistance. Photo by © Studio Marco Vermeulen.

SMV chose a bio-composite instead of a glass fiber-reinforced, petroleum-based plastic for the station's cladding because the bio-composite kept with the overall project goals. Initially, SMV wanted to make the composite panels out of waste material from sugar as the agricultural and food cluster has retention basins made from the same material; however, the material could not meet the fire-spread requirements needed for a gas receiving station. NPSP, a company located in the Netherlands, had developed a bio-based resin called Nabasco (shortened from "nature-based composites") that could be modified to meet the necessary code requirements. The composite's gelcoat—the first layer that is applied to the mold—is the fire-resistant coating and SMV Founder Marco Vermeulen estimates that the panels are approximately 60–70% bio resin.[1]

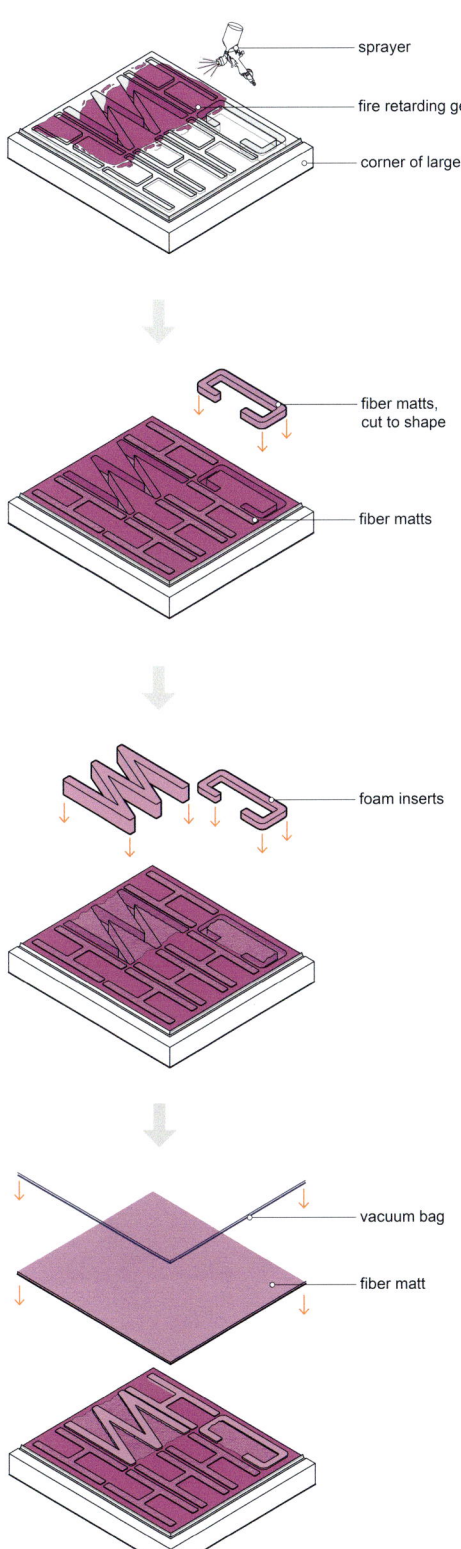

sprayer

fire retarding gel coat

corner of larger mold

fiber matts,
cut to shape

fiber matts

foam inserts

vacuum bag

fiber matt

Figure 3.2.6
A diagram of the case study process. Only a corner of
the mold and the panel is illustrated.

Figure 3.2.7
A detailed photograph of the panels. The VIP panels are mounted to the
building's doors, giving a seamless look. Photo by © Ronald Tilleman.

Figure 3.2.8
Image of the NPSP manufacturing facilities. Photo by © Studio Marco Vermeulen.

Figure 3.2.9
A close-up image of the aluminum mold surface.
Note the rounded corners and eased edges.
Photo by © Studio Marco Vermeulen.

Figures 3.2.10a and 10b

The cut, dry mats are being laid onto the mold and cut foam inserts are
used for the deep recesses. Photo by © Studio Marco Vermeulen.

SMV designed the panels so that they depict the chemical composition of gas in its ratio of hydrogen (H), carbon (C), and nitrogen (N) atoms. Vermeulen believes that this project was a constructive collaboration between SMV and NPSP. SMV used SketchUp for the panels' design and NPSP developed the panel with draft angles and rounded corners and restricted the panels' relief depth to 2.4 in (6 cm) to facilitate manufacturability.

NPSP provided the composite and manufactured the 102 panels. NPSP recommended VIP as it was the best solution to form the deep relief of the letters, because traditional contact molding would have been labor intensive and likely to result in voids in the recesses. To provide pressure into the deep recesses, NPSP placed foam blocks between the composite and the vacuum bag (see Figures 3.2.6 and 3.2.10b).

Figure 3.2.11
The VIP panel removed from the mold. Note that
the edges are rough and have not yet been trimmed.
Photo by © Studio Marco Vermeulen.

Figure 3.2.12
The VIP panels after post-production trimming. Photo by © Studio Marco Vermeulen.

For the manufacturing process, NPSP sprayed the initial gelcoat onto the custom CNC-milled, solid aluminum mold and waited for it to cure. Then NPSP used a stencil to cut the letters from hemp fiber matts and inserted them into the mold with other dry fiber matt shapes to cover the mold surface. NPSP added foam into the deep recesses where required. NPSP mixed the resin, and it was pulled through the dry fibers by the vacuum. NPSP produced one panel each day. Over ten years later, Vermeulen says that the building and its panels still look great.

Note

1. Vermeulen, Marco. *Personal Interview.* 17 May 2022.

Reference

Vermeulen, Marco. *Personal Interview.* 17 May 2022.

3.3

Filament Winding

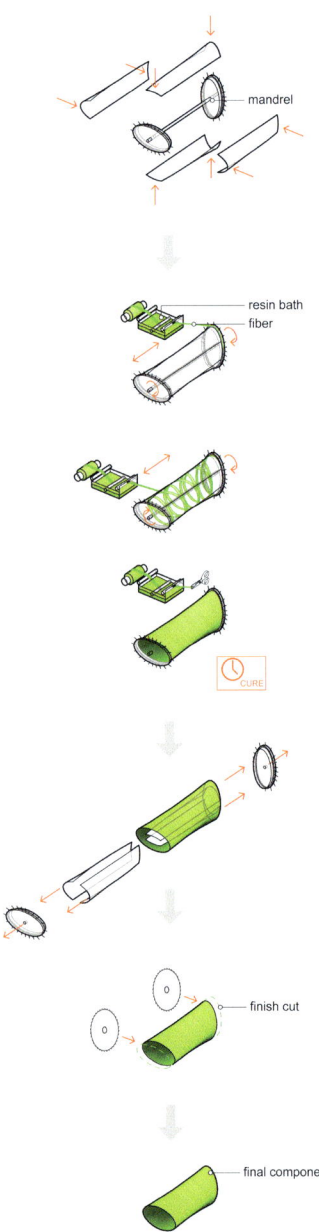

mandrel

resin bath
fiber

CURE

finish cut

final component

Figure 3.3.1
Filament winding process diagram. Drawing by author.
A version of this diagram originally appeared in
Manufacturing Architecture (Laurence King, 2018).

DOI: 10.4324/9781003299196-15

Filament winding is the manufacturing process that wraps plastic-impregnated reinforcing strands around a rotating mandrel. The plastic used is typically a thermoset plastic, although thermoplastic can be used. The common reinforcing strands are made of glass, but the process can use carbon or aramid fibers and the strands can be arranged as rovings, tow, or tape. This process makes hollow components in a range of sizes from golf-clubs to airplane fuselages. Today, computer-numeric control (CNC) equipment controls the placement of the strands and the mandrel's rotation. This process can manufacture irregular and complex shapes, such as T-shaped pipe fittings, oars, and drive shafts. Manufacturing pressures for this process are low; therefore, mandrels can be made of low-cost materials and capital costs are low. Generally, filament winding is time intensive and multiple molds may be used for parallel productions to reduce the overall production schedule.

In filament winding, the mandrel rotates as the carriage moves along the mandrel's axis placing the plastic-impregnated reinforcing strands on the tool. The fiber strands may be pre-impregnated, also known as prepreg, or may pass through a resin bath before placement. The carriage holds the winding eye that controls fiber placement and orientation and rollers to maintain proper tension in the fibers. As the mandrel spins, the carriage moves along the primary mandrel axis to place the impregnated fibers. Most filament winders are sophisticated in their movements, and the carriage moves in additional axes, including moving up and down, closer to and further from the mandrel, and rotating horizontally and vertically. After the fibers are placed, the fibers are cut, the mandrel and composite are removed from the filament winder, the plastic cures or hardens, and then the component is removed from the tooling. Any postproduction, such as trimming or finishing, is done as needed.

One option is to have the filament winding in full contact with the tooling. Tooling can be solid, and sacrificed during the winding process, or can come in multiple pieces that are deconstructed for demolding. The other option is to have the filament winding in partial contact with the tooling. Tooling can be frames or jigs used to support the filament windings only at key locations with the fibers' tension forming straight lines between supports. Tooling may be made of a range of materials, including cardboard, plastics or foam, wood, fiberglass, and metal. Generally, gentle convex and concave curves are easy to wind, but full molds are used to form sharp angles and interior corners. If not using prepreg strands, then the inside of the filament winding is often the components' finished face, as it is in direct contact with the tooling. If the exterior of the winding is to be finished, manufacturers may wrap the component in plastic film wrap before curing. This brings the resin to the outside surface to create a finished surface. After curing, the plastic wrap is removed.

Filament winding may also be known as *filament placement*.

Fiber-Reinforced Plastic (FRP)

Typically, filament winding uses thermoset plastics, such as epoxy or polyester, that have a low viscosity, so the fibers stay in place with little to no slipping during the winding process. Additives may be added to the resin to improve the plastic's resistance to fire, reduce its degradation to UV light, or to change its opacity and color. The specific thermoset plastic used will affect material strength, creep, and the curing process, including curing time, heat generated, warping, and wear on the mold.

Typically, filament winding uses glass fibers. Carbon fibers provide the best strength-to-weight ratio; however, they are expensive and are difficult to wind as they flake during winding, creating dust that prematurely wears the electronic circuitry of the winders.

Figure 3.3.2
Aerial photograph of the 2019 Bundesgartenschau (BUGA) Summer Island. The Fiber Pavilion
can be seen in the foreground and the Wood Pavilion in the background. Photograph by Nikolai Benner.

Figure 3.3.3
Photograph of the 2019 BUGA Fiber Pavilion. The pavilion is made from discrete custom fiber wound structural elements bolted together with a tensioned EFTE membrane for weather protection. Courtesy of FibRGmbH.

Figure 3.3.4
Detail image of the dome. Courtesy of ICD/ITKE, University of Stuttgart.

Figure 3.3.5
Detail image of the pavilion foundation. Courtesy of FibRGmbH.

BUGA Fiber Pavilion in Heilbronn, Germany

By University of Stuttgart Institute for Computational Design and Construction (ICD) and the Institute for Building Structures and Structural Design (ITKE)

The University of Stuttgart Institute for Computational Design and Construction (ICD) and the Institute for Building Structures and Structural Design (ITKE) designed two pavilions for the 2019 Bundesgartenschau (BUGA). BUGA is a national biennial horticultural show and was held from April 17 to October 6, 2019, in Heilbronn, Germany. The BUGA Fiber Pavilion, along with its sister, the BUGA Wood Pavilion, was sited in the show's Summer Island area, among the rose garden, climbing rocks, and beach play areas. The Wood Pavilion provided a small stage for presentations and activities, while the Fiber Pavilion hosted video installations. The Fiber Pavilion is made of discrete structural elements that are assembled with mechanical bolts to form a geodesic-like dome. Each element is made of filament wound glass and carbon fibers, with an epoxy resin. The pavilion spans 75.5 ft (23 m) with a maximum height of 22 ft 4 in (6.8 m) and covers 4306 ft² (400 m²) of floor area. Offset from the structure is a tensioned ethylene tetrafluoroethylene (ETFE) membrane that provides the necessary rain protection for its video installations. Both pavilions were disassembled after BUGA ended, and according to ITKE Head Jan Knippers, they are negotiating with BUGA to see if the Fiber Pavilion will be reinstalled for the 2023 event.[1]

Figure 3.3.6
Detail of the bolted connection between structural members. This allowed the building to be disassembled at the end of the event. (Photograph taken during construction, before the ETFE membrane was installed.) Courtesy of ICD/ITKE, University of Stuttgart.

Figure 3.3.7
Load testing of individual structural members and filament winding of prototype elements, completed at the University of Stuttgart. Courtesy of ICD/ITKE, University of Stuttgart.

Figure 3.3.8
Production filament winding done a FibR GmBH. Courtesy of FibRGmbH.

Many of the previous ICD/ITKE pavilions have looked to nature for design inspiration—such as sand dollars for the 2011 pavilion, a lobster's exoskeleton for the 2012 Pavilion, or a water spider's web for the 2014–2015 pavilion; however, the design for the BUGA Fiber Pavilion was required to be more pragmatic. First, the structure needed to be designed so that it could be disassembled

Figure 3.3.9
Image of FibR GmBH during production. Courtesy of FibRGmbH.

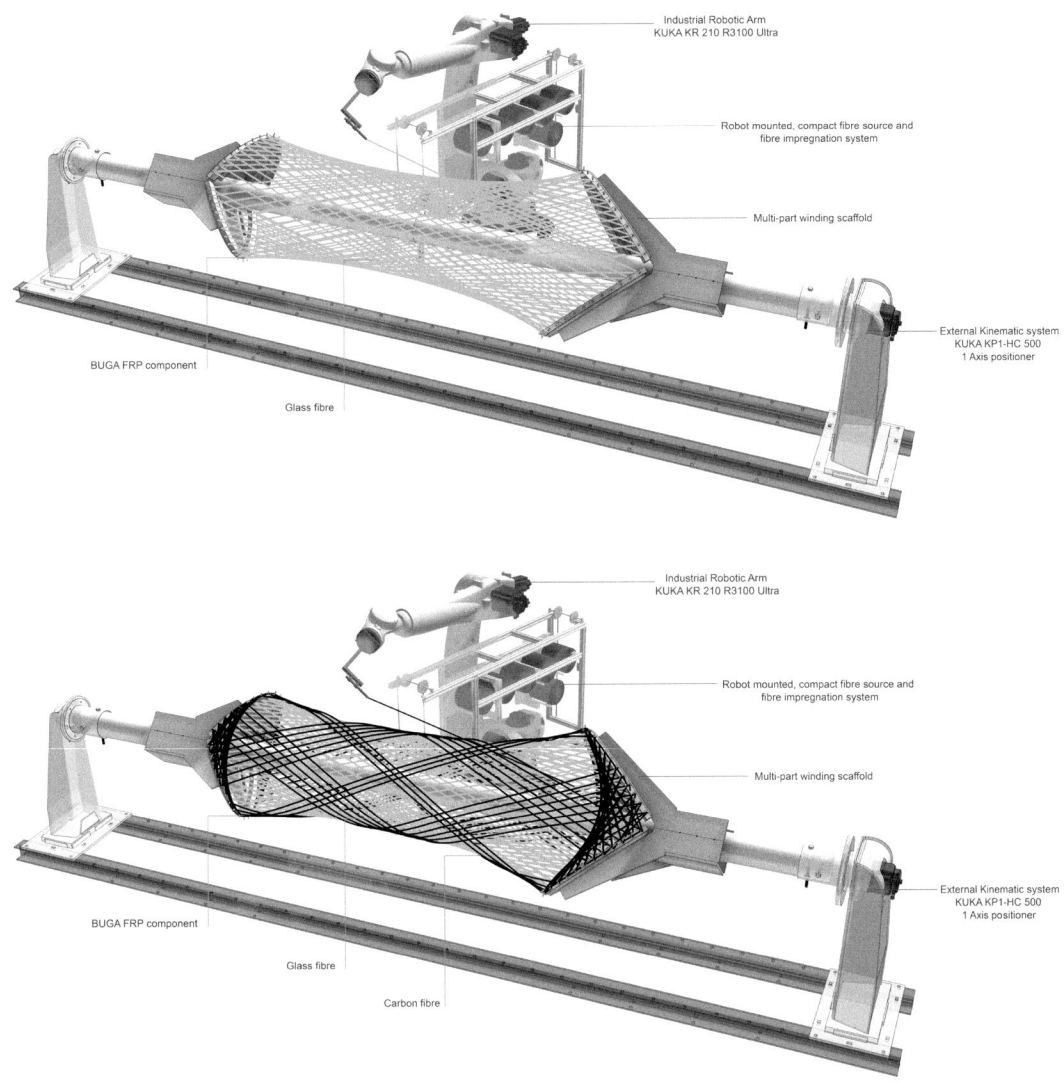

Industrial Robotic Arm
KUKA KR 210 R3100 Ultra

Robot mounted, compact fibre source and
fibre impregnation system

Multi-part winding scaffold

External Kinematic system
KUKA KP1-HC 500
1 Axis positioner

BUGA FRP component

Glass fibre

Industrial Robotic Arm
KUKA KR 210 R3100 Ultra

Robot mounted, compact fibre source and
fibre impregnation system

Multi-part winding scaffold

External Kinematic system
KUKA KP1-HC 500
1 Axis positioner

BUGA FRP component

Glass fibre

Carbon fibre

Figure 3.3.10

Glass fibers are wound first to provide the scaffolding for the carbon fiber placement. Courtesy of ICD/ITKE, University of Stuttgart.

after the event. Second, ICD/ITKE needed to demonstrate to BUGA that the pavilion would stand for the six-month event. Third, because the structure was not going to be on the University of Stuttgart's campus like the other pavilions, the design team needed to go through the full building approval process. By designing the pavilion with smaller individual structural elements, the design team could physically test the structural integrity of each component.

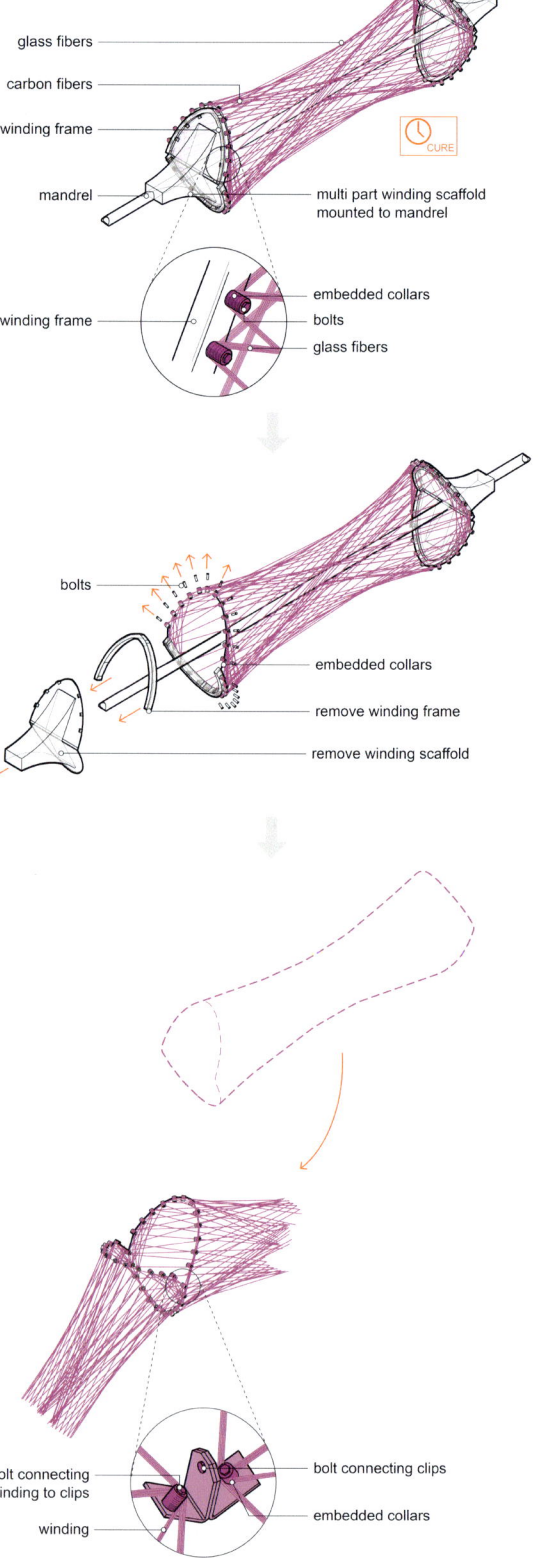

ICD/ITKE filament wound the initial prototypes of the structure components and did their strength testing. FibR GmBH, a private filament winding company that started out of ICD/ITKE projects, manufactured the components used in the Fiber Pavilion. The structural components are wound, using a variant technique of filament winding that ICD/ITKE states it developed, called coreless filament winding (CFW).[2] In this process, two end boundary frames are mounted on a spinning axle, and the filament winder carriage moves back and forth, as the axle and frames spin. For the Fiber Pavilion, glass fibers impregnated in an epoxy resin bath were wound first, then carbon fibers impregnated in an epoxy resin bath were wound over the glass fibers.[3] Each component took 4–6 hours to wind. After winding, the axle and the mounted boundary frames were then put into an oven to cure the epoxy. After the epoxy set, the boundary frames were removed, leaving behind metal collars. Bolts would connect the collars to steel brackets to join the wound elements to one another (see Figure 3.3.11).

Figure 3.3.11
Metal collars were wound into the structural components. These collars were used to connect the elements to one another during construction.

The Fiber Pavilion has 60 filament wound components, in seven different configurations with each configuration having its own boundary frame and winding pattern.[4] The configurations have different lengths, depending on their placement in the structure, with the longest being at the bottom of the dome and having length of 15 ft 9 in (4.8 m). The expensive carbon fibers are the load-bearing strands of the components, and they are only placed where needed structurally. The components with more carbon fibers are those at the bottom of the pavilion, where the loads are the highest, and the components with less carbon fibers are at the top of the dome where the loads are the least.

Since the filament wound components are coreless, the precise fiber placement depends on the fibers' tension as the carriage moves back and forth.[5]

Figure 3.3.12
Detail image. Courtesy of ICD/ITKE, University of Stuttgart.

ICD/ITKE has written that there is an inherent lack of precision with the final fiber placement in CFW,[6] but Knippers acknowledged that the team has not measured the accuracy of the winding against the digital model; therefore, tolerances are not yet clear for this project.

fibers rather than the glass fibers. This in essence made the glass fibers redundant, except that they acted as winding core for the carbon fibers.

4. Two of the configurations—the components located at the dome base—are mirrors of one another. The five others are completely different.

5. Gil Pérez et al.

6. *Ibid.*

Notes

1. Knippers, Jan. *Personal Interview.* 8 July 2022.
2. Gil Pérez, Marta, et al. "Structural Design Assisted by Testing for Modular Coreless Filament-Wound Composites: The BUGA Fibre Pavilion." *Construction & Building Materials*, vol. 301, 2021, pp. 124–303.
3. The structural members were designed so that all the structural loads were to be carried by the stronger carbon

References

Gil Pérez, Marta, et al. "Structural Design Assisted by Testing for Modular Coreless Filament-Wound Composites: The BUGA Fibre Pavilion." *Construction & Building Materials*, vol. 301, 2021, pp. 124–303.

Knippers, Jan. *Personal Interview.* 8 July 2022.

CHAPTER
3.4

Rotational Molding

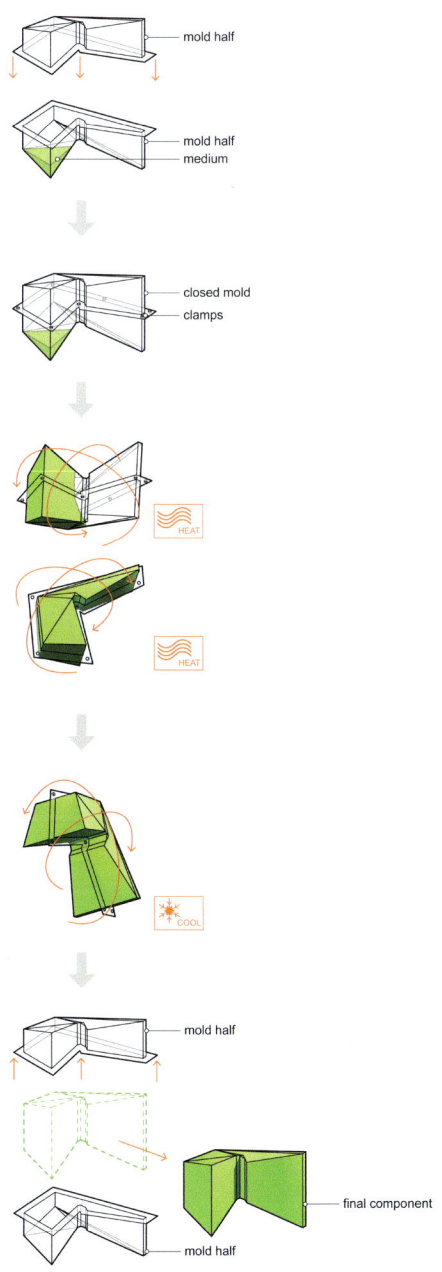

Figure 3.4.1
Rotational molding process diagram.

DOI: 10.4324/9781003299196-16

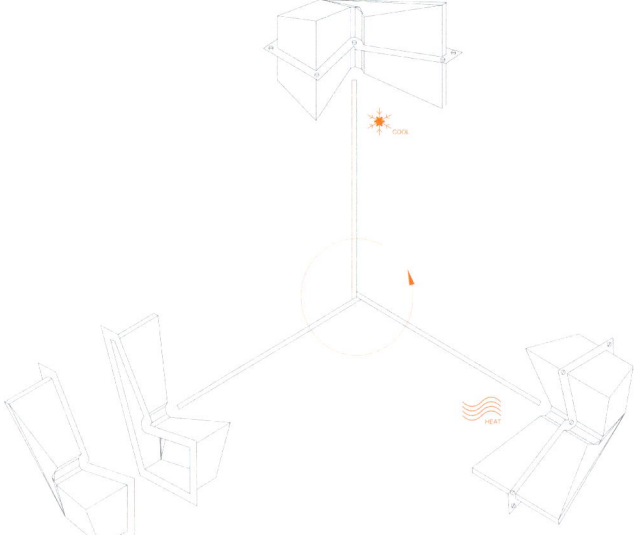

Figure 3.4.2
Most rotomolders use a carousel to rotate multiple molds through three stations. Station 1 loads the plastic and demolds the final component. Station 2 heats the mold and the medium. Station 3 cools the mold and the medium.

Rotational molding, or *rotomolding* for short, is the manufacturing process that uses a slowly rotating mold and gravity to coat the mold's interior surface with a medium in its liquid state. Typically, rotomolding uses thermoplastics, which soften when heated, but can include thermoset plastic or fiber-reinforced plastic (FRP). Rotomolding is used to make hollow, closed components, in which the component's interior can only be accessed through post-production cutting or drilling. Shapes can range from simple cubes or spheres to complex shapes such as the doubled walls for a plastic insulated cooler. It is better to form complex curves with this manufacturing process, as large flat surfaces tend to warp during cooling. The rotation of the mold is slow; therefore, the molds are not structured to resist high pressures. However, the molds must withstand the heating and cooling processes necessary to liquify and solidify the plastic. Generally, rotomolding is time intensive and often multiple molds are used for parallel productions.

In rotomolding, a predetermined amount of plastic is placed into the cavity of a closed mold. The mold and plastic are then heated to liquify the plastic as the mold slowly turns in multiple directions, essentially evenly coating the inside of the closed mold with the liquid plastic. Once the mold is coated, it is cooled while continuing to be rotated. Once the plastic is solid enough to hold its shape, the mold is opened, and the component is demolded. Any residual flashing is removed, and then the component is ready for optional post-production processes. To speed up manufacturing, most rotomolders use a rotary-arm machine, also known as a carousel, which rotates multiple molds through three stations: station (1) loading in the plastic and demolding the component, station (2) heating, and station (3) cooling. The carousel can either accommodate multiples of the same mold for parallel productions, or different molds with similar cycle times for concurrent productions.[1]

Tooling in rotomolding includes the mold and the mold frame that attaches the mold to the rotator. Molds need to conduct the heating and cooling temperatures to the medium inside the mold cavity. Production molds are typically made from aluminum or steel and can be fabricated by casting, machining, or shaping sheet metal. For prototypes, molds can be made from FRP via contact molding (Chapter 3.1), vacuum infusion process (VIP) (Chapter 3.2), or filament winding (Chapter 3.3). Typically, molds are made of two parts: however, molds can be made of more parts for components with complex shapes, interior holes, or undercuts.

Thermoplastic

The long cycle times and relatively high temperatures (approximately 400°F or 204°C) limit the specific types of thermoplastics that can be used. Rotomolded plastics need to have thermal stability, low shear melt viscosity, and good heat transfer. Depending on the plastic, it is loaded into the mold cavity in a powder, pellet, or liquid form. Reinforcement fillers can be added and can include glass and carbon fibers, glass spheres, wood chips, and metal powders. The fillers should be a similar size and density to the plastic so that natural separation does not occur.

Figure 3.4.3
Photograph of the exterior of the Beijing Tea House, with its translucent structure.

Figure 3.4.4
Photograph of roof terrace with views of the Forbidden City.

Figure 3.4.5
Photograph of tearoom.

Beijing Tea House in Beijing, China
By Kengo Kuma and Associates

Kengo Kuma and Associates (KKAA) renovated a historic, siheyuan-style building into a modern teahouse. The building is in a prime location, in the center of Beijing, near the Palace of the Forbidden City's East Gate. The building has a simple floor plan with a public tearoom, two private tearooms, and a roof terrace with views of the Forbidden City. The historic building's street-front roof and structural masonry remained with KKAA inserting a new structural system made of custom, rotationally molded blocks made from white translucent polyethylene. According to Yoshihiko Seki, the person at KKAA in charge of the Beijing Tea House, the design of the rotomolded structure is like traditional Hutong buildings but unlike those building's dark structure, the rotomolded pieces offer the passage of light because of their translucent nature.[2]

The rotomolded components were custom made at a Shanghai factory. The Beijing Tea House's structural system used four different mold types: (1) a main H-shaped module that made up the walls; (2) a vertical, column module; (3) a horizontal, beam module; and (4) a wide module placed in the building center for lateral stability. The four mold types stack together like a modern version of masonry construction.[3] The thermoplastic modules mechanically connect through integrated mortise and tenons, similar to Legos snapping together, and are bonded with a special glue from 3M. A thick, standard-profile polycarbonate panel slots into the structural module, providing the building's enclosure. KKAA worked with a structural engineer and the rotomolder from the project's beginning, collaborating on the building's system with many research and experimentation attempts.[4] According to Seki, because the building was a private project, additives did not need to be added to the plastic to meet legal building requirements.

The Beijing Tea House is not the first project that KKAA has used stackable, hollow plastic blocks as a building structure. KKAA investigated this building material through inhabitable installation projects called the Water Branch House. The first iteration of the Water Branch House was Water Block I and it was exhibited nationally and internationally. The module for Water Block I could

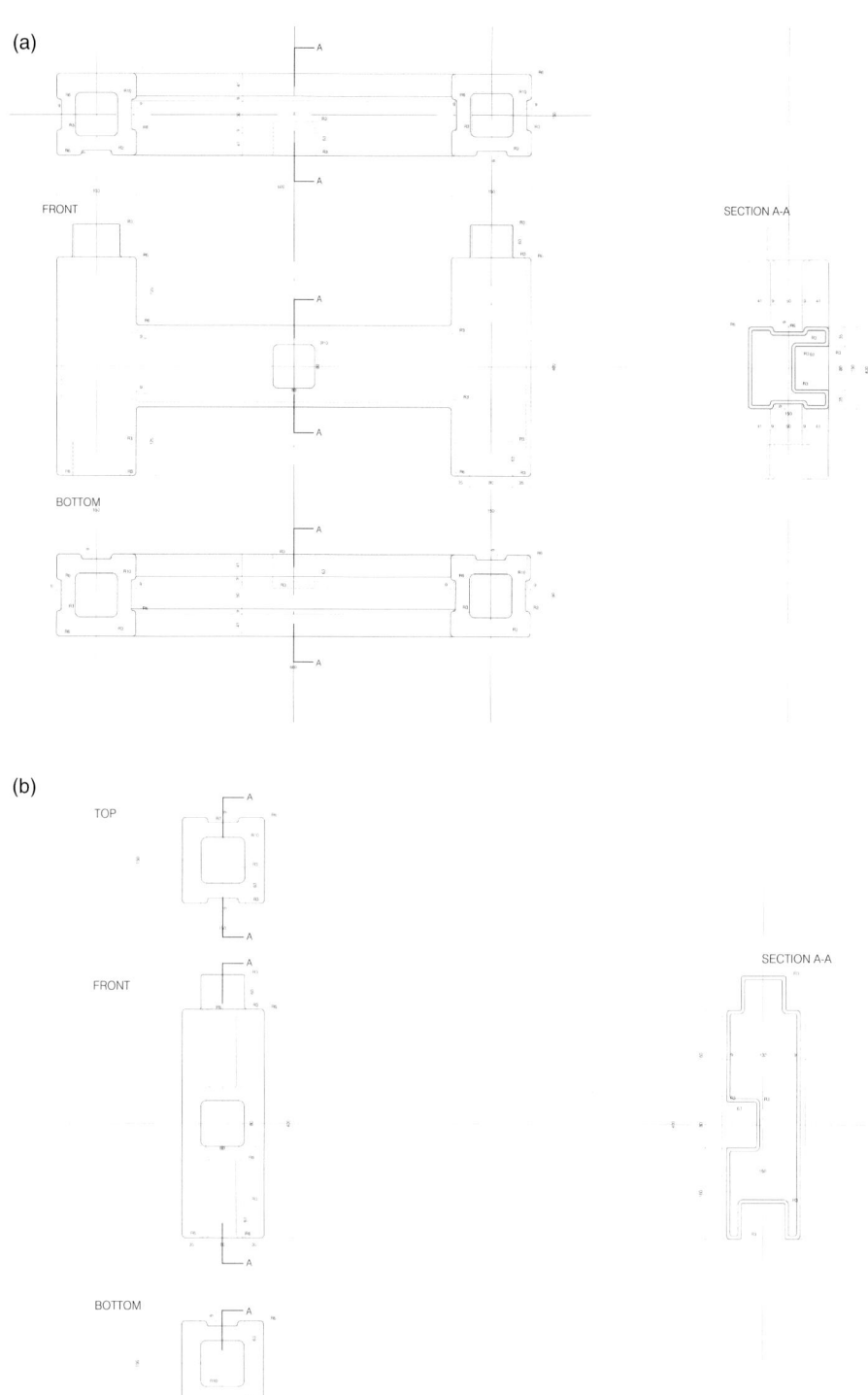

(a)

FRONT

BOTTOM

SECTION A-A

(b)

TOP

FRONT

BOTTOM

SECTION A-A

Figure 3.4.6
Plan, section, and elevation of the four different component shapes. (a) Main, H-shaped module,
(b) column module, (c) beam module, and (d) wide module for lateral stability. (*Continued*)

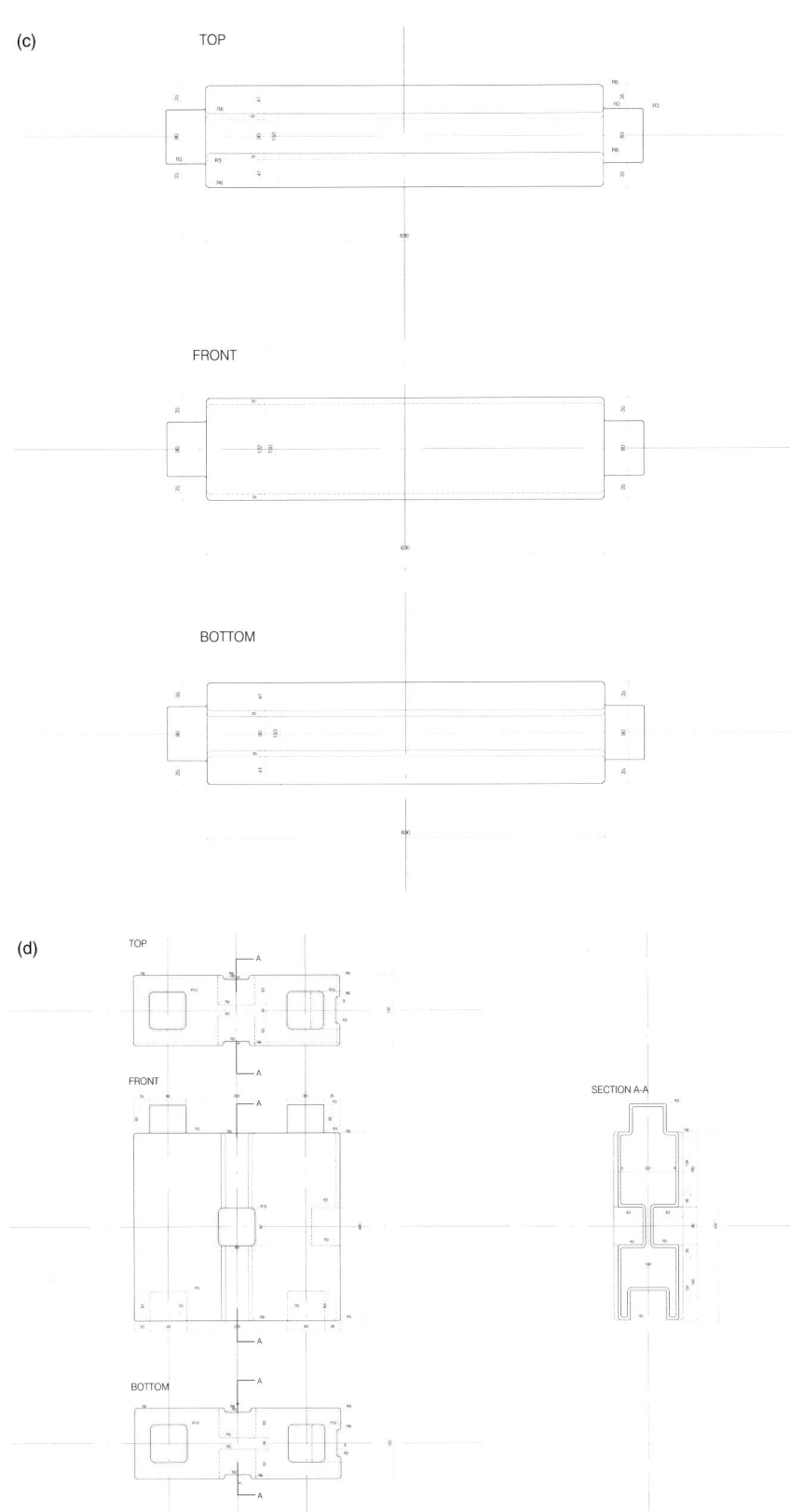

Figure 3.4.6
(*Continued*)

Figure 3.4.7
Ceiling structure of tearoom.

form building walls, but not a roof; KKAA's second iteration, Water Block II, could be used for both walls and roofs. The Water Branch House's Water Block II was produced and displayed in 2008 at New York's Museum of Modern Art's *Home Delivery* exhibition. The Water Branch studies demonstrated to KKAA that plastic is viable as a stackable building structure and that like Legos, small, repetitively stacked units offer a lot of design flexibility. It was likely that these investigations led them to using the process and structural system for Beijing Tea House.

Notes

1. A three-armed rotary machine means that a component with a 30-min cycle time could be demolded every 10 min, speeding up production.
2. Seki, Yoshihiko. *Personal Email*. 11 May 2022.
3. "Beijing Tea House." *The Japan Architect*, Spring 2018, pp. 164–165.
4. Seki.

References

"Beijing Tea House." *The Japan Architect*, Spring 2018, pp. 164–165.

Seki, Yoshihiko. *Personal Email*. 11 May 2022.

CHAPTER 3.5

Spin Casting

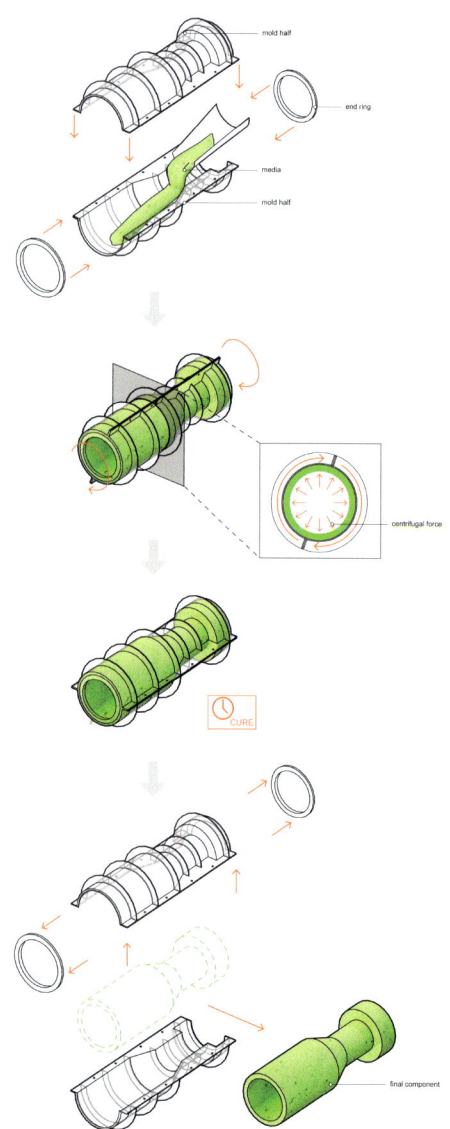

Figure 3.5.1
Spin casting process diagram. Drawing by author.
A version of this diagram originally appeared in
Manufacturing Architecture (Laurence King, 2018).

 DOI: 10.4324/9781003299196-17

Spin casting is the manufacturing process that spins a cylindrical-shaped mold filled with a medium at high speeds so that a centrifugal force pushes the medium against the inside of the mold surface. For customized architectural applications, spin casting uses concrete. However, metal can be spin-cast; the process is the same and is referred to as *centrifugal casting*. Generally, for architecture, spin casting is used to make hollow concrete columns. Shapes can include tapers or entasis; a variety of cross-sectional shapes such as circles, ovals, octagons, squares, and rectangles; and components may have surface textures. Spin casting includes reinforcing steel that can be prestressed and can include high early strength concrete, ultra-high performance concrete (UHPC), or polymer-based concretes. Exterior surface quality for spun cast components is excellent, as the centrifugal force eliminates air pockets, or bug holes, on the surface. Large and specialized equipment is needed for spin casting, and this process requires specialized manufacturers. Molds are typically made from steel and often multiple molds are required to reduce the overall production schedule.

In spin casting, the cylindrical-shaped mold is partially filled with concrete. The mold is closed and then rotated at a high speed, creating a centrifugal force to evenly distribute the concrete along the mold. Then the rotational speed is increased; this compacts the concrete and any excess water, not needed for curing the concrete, is driven to the component center, and is expelled from the mold. The top spinning speeds are maintained for 5–10 min, depending on the component's wall thickness. The component is left in the mold for curing and then is demolded and allowed to continue curing until it reaches full strength. If specified, concrete components can be grit blasted, acid washed, ground, or polished, like other concrete finishing methods.

Tooling for spin casting includes the mold and the end ring. Molds for spin casting are fabricated from sheet steel and are reinforced with stiffener plates or steel angles. To eliminate joint lines between sheets, joints are often welded and ground. The finish on the inside mold face can range from grit to polish and affects the resulting concrete surface finish. If complex surface textures are desired, then rubber formliners can be added to the inside of the mold. Typically, spin-casting molds are made in two parts but can be made from additional parts if necessary. The parting line will leave a mark on the component, and any resulting flashing will be removed during handling. The end rings are made from steel plates and are in the shape and wall thickness of the spun cross section. The end rings stiffen the mold and keep the concrete inside of the mold during spinning, while allowing excess water to escape.

Concrete

The concrete used for spin casting is different than that of conventional, wet-cast concrete (Chapter 4.1). With to the centrifugal force, the concrete mix must not separate during spinning and therefore additives such as fly ash are needed to reduce the segregation of ingredients. Since water is expelled during the process, high early strength concrete should be used. This means that components can be demolded sooner and cycle times shorter than traditional precast concrete. Generally, spun concrete components are less porous and have greater strength, density, and modulus of elasticity when compared to traditional precast concrete, thus making spun columns suitable for withstanding harsh exterior environments.

Figure 3.5.2
View of southeast corner and main building entrance.

Figure 3.5.3
Site plan with roof.

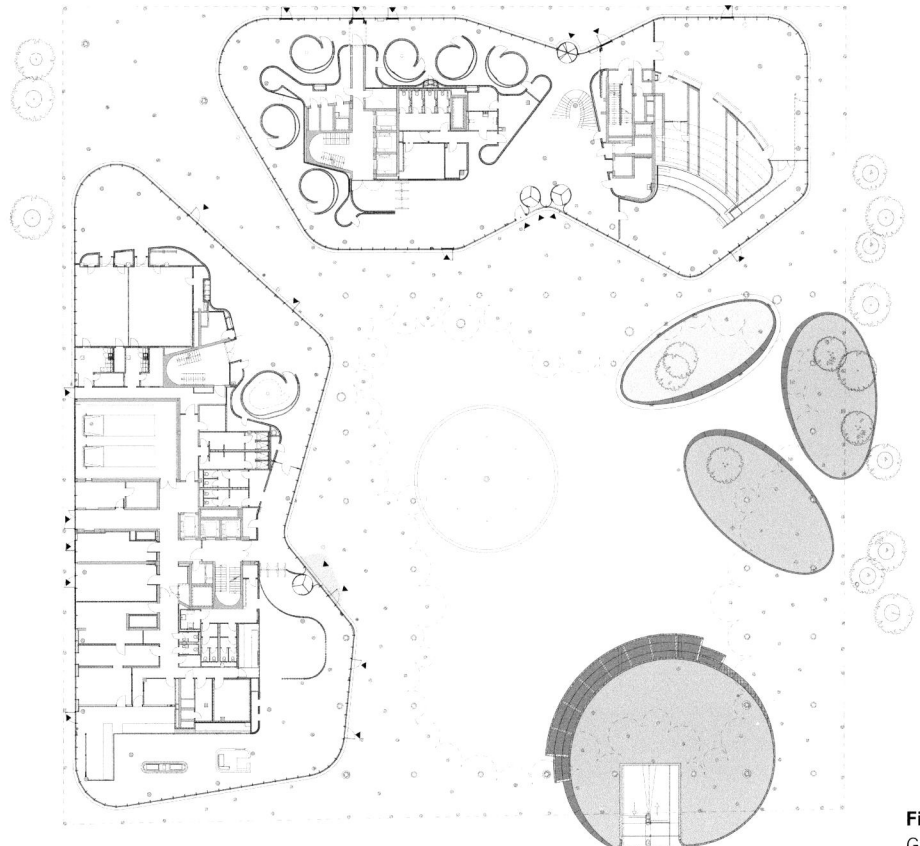

Figure 3.5.4
Ground-floor plan.

Figure 3.5.5
Public courtyard with custom spun columns made with white cement and color additive.

Sächsische AufbauBank (SAB) in Leipzig, Germany
By ACME

The Sächsische AufbauBank (SAB) is a government-managed bank of Saxony and its primary function is to serve the public by providing grants, subsidies, loans, and guarantees to businesses, investors, developers, and homeowners. The building is 242,188 ft² (22,500 m²) and accommodates approximately 600 staff. In addition to its banking function, the building has a conference center with auditorium, cafeteria, underground parking, and space dedicated to public use. The SAB is built just north of the historic Leipzig city center and is just two blocks east from the city's train station. In 2013, the client held an invited international design competition for the SAB and ACME won. According to ACME Associate Director Heidrun Schuhmann, it was critical that the design team thought about how to represent the public-serving function of the building in their design.[1] The resulting massing is modest and similarly scaled to its neighboring buildings. Only about a third of the site is taken up with building program at ground level, and the rest is left open as public space.

Figure 3.5.6
Interior columns are made from traditionally cast precast concrete with Portland cement.

Figure 3.5.7
There are three different diameter columns. The two smaller diameter columns support the roof and canopy; the largest diameter columns vent the underground parking garage.

SAB's design is dominated by round concrete columns. Schuhmann stated that the design plays on the historic typology of banks and their formal language of columns organized on a grid. The projects' interior, structural columns—for both the building and the below-ground parking garage—are standard precast concrete (Chapter 4.1) with gray, Portland cement. The projects' exterior columns are spun cast with white cement and added pigment. Knippers Helbig, the structural and facade engineer for the project, directed ACME to FUCHS Europoles, as Europoles had recently supplied the custom octagonal and partially ornamented spun columns for the Great Mosque in Algiers.[2]

The SAB has 299 spun columns, made in three different diameters: 15 ¾ in (400 mm), 27 ½ in (700 mm), and 43 ¼ in (1100 mm). The two smaller diameter columns are 66 ft (20 m) tall and support the concrete roof or a fabric shade structure. Those columns provide acoustic and visual separation from the neighboring highway to the south, shading for the public space and building's floor-to-ceiling windows, and in some cases, vertical chases for roof drainage pipes. There are only four of the large diameter columns, and their primary role is the vent the underground parking garage. Due to weight-carrying limits of the erection crane, the large diameter columns' heights were limited; they were spliced together with integrated cast steel base and top plates to reach the full sixty-six-foot height. The spun columns use pretensioned reinforcing strands, which allow the columns to be thinner and taller than conventionally reinforced concrete. The bottom of each column has an embedded steel anchor plate that is bolted on site onto a base plate with a reinforcing cage tied into the site-cast concrete slabs.

According to Europoles Key Account Manager Jürgen Hegel, the column diameters for this project were standard diameters for Europole to produce; however, ACME designed the columns with a custom flare. This visually transitioned the column shaft to its mushroom-style capital.[3] Europoles modified the molds to add in the custom portion of the tooling made from solid, CNC-milled steel.

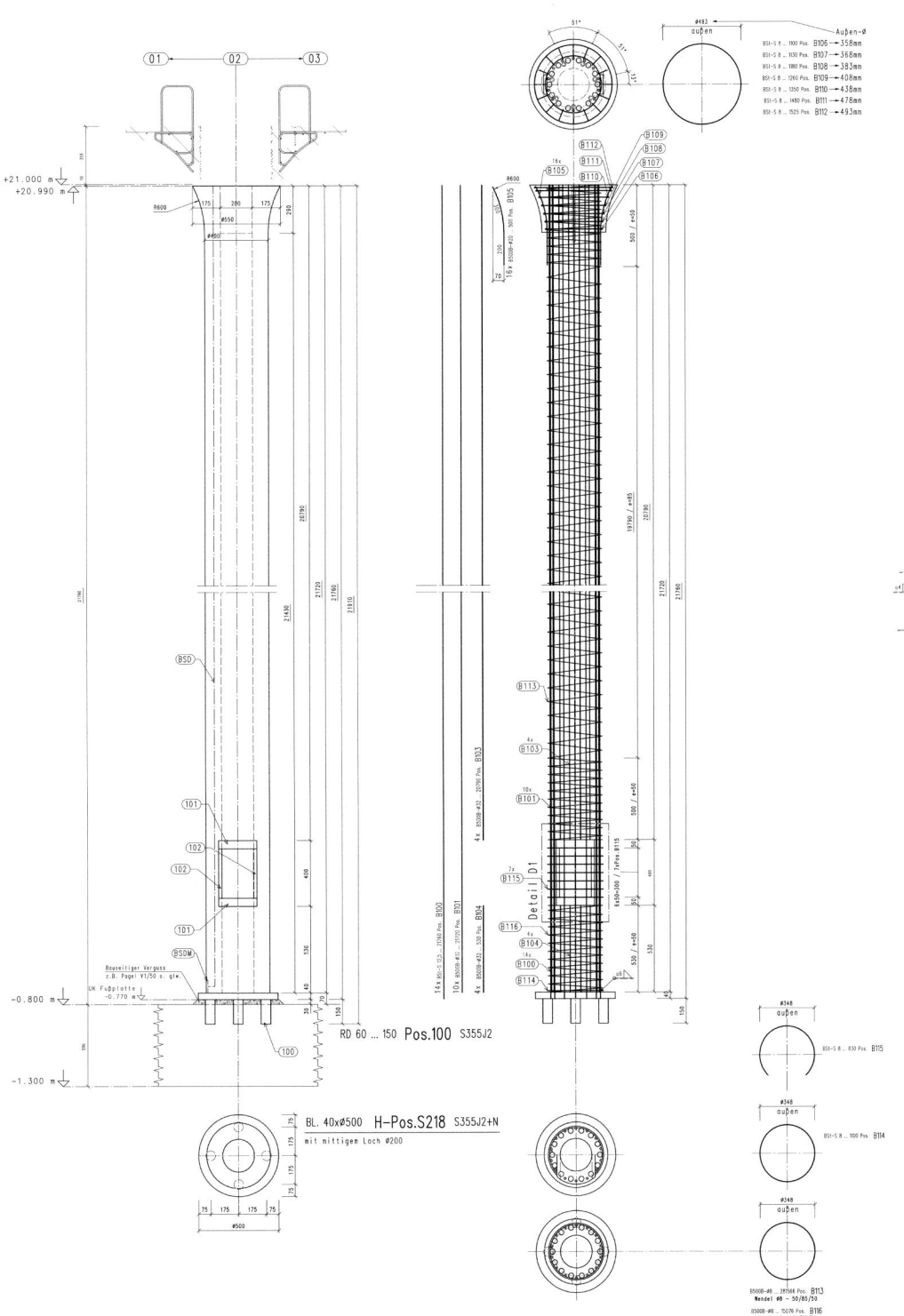

Figure 3.5.8
FUCHS Europoles shop drawings of the spun cast columns for SAB. Drawing by FUCHS Europoles Germany.

Figure 3.5.9
Detail image of spun cast column surface, taken from second level balcony.

Figure 3.5.10
Image of mold half with reinforcing steel cage at Europoles. The end ring can be seen on the wall, behind the mold. Photo by FUCHS Europoles Germany.

Figure 3.5.11
A standard Europoles mold was modified with
CNC-milled steel to provide the flare at the top of
the column. Photo by FUCHS Europoles Germany.

Notes

1. Schuhmann, Heidrun (ACME), Tim Laubinger (ACME),
   and Jürgen Hegel (FUCHS Europole). *Personal Interview.* 29 June 2022.
2. KSP Engel. *Djamma el Djazaïr.* Algiers, Algeria. 2019.
3. Schuhmann et al.

Reference

Schuhmann, Heidrun (ACME), Tim Laubinger (ACME), and
   Jürgen Hegel (FUCHS Europole). *Personal Interview.*
   29 June 2022.

Slip Casting

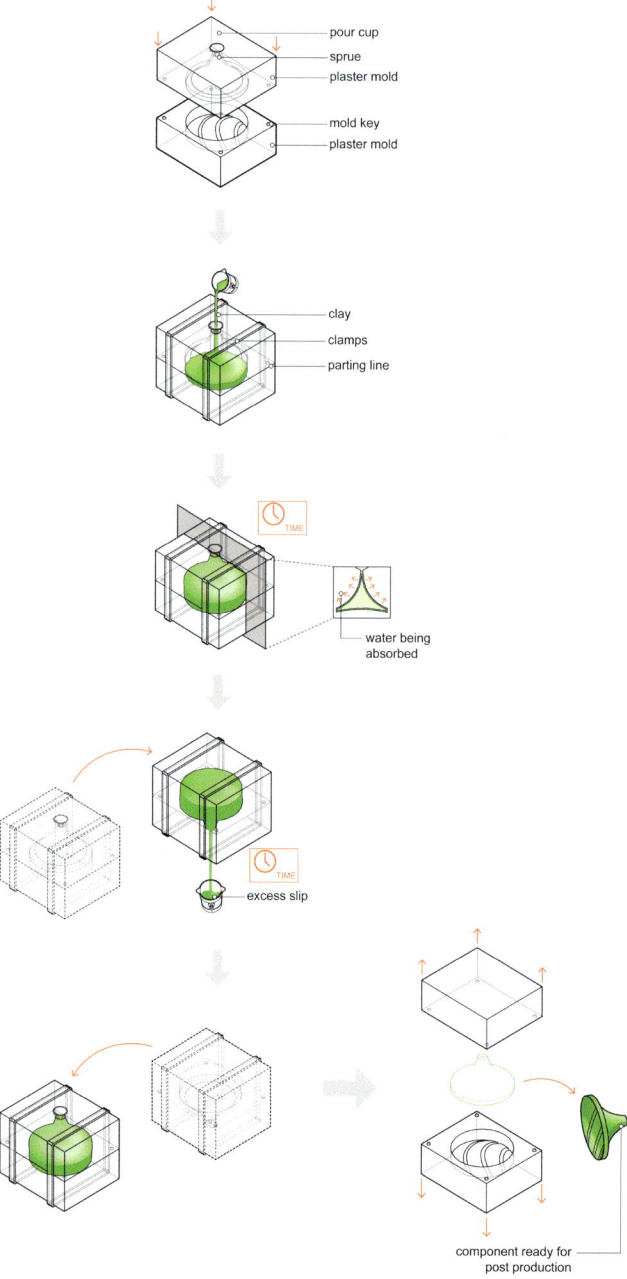

pour cup
sprue
plaster mold

mold key
plaster mold

clay
clamps
parting line

TIME

water being
absorbed

TIME
excess slip

component ready for
post production

Figure 3.6.1
Slip casting process diagram.

DOI: 10.4324/9781003299196-18

Slip casting is the manufacturing process in which a liquid clay or porcelain slurry, called slip, is poured into a closed, semi-closed, or open plaster mold. The plaster absorbs the water from the slip, leaving the clay or porcelain behind. Slip casting is generally used to produce relatively small, hollow components and can produce shapes with complex curves, fine surface details, and crisp edges. Since slip casting relies on plaster pulling water from the slip, molds must be made from plaster; however, plaster molds are relatively inexpensive, are easy to form, and therefore capital costs are low. The component's wall thickness will depend on the water-to-clay ratio of the slip and the length of time the slip is left in the mold. Generally, slip casting is time intensive and often multiple molds are used for parallel productions to reduce the overall production schedule.

In slip casting, the multipart plaster mold is closed and clamped together typically with band-style clamps. Casting slip is poured into the mold and water from the slip is absorbed by the plaster, leaving a film of clay or porcelain on the mold surface. The manufacturer leaves the slip in the mold for 5–30 min, depending on the component size and detail, the amount of water in the slip, and the desired wall thickness. Once the time limit is reached, excess slip is poured out of the mold and is allowed to drain. After draining, the casting continues to sit in the mold for 1–24 hr, until the component is stiff enough for demolding. The sprue or any excess is trimmed before the mold is opened and the component is removed. When removed from the mold, the component is ready for any necessary post-production processes and will then be allowed to harden until it is ready for glazing or firing.

Molds for slip casting are made of plaster and can be closed, semi-closed, or open.[1] This process is best suited for closed molds or semi-closed molds that have small component openings relative to the component's depth (e.g. vases). The molds are fabricated in at least two parts, but it is not unusual for molds to be made in three to five parts. The number of parts depends on the complexity of the component's design and having more mold parts increases mold and labor costs. The plaster for the mold can be cast against a pattern—this is often done when multiple molds are needed for parallel productions—or can be CNC-milled from a solid cast plaster block.

Slip

The slip used for slip casting is a mixture of clay or porcelain, water, and additives (e.g. sodium silicate, soda ash, and polymers). The additives make the casting slip more fluid than traditional slips used in clay working and the consistency of the casting slip should be like thick cream.

Figure 3.6.2
Restored front facade of the Eastside Townhouse. Its hand-pressed terracotta components are indistinguishable from the original carved brownstone elements. MKCA had to obtain authorization from the NYC Landmarks Preservation Commission to recreate the townhouse's neo-Grecian, brownstone front facade with terracotta instead of stucco or precast concrete. Photography by Alan Tansey.

VARIEGATED TERRA COTTA RAINSCREEN

VERTICAL GARDEN

ROOF TERRACE//ENTERTAINING AND FAMILY SPACE

LIBRARY//ENTERTAINING AND FAMILY SPACE

BEDROOMS//PRIVATE SPACE

MASTER SUITE//PRIVATE SPACE

LIVING AND DINING//ENTERTAINING SPACE

Figure 3.6.3
Building section by Michael K. Chen Architecture (MKCA), highlighting new rear facade with custom extruded terracotta rainscreen and slip-cast planters for the vertical garden wall. Drawings courtesy MKCA.

Eastside Townhouse in New York City, New York
By Michael K. Chen Architecture

The client for the Eastside Townhouse was an international family that needed a residence for when they were in New York and could be used for entertainment, meetings, and visiting family members. Completed as a renovation, the original house was a 1879 single-family home that had been transformed to single room occupancy (SRO) housing. The building's back facade had been covered with a tar and asbestos matrix and was so degraded that it crumbled when touched. For the renovation, the rear facade was demolished, the house footprint expanded toward the back, and a new facade was added. The final building is seven stories, including a cellar and a roof garden, and is over 9500 ft² (883 m²).

Figure 3.6.4
Aerial photograph of new rear facade and garden. Landscape Architecture by Local Office Landscape Architecture. Photography by Alan Tansey.

Figure 3.6.5
Photograph of new rear facade and vertical garden wall. Photography by Alan Tansey.

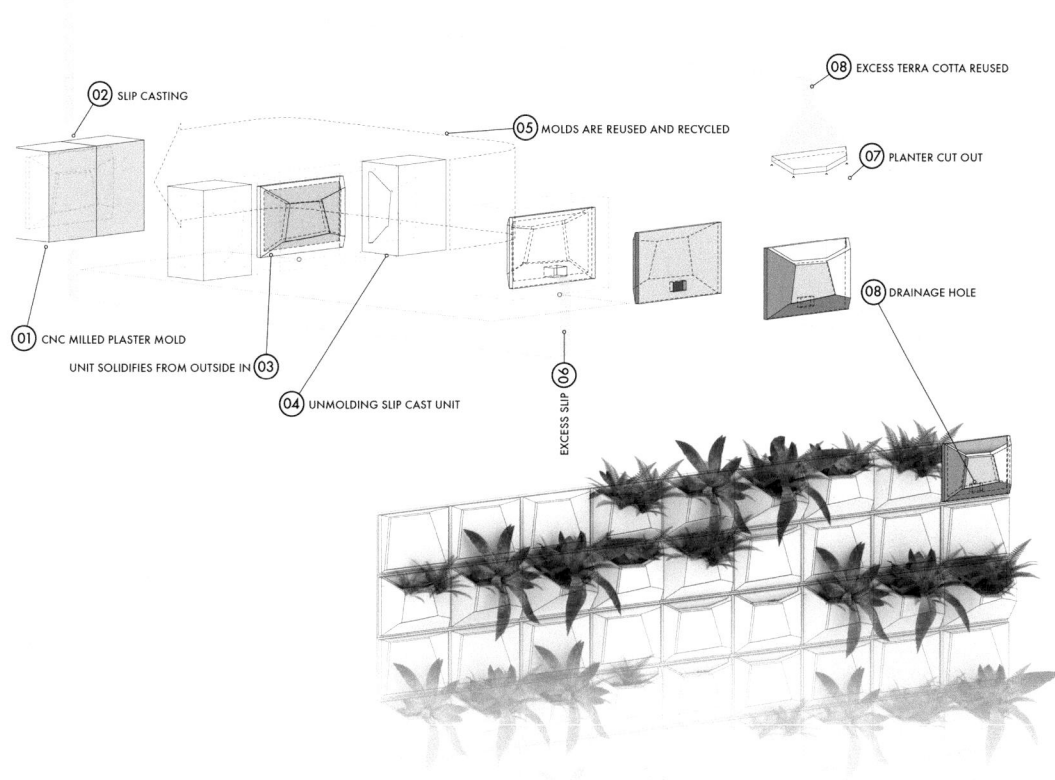

Figure 3.6.6
MKCA diagram of slip casting process used by Boston Valley Terra Cotta. Diagrams courtesy MKCA.

Michael K. Chen Architecture (MKCA) chose terracotta for the new rear facade because of the building's front facade. The front facade needed restoration and was made of a local brownstone that is no longer available. In New York, most brownstone restorations use stucco or precast concrete; however, because the original facade had sharp V-grooves between the brownstone blocks, founding MKCA Principal Michael Chen believed that the block details indicated machined rather than hand-craft aesthetic.[2] With the mechanized look, MKCA felt that terracotta would be more period appropriate. In consultation with Buro Happold, the project's structural, MEP, and facade engineering consultant, MKCA connected with Boston Valley Terra Cotta, located in Orchard Park, just outside of Buffalo, New York. To get to know the manufacturer and their capabilities, MKCA visited Boston Valley's facilities. Through their visits and discussion, MKCA realized that Boston Valley would be a great collaborator and well suited to make the components for the building's rear facade. As Chen noted that once the firm finds a partner, "they tend to go all in."

Figure 3.6.7
Photograph of the slip-cast elements during production at Boston Valley: (a) slip-cast element, after demolding; (b) planter being hand trimmed for opening; and (c) fully trimmed planter. Note the difference between the interior and exterior surface quality of the slip casting. Process imagery courtesy Boston Valley Terra Cotta.

Figure 3.6.8
MKCA design rendering of ground level, looking toward the backyard. Both the custom hydraulic-pressed interior tile and custom slip-cast planters are illustrated. Drawings courtesy MKCA.

On their tour of Boston Valley, MKCA saw the different ways that the manufacturer works with terracotta, including hand pressing, hydraulic pressing, extrusion, and slip casting. MKCA customized all of Boston Valley's processes for the Eastside Townhouse. The historic pieces on the building's front facade used hand pressing, an interior double-height feature wall used hydraulic-pressed clay tile, the rear facade used extruded rainscreen tiles, and backyard vertical garden wall used slip-cast planters. Slip casting was ideal to make the planters because it inherently forms hollow components that could hold the soil and plants (see Figure 3.6.7).

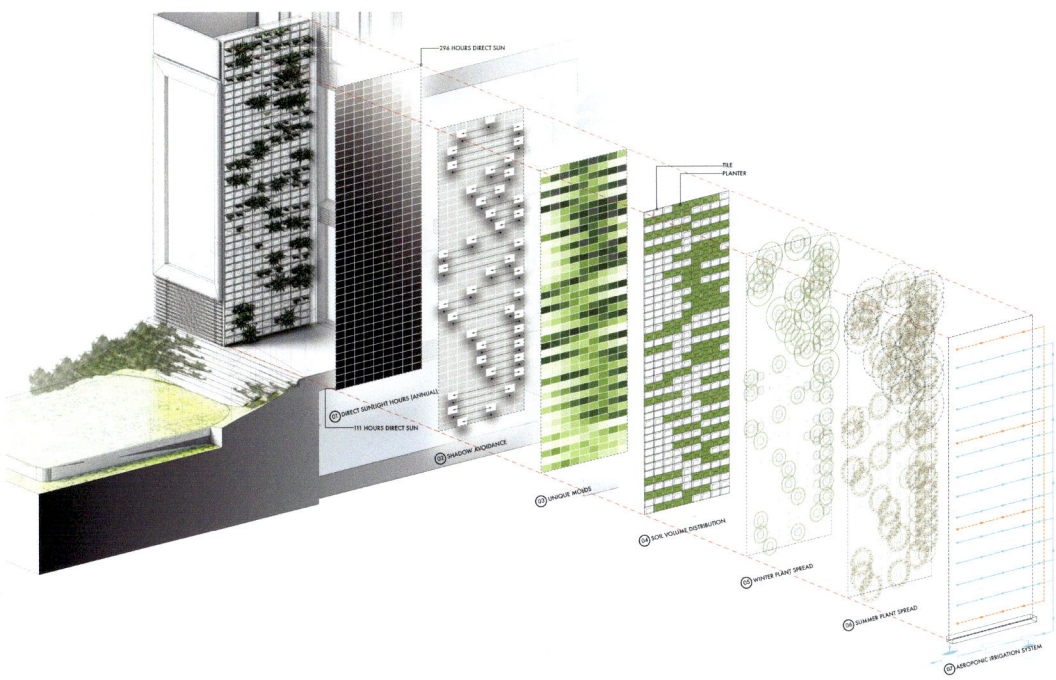

Figure 3.6.9
MKCA studies of the vertical garden. Succulents are at the garden top where they receive the most direct light, and ferns and mosses are toward the bottom, where they will be shaded more. The vertical garden uses a mist-driven irrigation system, located between the rainscreen mounting track and the planters, so that it is hidden from view. Diagrams courtesy MKCA.

MKCA designed the vertical garden in collaboration with Local Office Landscape Architecture and State University of New York (SUNY) conservation botanists. The design team obtained permission from the United States Fish and Wildlife to use some of the planters to propagate endangered ferns from New York State Hudson Valley's cliff faces. In an urban location, because of the narrow building lot, access to the sun was precious and MKCA wanted to limit self-shadowing as much as possible. The planters' sizes, depths, and locations on the wall are determined by the daylight needs of the individual plants, and shadows from the deeper units do not cast on the plants below when not desired.

Boston Valley used 11 different molds to make the vertical garden tiles' 11 different profiles, with only three to four of those profiles being deep enough to be made into planters. All the components were slip cast on a CNC-milled, two-part plaster mold (see Figure 3.6.6). The sprue hole, used for pouring and draining the slip, stayed after casting and became either the drainage hole for the planter units or a hole for air exchange for the hollow, non-planter units. After demolding, the non-planter units would continue to dry, whereas the planter units would be hand trimmed at their top, creating the planter opening for plants and soil (see Figure 3.6.7). Boston Valley cut slots into the bottom edge of the slip-cast components, so that the components could be supported by their standard rainscreen track system. The dark and mottled glaze on the slip-cast elements gave them a handmade look and provided a contrast to the plants' bright green color and the lighter color of the rear facade's extruded profiles.

Figure 3.6.10
Boston Valley carved slots into the bottom of the slip-cast components. The slots with additional rubber friction gaskets to hold the units in place on a Boston Valley standard rainscreen track. Photography by Alan Tansey.

Figure 3.6.11
Detail image of the bottom of the garden wall. Photography by Alan Tansey.

Notes

1. The component shapes achievable with an open mold most likely would be better suited for other manufacturing processes such as hydraulic or hand pressing (Chapter 4.7) as the cycle times will be shorter and the labor costs lower than slip casting.
2. Chen, Michael K. *Personal Interview*. 2 May 2022.

Reference

Chen, Michael K. *Personal Interview*. 2 May 2022.

Forming Solid

4

This part includes those manufacturing processes that form solid components in either open or closed molds. Manufacturing processes include Chapter 4.1, Casting Concrete; Chapter 4.2, Casting Metal; Chapter 4.3, Casting Glass; Chapter 4.4, Vibration Press Casting; Chapter 4.5, Vibration Tamping; Chapter 4.6, Pressing; and Chapter 4.7, Injection Molding. Materials in this part are varied and include wet-cast concrete mixes, such as traditional Portland and white cement concrete, ultra-high-performance concrete (UHPC), glass fiber-reinforced concrete, and polymer concrete; dry earth-moist concrete; metal; glass; clay and stiff mud; and plastic. With open molds, surfaces in contact with the mold face are significantly better than surfaces not in contact with the mold face.

DOI: 10.4324/9781003299196-19

Casting Concrete

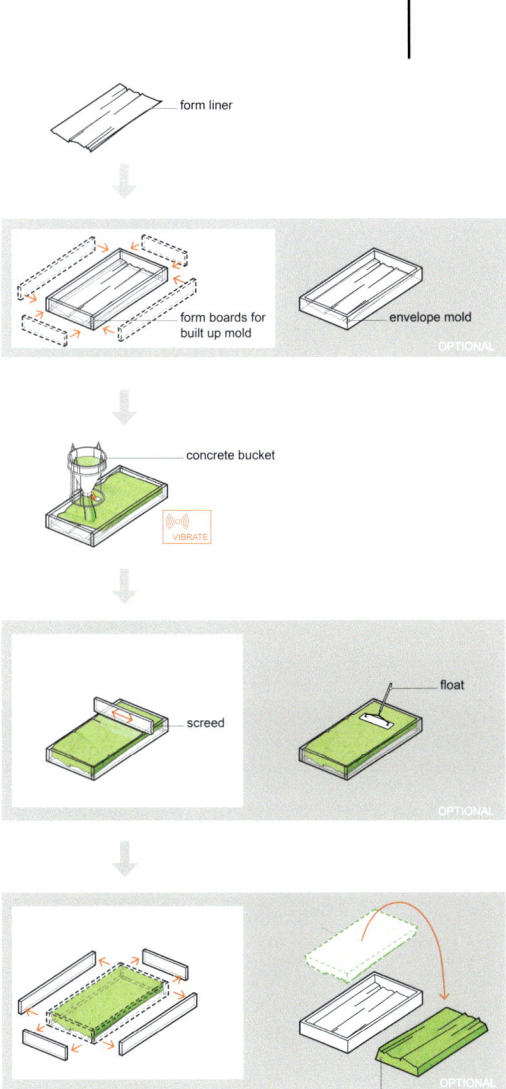

Figure 4.1.1
Casting concrete process diagram. Drawing by author. A version of this diagram originally appeared in *Manufacturing Architecture* (Laurence King, 2018).

Casting concrete is the manufacturing process that pours a wet mixture of concrete into a mold; it is the combination of gravity, water, additives, and vibration that help fully fill the mold with the concrete. Architectural wet-cast precast concrete, known simply as *architectural precast*, is designed and manufactured to be exposed and part of the aesthetic expression of the building. Architectural precast can be decorative and non-load bearing or structural. Casting concrete can produce complex geometries, simple shapes, sharp corners, and fine surface details. Component sizes are typically limited by shipping size and weight regulations (e.g. United States Department of Transportation and additional state regulations), site access issues, and erection crane limits. The molds for cast concrete must be able to support the weight of the wet concrete and can be made of a wide range of materials. Manufactures for architectural precast are specialized, as they require large equipment such as gantry cranes or mobile cranes to move the components, large yards for storage, and access for concrete mixing; however, small concrete manufacturers do exist but will be limited to small components that may not be structural. Generally, casting concrete is time intensive and often multiple molds are used for parallel productions to reduce the overall production schedule.

In casting concrete, the typical mold is open with the component cast finish face down. The mold may be an envelope mold in which it is solid on all five sides, or it may be a built-up mold in which some or all the sides are removed during demolding. Envelope molds require draft angles on the mold edges so that the component releases from the mold without damage to the mold or component; whereas built-up molds can have no draft angles as the sides are removed before demolding. The mold surface is sprayed with a release agent so that the concrete does not bond to the mold. The concrete is then poured into the mold and vibrated as needed. The open face of the concrete is screeded and may be floated or finished if required. The concrete is left in the mold for 12–24 hr so that it is set enough to be demolded. The concrete continues to cure outside of the mold until it reaches its needed strength.[1] The components can go through post-production processes such as grit blasting, acid washing, grinding, or polishing.[2]

Although typically cast in open molds, concrete may be cast in closed or semi-closed molds. These alternative molds are often done to form complex shapes or to create panels with two broad finished faces. During vibration, air bubbles release easily from horizontal or near-horizontal surfaces but tend to get trapped on vertical or upper surfaces, forming bug holes. When using closed molds, many large bug holes may form on the upper mold surface, often necessitating patching on these surfaces to make their finish acceptable.

To keep water out of the joints in a built-up or multipart mold, the seams between the elements often need to be caulked before casting. Caulking increases the cycle time and labor costs of built-up molds when compared to envelope molds. Concrete molds can be made from a range of materials, such as foam; thermoformed plastic (Chapter 1.2); wood, plywood, or MDF; contact molded (Chapter 3.1), vacuum infusion process (VIP) (Chapter 3.2), or 3D-printed FRP; rubber or elastomeric; and steel.

Other terms used to describe this manufacturing process can include *precasting concrete*, *precasting architectural concrete*, and *wet-casting concrete*. Unlike the other processes in *Custom Components in Architecture* both the mold material (e.g. plywood, foam, or FRP) and the specific concrete mixture (e.g. Portland cement, brick and concrete composite, or ultra-high-performance concrete [UHPC]) will impact the component's design. Therefore, this book includes the descriptions and case studies of a range of different mold materials and concrete mixes.

[Mold] Foam

Foam is used for its ease of shaping as it can be shaped quickly by hand or by CNC equipment such as hot wire cutters and millers. The benefits of foam are balanced by its low durability, as manufacturers often dispose of foam molds after a single cast, although multiple pours can be done if using a high-density foam with greater draft. In addition, the insulating properties of foam will trap heat generated by the concrete during the curing process; this in turn may negatively affect the concrete's surface quality.

[Mold] Wood

Included with wood are wood products, including molds made from plywood, MDF, melamine particle boards, or any mixture of those. Wood molds are typically built up by hand, with only key elements or specialty items fabricated by CNC mills. To keep water from being absorbed by wood molds, the inside face of the mold may be treated with wax, polyurethane, a thin sheet of fiberglass, or a self-leveling resin. Joints between mold parts are caulked between pours. The caulk keeps concrete from seeping into the joints and lengthens the life of the mold. Often the caulk is stripped as the component is demolded, and

therefore, wood molds must be recaulked between pours, increasing cycle times and labor costs.

[Mold] 3D-Printed FRP

Recent developments with 3D printing have led to using large 3D-printed carbon fiber–reinforced plastic molds for precast concrete. Once the printed FRP cures, its casting surface must be smoothed to remove traces of the printed layers. This can be done by hand but is best done by a five-axis CNC miller. The smooth surface eases demolding, reduces stress on the mold, and increases the mold's durability. 3D-printed FRP molds are expensive but have been proven to be extremely durable and can be reused for 100 pulls or more. Unlike built-up wood molds, 3D-printed FRP molds can be made seamless at corners, reducing or eliminating the need for caulking between pours and thereby decreasing cycle times and labor costs.

[Mold] Rubber or Elastomeric

Rubber or elastomeric molds are fabricated by casting a liquid latex onto a pattern to form the mold. The pattern can be made from a range of materials, including MDF and high-density foam. Rubber molds are expensive because of this two-step fabrication costs (i.e. making the pattern and then casting the mold) and because the rubber is costly. Rubber is durable and lasts 50–100 pulls. Rubber molds are flexible and are best suited for components with sharp corners, steep or negative draft angles, or fine surface details.

[Mold] Steel

Steel is typically used to manufacture structural precast (e.g. single Ts, double Ts, and hollow core). It is extremely durable and is suitable for high-volume productions over a long period of time. Unlike molds made from wood, FRP, or rubber, steel molds will not degrade over time and the first casting will be nearly identical to a casting done on the same steel mold five years later. Metal is expensive and difficult to shape; therefore, it is best suited to produce a lot of components with simple details.

[Material] Portland or White Portland Cement-Based Concrete

Portland or white Portland cement-based concrete is a traditional concrete that mixes cement with large and small aggregates, water, and additives. Portland and white cements are chemically similar and can be used to make concretes to the same ASTM standard; however, white cement is finer texture, sets slightly faster, and more expensive than Portland cement. The color of Portland cement can vary, resulting in concrete that ranges from grey to buff, depending on the manufacturer, the batch, and the curing conditions. If color consistency is important, then consider using white cement and a color additive. Due to the manufacturing process, precast concrete components made with Portland or white cement are less porous and have greater strength and density than site-cast concrete, making precast concrete a suitable exterior building material.

[Material] Composite

Concrete panels can be composite, with different materials used for the component's finished face than the back of the panel. A finished face may use a white cement-based concrete, while a less expensive Portland cement is used for the back of the panel. A finished face may use non-cement-based materials such as brick or ceramic tiles with concrete as a backup material. For composite panels, the face material must properly bond to the backup materials. If forming a composite concrete panel, the back face of the first layer of concrete is often scored to create a mechanical connection between the composite layers. If the face material is brick or ceramic, then the manufacturer may rely on a chemical connection between the brick or ceramic and the concrete; however, in regions that experience freeze-thaw cycles, the brick or ceramic tiles may also rely on a mechanical connection such as scoring or dovetails to bond to the backup concrete.

[Material] Sandwich Panels

Sandwich panels are precast panels made with two wythes of concrete separated by a layer of solid insulation. The insulation is continuous and goes to the edges of the concrete, creating a full thermal break between the wythes. Plastic or FRP ties connect the wythes, through the insulation. The exterior wythe is 2 in or thicker, depending on the surface relief and performance, whereas the interior, or backup, wythe is 6 in or thicker. The backup wythe can be designed to be load bearing.

[Material] Polymer-Based Concrete[3]

Polymer-based concrete includes small and large aggregates and a polymer-based epoxy, or some other thermoset plastic, as the binding agent instead of cement. Polymer-based concrete can use decorative aggregate and can be reinforced with steel. Polymer-based concrete is more chemically and water resistant than cement-based concrete. For casting, polymer-based concrete flows easily and it is suitable for components with thin walls and decorative surfaces. Polymer-based concrete is more expensive, more prone to creep, and may perform worse under fire conditions than cement-based concrete.

[Material] Wet-Cast Fiber-Reinforced Concrete (FRC)

Wet-cast fiber-reinforced concrete (FRC) can also be known as *direct cast FRC* and is different than the layered approach associated with spray- or trowel-applied, contact-molded FRC (Chapter 3.1). Portland cement is typically used as the binding agent and fibers can be glass (GFRC), plastic, steel, basalt, or any combination of those. Generally, in wet-cast FRC, the aggregate sizes are kept small and the fiber length short so that the fibers do not tangle during the mixing and casting process.

[Material] Ultra-High-Performance Concrete (UHPC)

UHPC can also be known as *reactive powder concrete*. UHPC uses Portland or white cements, mixed with other fine powders (e.g. silica fume and quartz flour), sand, fiber reinforcement, water, and additives. The additional powders are so fine that the sand essentially becomes the large aggregate of the concrete mixture. UHPC is a type of FRC, typically using steel fibers for reinforcement, but could use plastic or glass fibers. UHPC has a very low cement-to-water ratio and includes admixtures and plasticizers to increase the concrete's workability. UHPC costs substantially more than traditional concrete; however, it is significantly stronger and often does not use traditional steel reinforcement (e.g. rebar and welded-wire fabric). UHPC is very flowable and is well suited to components with fine details, thin walls, and vertical pours. Since little water is used in the UHPC mixture, components often are steamed cured.

Carapace Pavilion Joshua Tree National Park, California, United States
By University of Southern California School of Architecture

The Carapace Pavilion is a 12-ft wide by 42-ft long (3.66-m × 12.8-m) shade structure in a volunteers' campsite at Joshua Tree National Park. The pavilion's

Figure 4.1.2
Carapace Pavilion by University of Southern California School of Architecture.

tube shape and square-shaped holes in the roof are inspired by the skeleton of a cholla cactus, and its dusty-pink color is inspired by the desert sunsets.[4] Its name, *Carapace*, comes from the hard outer shell of desert tortoises. The Pavilion was part of an undergraduate design-build studio at the University of Southern California School of Architecture, taught by Professor Douglas Noble and sponsored by the Precast/Prestressed Concrete Institute (PCI) Foundation.[5] Noble has been working with Joshua Tree for over a decade, using it as a site for fictional park programs like visitors' centers. The park is seismically active and has extreme climate conditions, including hot temperatures during the summer, large diurnal swings, and only around 5 in (127 mm) of annual rainfall. The park is remote, with dirt roads providing access to many parts. Also, it is a native culture sensitive area with historic artifacts buried throughout the park. After discussions between Noble and Joshua Tree, spurred by their past collaborations, USC proposed designing and building a new toilet building type, related to the beauty of the park, that could be used in place of the park's ordinary precast concrete vault toilet buildings.

Before the semester started, Noble established initial parameters for the design-build studio. First, the project would use precast concrete. The benefits of precast concrete compared to site-cast concrete are that it reduces the disturbance to the land and the need for onsite labor to access the remote site. Second, the project would limit the number of molds needed to cast the pavilion so that only one mold would be used to cast all the concrete components. Third, working with the National Park Service, Noble had done an initial sketch with the rough size of the pavilion, and the students used this to start the design. Fourth, Noble was advising a USC Master of Building Science student, Ivan Monsreal, who was interested in developing a software tool—first in Grasshopper and later in Dynamo—to establish design parameters.[6] Once the studio started, Monsreal continued to work with Noble and the students to update the tool with new parameters as requested by the studio. The design phase for this project was approximately five weeks, and then the students moved into fabricating the mold.

Figure 4.1.3
All of the panels were cast on the same mold and are a complex curve.

Figure 4.1.4
The two-part mold at Clark Pacific.

Figure 4.1.5
The bottom half of the mold with high-density CNC-milled foam.

The design-build course worked with Walter P. Moore Engineers for the structural analysis. Originally the pavilion was going to be made from typical, Portland cement concrete, but as the design developed, the loads were too high, the material too thick, and the weight too heavy. The studio

Figure 4.1.6
Students added a layer of resin and glass fiber mats on top of the foam. The seams between the fiber mats were offset from the foam panel seams. Students prepared the surface by sanding smooth. The plugs for the apertures had gelcoat only, without fiberglass and resin.

Figure 4.1.7
The five panels in fabrication yard at Clark Pacific. The two side panels have been
attached to the bottom and the two roof panels are still unattached.

transitioned to using UHPC. UHPC eliminated any
traditional reinforcement (e.g. rebar or welded wire
fabric). Since the panels were double curved, fabricating the reinforcement to the pavilion geometry
would have been extremely difficult for the students. The UHPC strength and internal steel fibers
allowed the pavilion to be made with a 2-in (50-
mm) cross section at its thinnest parts and allowed
for the pavilion's significant double cantilever.
UHPC is more durable than typical concrete, allowing the pavilion to last 100 years or more with no
maintenance.[7] Since UHPC is more flowable than
typical concrete and will seek its own level during
casting, the double-curved, saddle shape of the

panels required that the UHPC be cast in a closed,
two-part mold.

The students fabricated both halves of the
mold. The bottom mold half was a plywood egg
crate, with a high-density foam and glass fiber–
reinforced plastic (GFRP) casting surface. Using
the digital files created during the design process, students CNC-milled sixteen and a half foam
panels that measured approximately 3 ft by 8 ft
(.91 m × 2.44 m). The plywood and foam panels
were brought to Clark Pacific, a precast concrete
manufacturer with a plant in Fontana, California.[8]
After assembling the plywood and foam panels at
the plant, students applied resin and glass fibers

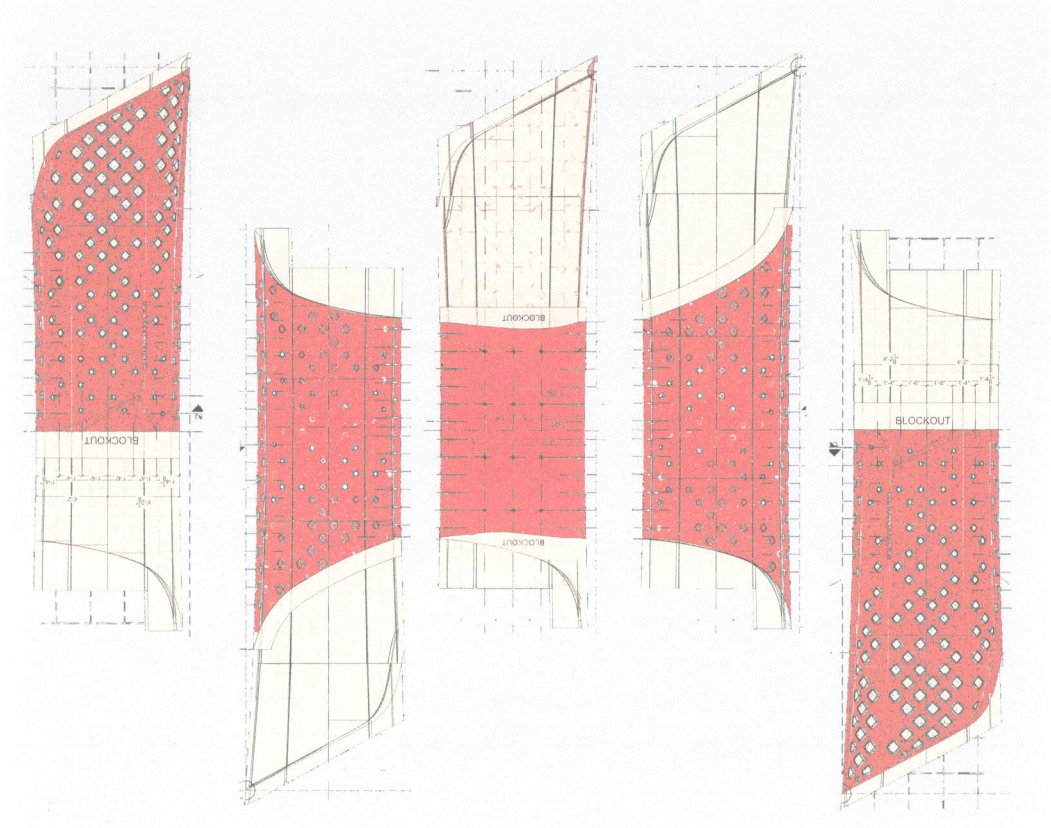

Figure 4.1.8
An unrolled plan of all five panels as they appear on the pavilion, with the mold repositioned for casting. From left to right, one roof panel, one side panel, the bottom panel, one side panel, and the other roof panel.

to the foam surface, sanded, applied a finished gelcoat, re-sanded, and applied a final gelcoat to the mold surface. Students cast five elements—one panel for the base, two panels for each side, and two panels for the roof. Students cast the components on different portions of the mold and used blockouts, or dams, to constrain and shape the panels' ends (see Figure 4.1.8). The bottom mold needed minor repairs between casts; Noble recalled that it took less than a day of work to repair the mold for the next cast.

The roof panels were the only two panels that were cast against the straight edge of the mold. The roof apertures at this edge were integrated with the CNC milling of the foam base. For the portions of the roof that overlapped with the side and base panels, the foam plugs for those apertures were installed after casting the side panels and before casting the roof panels. All the plugs were gel coated, but they did not have glass fiber reinforcing, as their protruding geometry made it too complex the lay the glass fiber mats without folds or bunches. According to Noble, since the apertures would only appear on the two roof panels, Clark Pacific was confident that the plugs with the gelcoating would be durable enough for two pulls. There are 99 apertures in each roof panel and each aperture is a unique shape. Noble estimates that half of plugs for the apertures were added between casting the side and roof panels. All the plugs have a draft angle that is measured against a vertical axis rather than perpendicular to the curved mold surface. The plugs point straight up as Clark Pacific demolded the panels by lifting them vertically from the mold. Only 18 of the 99 plugs stuck to the UHPC during demolding, and the team was able to recover 17 of those for reuse in the second roof panel casting.

Figure 4.1.9
Clark Pacific demolded the panels by lifting them vertically from the mold; therefore, all of the plugs for the apertures pointed vertically, with the draft angles measured off of the vertical angle.

Figure 4.1.10
The top portion of the mold half had two layers of thin plywood, with the seams offset from one another.

Figure 4.1.11
Casting the UHPC from a crane bucket.

Figure 4.1.12
The students applied a skim coat of UHPC to the outside of the pavilion surface to fill the bug holes that formed on the upper part of the closed mold.

The top portion of the mold half was a plywood egg crate with two layers of thin plywood, with the seams offset from one another. The two high points of the mold cavity had funnels. Students used one funnel to pour the UHPC into the mold. Since the mold was closed, when casting students could not see the level of the UHPC; instead, they poured the UHPC in one funnel and relied on the batter-like consistency of the UHPC to self-level and rise out of the funnel on the other side. By design, the foam apertures were to sit tight against the upper mold surface; however, only one or two touched the upper mold surface to create the through openings. For some apertures, only a little UHPC got between the mold parts and that would be removed with a hammer after demolding. For other apertures, there

Figure 4.1.13
A crane lifts the assembled pavilion off of a low-boy trailer to install on site.

Figure 4.1.14
Additional dirt was placed on the site to ensure any native artifacts are not disturbed.
The installation team used hand shovels to place dirt inside the pavilion.

was approximately ¾ in (19 mm) of UHPC covering the opening. Students used a rotary hammer drill to remove the UHPC for those openings. A substantial number of air bubbles, or bug holes, formed against the surface of the upper mold. The students applied a skim coat of UHPC to the outside of the pavilion surface to fill the bug holes.

The studio assembled the five cast panels at the Clark Pacific plant before placing the pavilion onto a low-boy style truck trailer for shipping. A small boom crane lifted the pavilion from the truck and placed it on site. Installation took less than 5 hr. To not disturb the existing landscape

and any possible existing artifacts, the team raised the level of the grade beneath the pavilion. The pavilion's foundation uses earth anchors, and dirt was backfilled to provide the inside floor. The pavilion is designed to accommodate two back-to-back toilet rooms in the future. The National Park Service has been happy with the structure but has no future plans to turn the Carapace Pavilion into a toilet building, as their needs have changed. Noble is working with the park for USC School of Architecture to design, build, and install tiny homes to solve a housing need for Joshua Tree Park rangers.

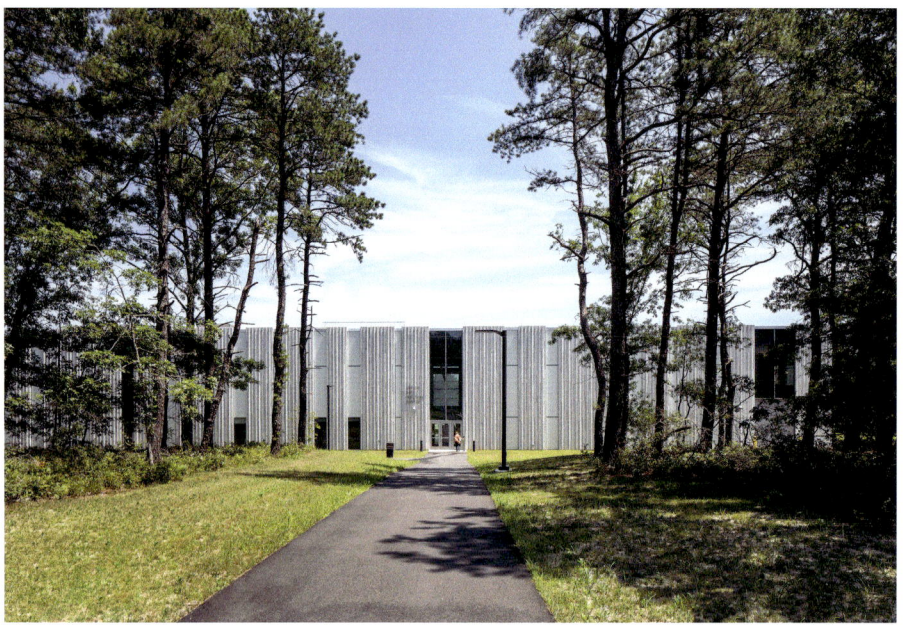

Figure 4.1.15
West entrance of the Health and Wellness Center, as seen through a stand of pine trees.
Image credit: ikon.5 architects.

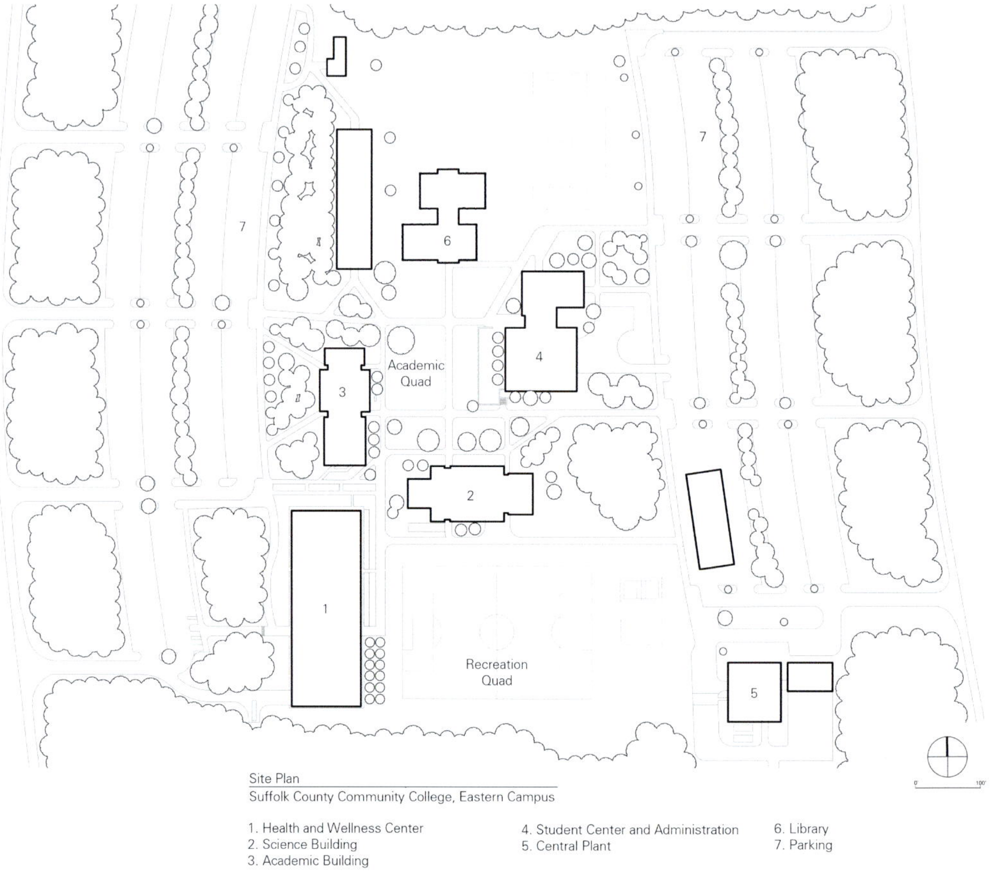
7
6
Academic Quad
3
4
2
1
Recreation Quad
5
7
Site Plan
Suffolk County Community College, Eastern Campus
1. Health and Wellness Center
2. Science Building
3. Academic Building
4. Student Center and Administration
5. Central Plant
6. Library
7. Parking

Figure 4.1.16
Campus plan, including the new building at the southwest corner.
It forms the west edge of the new Recreation Quadrangle. Image credit: ikon.5 architects.

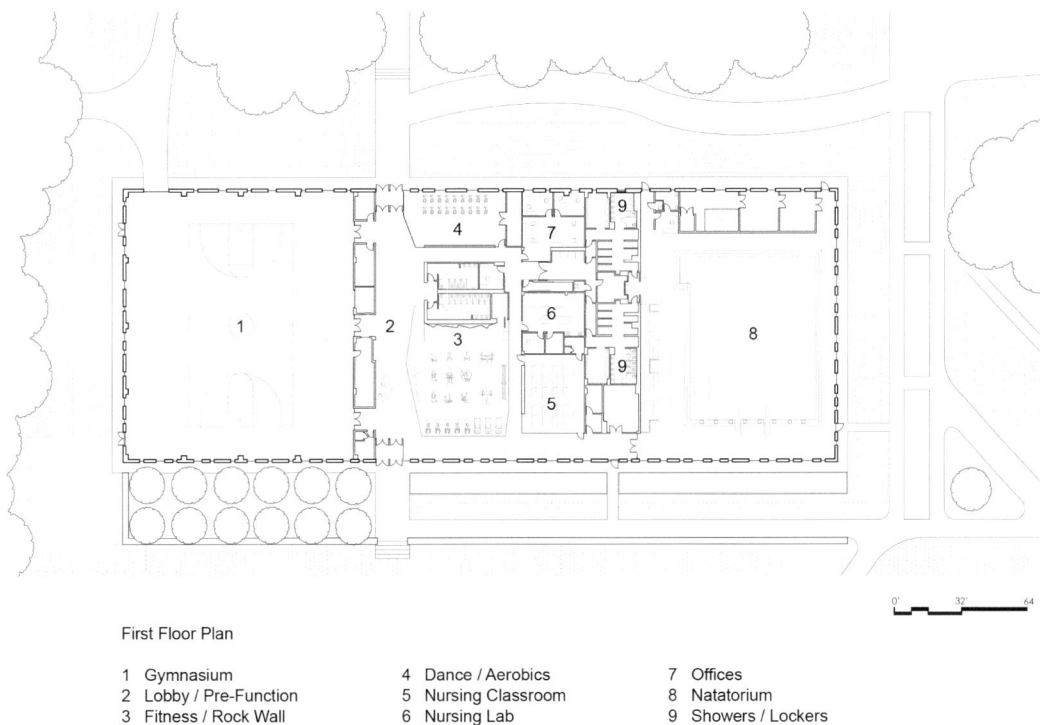

First Floor Plan

1 Gymnasium	4 Dance / Aerobics	7 Offices
2 Lobby / Pre-Function	5 Nursing Classroom	8 Natatorium
3 Fitness / Rock Wall	6 Nursing Lab	9 Showers / Lockers

Figure 4.1.17
Building ground-floor plan. Image credit: ikon.5 architects.

Health and Wellness Center for Suffolk County Community College in Riverhead, New York, United States
By ikon.5 architects

Suffolk County Community College is in Eastern Long Island, in an undeveloped area near Long Island State Pine Barrens Preserve. The Health and Wellness Center creates the western edge of the new Recreation Quadrangle for the community college. It is a 45,000-ft^2 (4181-m^2) facility that has academic classrooms for the Departments of Nursing and Physical Education and athletic facilities, including a natatorium, gymnasium, and locker rooms for students, faculty, and the neighboring community. The building's exterior has precast concrete insulated sandwich panels, window glazing, and a translucent panel system. Although precast sandwich panels can be used in a load-bearing application, the building's vertical structure uses steel columns. According to ikon.5 architects Principal Charles Maira, using structural steel columns instead of bearing on the precast panels, reduced the coordination between the trades, and its construction could be sequenced smoothly and quickly.[9]

Figure 4.1.18
Image of the short end of the building. Notice the detail of the butt joint with the reveals as the panel turns the corner. Image credit: James Ewing.

Figure 4.1.19
Image of the building exterior at night, showing the precast panels, vision glass, and translucent panels. Image credit: James Ewing.

Figure 4.1.20
Ikon.5 fabricated full-sized models made of painted foam. The small sample on the left was from a precast manufacturer that illustrated a stock design that the manufacturer proposed could be used instead. According to Tattoni, the scale of the sample's reveals was too shallow and texture too small to fit with the needed scale of the building. Image credit: ikon.5 architects.

Ikon.5 Architects Principal Joseph Tattoni stated that the design team chose precast concrete for several reasons. First, it provided a durable finish for the building's exterior and interior, particularly in relation to the chlorinated water of the natatorium and the impacts associated with the gymnasium. Second, it was a system that could be erected quickly. Third, its thermal mass would be beneficial in reducing heating and cooling loads. With the community college's location next to the Pine Barrens Preserve, and the building's site being right next to an existing stand of trees, Ikon.5 wanted to honor the site by designing the exterior surface of the precast panels as an abstracted representation of the surrounding tall stands of pines.[10] The design team digitized a photograph of a stand of pine trees, made that image into a drawing, and from the drawing made a 3D digital model in Rhinoceros. Ikon.5 made full-sized models from painted extruded polystyrene to evaluate.

Figure 4.1.21
CoreSlab's plywood mold with hardwood and plastic reveals. Image credit: ikon.5 architects.

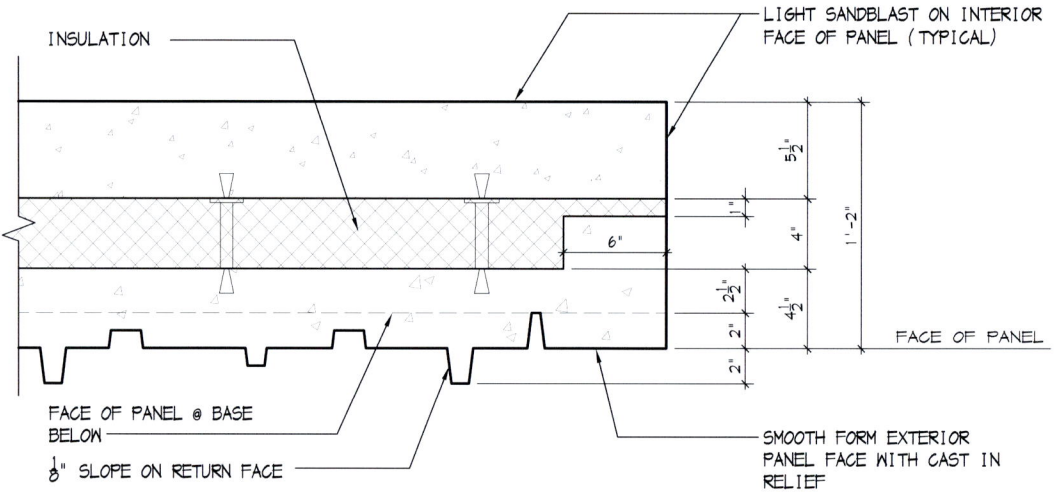

INSULATION

LIGHT SANDBLAST ON INTERIOR
FACE OF PANEL (TYPICAL)

FACE OF PANEL

FACE OF PANEL @ BASE
BELOW

⅛" SLOPE ON RETURN FACE

SMOOTH FORM EXTERIOR
PANEL FACE WITH CAST IN
RELIEF

Figure 4.1.22
Ikon.5 plan detail for typical panel edges, where the panel meets a window or translucent panel.
Image credit: ikon.5 architects.

Figure 4.1.23
Interior image of the gymnasium. Note the structural steel and light
sandblast finish of the precast. Image credit: James Ewing.

The Wellness Center was a publicly bid, design-bid-build project delivery method, with the contracts going to the lowest bidder. The general contractor for the Wellness Center selected Core-Slab Structures as the project's precast manufacturer. CoreSlab's plant in Thomaston, Connecticut manufactured the panels, and they traveled by truck 150 mi (241 km) to the site.[11] The sandwich panels have solid insulation that go to the panels' edges to provide a full thermal break. CoreSlab used Thermomass fiberglass ties that crossed the insulation layer and connected the concrete wythes together. On the exterior finish face are ribs that are either proud or recessed up to 2 in (50 mm) from the main surface. The exterior concrete wythe is a minimum of 2 ½ in (63.5 mm) thick, with 4 in (100 mm) of solid insulation. The interior concrete wythe is 5 ½ in (140 mm) thick. There is a 3-in (75-mm) concrete return at the panels' exterior edges, reducing the insulation from 4 in to 1 in (25 mm) and allowing the building openings to be set back from the outside panel face. The panels' concrete mix used Portland cement with an integral color for color consistency between the panels. The inside finish of the panels has a light sandblast, while the outside of the panels was cleaned but retained its mold finish.

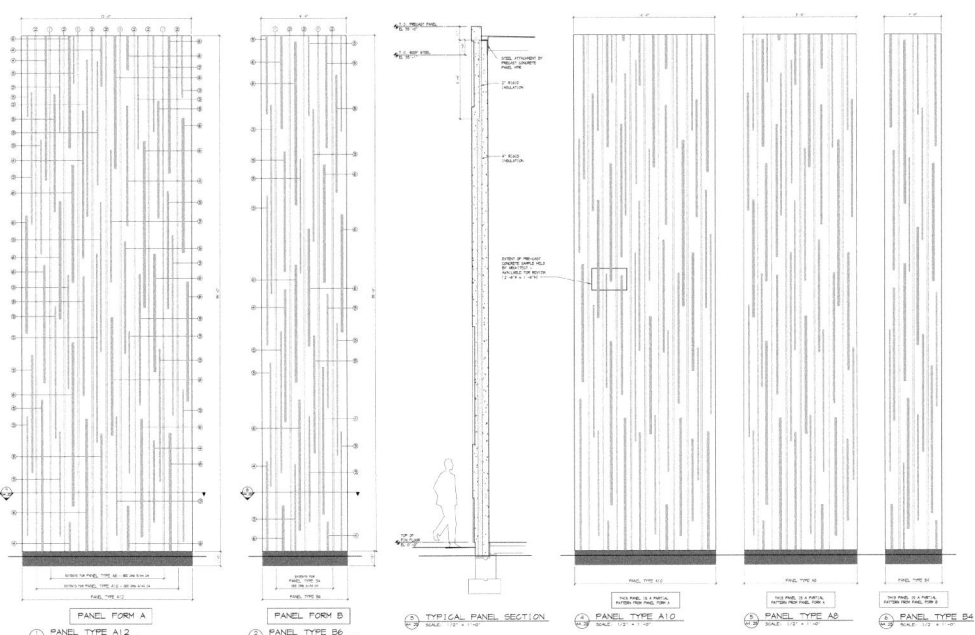

Figure 4.1.24
CoreSlab manufactured the five different panels on two different molds. CoreSlab used
Form A to make A12, A10, and A8 and Form B to make B6 and B4. Image credit: ikon.5 architects.

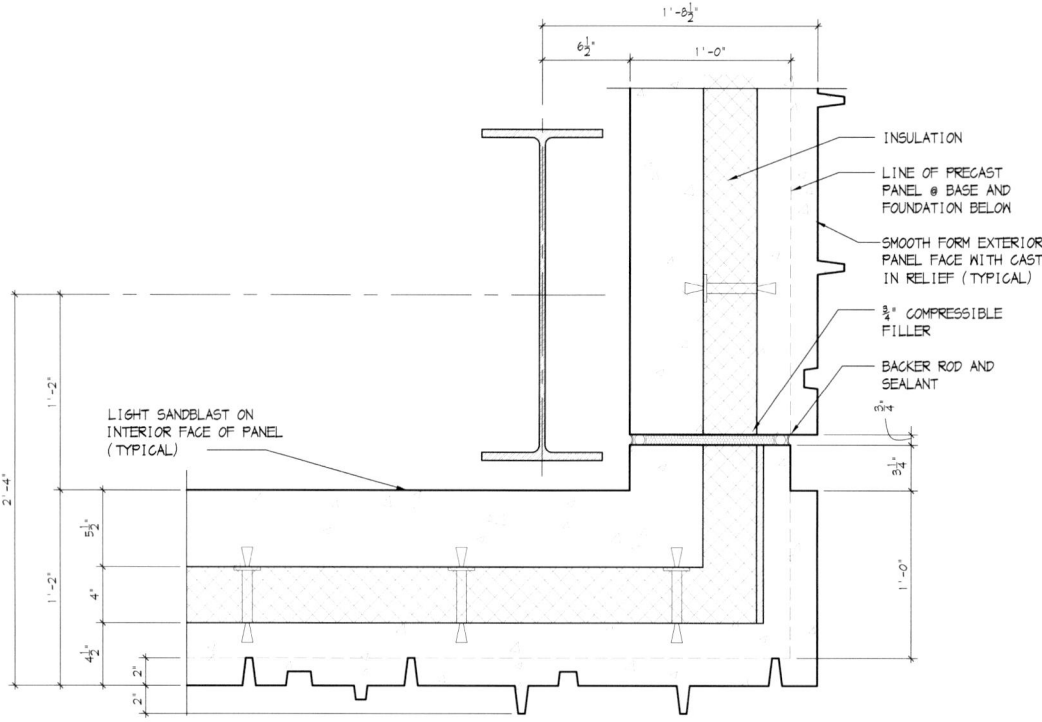

INSULATION

LINE OF PRECAST
PANEL @ BASE AND
FOUNDATION BELOW

SMOOTH FORM EXTERIOR
PANEL FACE WITH CAST
IN RELIEF (TYPICAL)

¾" COMPRESSIBLE
FILLER

BACKER ROD AND
SEALANT

LIGHT SANDBLAST ON
INTERIOR FACE OF PANEL
(TYPICAL)

Figure 4.1.25
Ikon.5 plan detail for building corners. Note the joint hidden by the reveal
requires that the reveal wood be removed prior to demolding. Image credit: ikon.5 architects.

The panels are 35 ft (10.7 m) tall and CoreSlab manufactured the five different panels on two different molds: Panel Form A and Panel Form B (see Figure 4.1.24). The molds were birch plywood with hardwood and plastic strips to form the ribs. According to CoreSlab Vice President and General Manager Robert Del Vento Junior, wood reveals were used for the longer projections as the plastic is less dimensionally stable with the temperature changes associated with casting concrete; however, CoreSlab uses plastic in its molds because it lasts longer than the wood. The edges of the mold are adjustable so that five different widths could be made on the same two molds. Form A was 12 ft (3.7 m) wide and formed panel A12. CoreSlab dammed Form A on both sides to make A10, a 10-ft (3-m) wide panel, and A8, an 8-ft (2.4-m) wide

panel. Form B was 6 ft (1.8 m) wide and formed B6. CoreSlab dammed Form B on both sides to make B4, a 4-ft (1.2-m) wide panel.

Ikon.5 distributed the panel types and widths across the facade, so that they appear random without an overall pattern emerging.[12] Two precast panels are always separated by a short section of curtain wall, except at the building corners. At the corner is a short return that allows for continuous insulation but appears as a butt joint on the building's less important short sides[13] (see Figure 4.1.25). The joint between the corner panels is hidden in the shadow of a reveal. CoreSlab would have to remove the reveal's wood molding before demolding. This was the first project that Ikon.5 has used precast insulated sandwich panels and since then has used them for later projects.

Figure 4.1.26
The building corner at the main facade, seen through a stand of pine trees. Image credit: James Ewing.

Figure 4.1.27
The parking garage from the corners of Tenth and Wyandotte Street in downtown Kansas City.
The renovated Crossroads Academy can be seen down Tenth Street, at the far end of the garage.
Photograph by Michael Robinson.

Figure 4.1.28
The main stairs on Tenth Street that take you to the second-floor level. Photograph by Michael Robinson.

Figure 4.1.29
The south face of the parking garage. Photograph by Michael Robinson.

Tenth and Wyandotte Parking Garage in Kansas City, Missouri, United States of America
By BNIM

The Tenth and Wyandotte Parking Garage in Kansas City, Missouri is 100,412 ft² (9329 m²) and has 300 parking spaces. The garage replaced a 30-year-old parking garage on the same site that was in disrepair. BNIM was working with developers on the renovation and addition of a neighboring office building for a new downtown charter school, Crossroads Academy Elementary, when the developers approached them about the parking garage. According to BNIM Associate Elvis Achelpohl, BNIM wanted to create cohesion with the two projects that addressed downtown Kansas City's eroded urban fabric due to 1970s and 1980s urban renewal.[14] BNIM designed

the new garage to form a small pocket park between it and the neighboring school for the school's protected use. Achelpohl said that the project developers viewed themselves as stewards of this area of downtown and supported the project to be more than just a parking garage. There were early discussions about making the parking garage mixed-use, but the client had concerns about project budget and property management. BNIM developed the idea to make the garage a public art piece. Independently, Achelpohl and Kansas City ceramic artist Andy Brayman had discussed the possibility of collaborating on a project. When BNIM realized that the garage may have an art component, Achelpohl invited Brayman to collaborate during conceptual design to make a proposal to the client. They made sure that their options fit in the anticipated project budget and could be produced by Brayman in his studio.

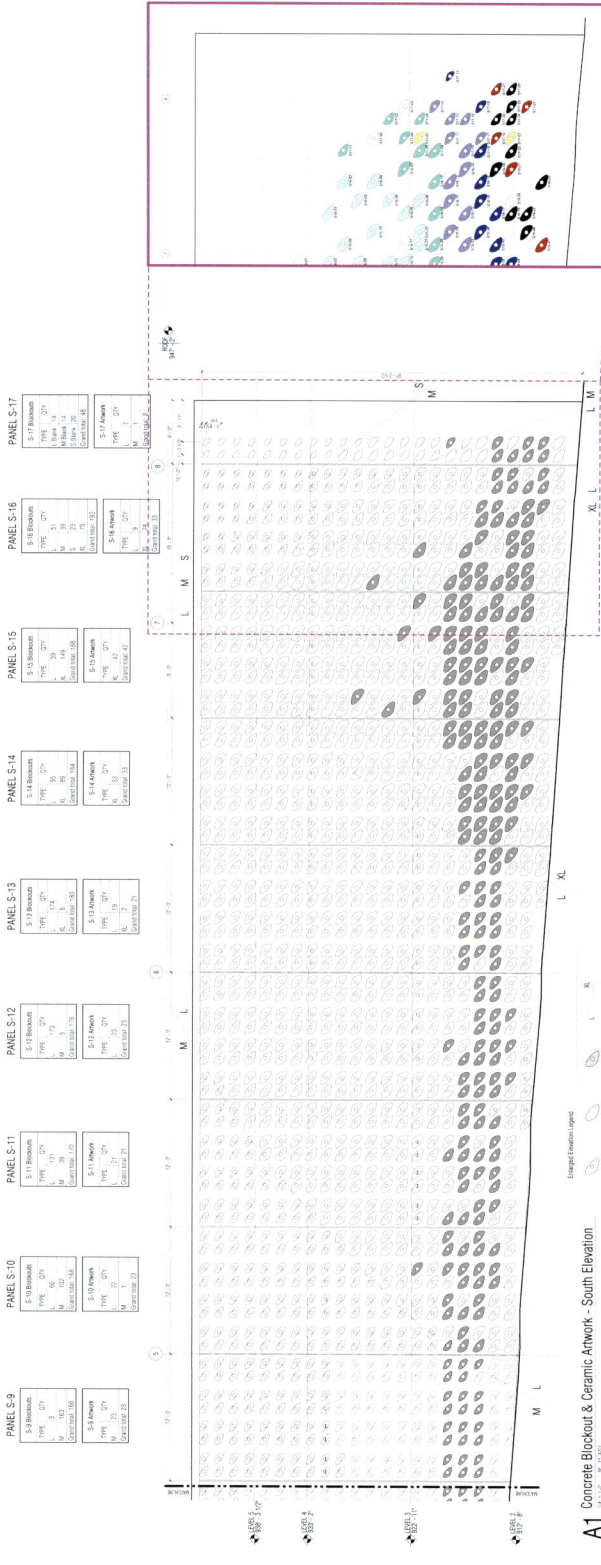

A1 Concrete Blockout & Ceramic Artwork - South Elevation

Figure 4.1.30

Partial South Elevation. The figure shows the location and size of bas reliefs, location of ventilation holes, and ceramic artwork. In addition, BNIM made a detailed elevation that depicted the location and color of every ceramic piece. (Partial drawing on the right.)

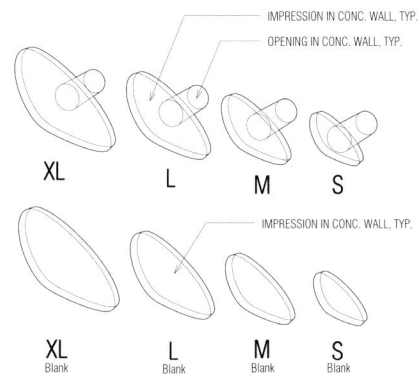

IMPRESSION IN CONC. WALL, TYP.
OPENING IN CONC. WALL, TYP.

XL   L   M   S

IMPRESSION IN CONC. WALL, TYP.

XL   L   M   S
Blank   Blank   Blank   Blank

Concrete Blockout Isometric Diagram
3/4" = 1'-0"   RE:

Figure 4.1.31
Drawing details of the blockouts.

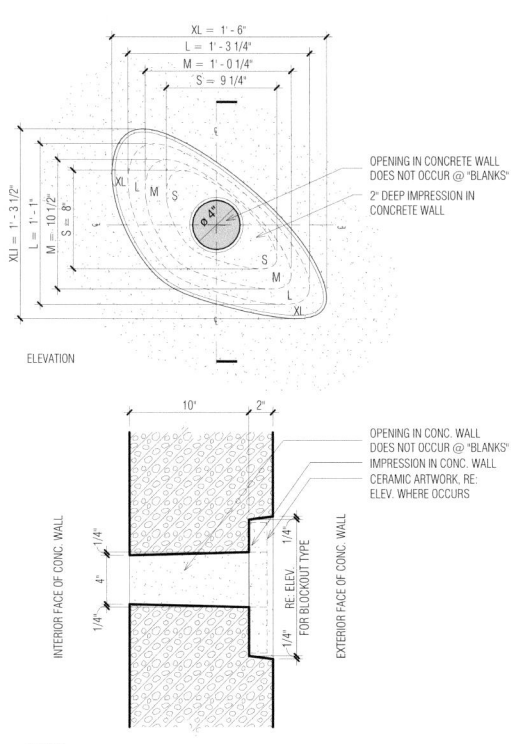

XL = 1' - 6"
L = 1' - 3 1/4"
M = 1' - 0 1/4"
S = 9 1/4"

XLH = 1' - 3 1/2"
L = 1' - 1"
M = 10 1/2"
S = 8"

Ø 4"

OPENING IN CONCRETE WALL DOES NOT OCCUR @ "BLANKS"
2" DEEP IMPRESSION IN CONCRETE WALL

ELEVATION

10"   2"

INTERIOR FACE OF CONC. WALL
1/4"   4"   1/4"
1/4"   1/4"
RE: ELEV FOR BLOCKOUT TYPE
EXTERIOR FACE OF CONC. WALL

OPENING IN CONC. WALL DOES NOT OCCUR @ "BLANKS"
IMPRESSION IN CONC. WALL
CERAMIC ARTWORK, RE: ELEV. WHERE OCCURS

SECTION

The design team selected precast concrete as it could serve as the garage's horizontal and vertical structural elements and addressed the project's need for cost effectiveness and efficiency. The Wyandotte Garage has 12-ft (3.66-m) wide, 12-in (30-cm) thick, load-bearing precast panels that showcase Brayman's work and provide a visual barrier to the cars. Within their 12 in depth, the panels have 2-in (5-cm) deep, ovoid-shaped bas reliefs in four different sizes—small, medium, large, extra-large, ranging in widths from 9 ¼ in to 1 ft 6 in (235–457 mm). When not interrupted by structure or service space (e.g. mechanical room), each bas relief has a 4-in (10-cm) diameter tapered hole that goes through the panels and provides ventilation to the garage behind. There are 5302 bas reliefs in the panels. Brayman designed and manufactured custom pressed ceramic tiles for approximately 10% of the bas reliefs. Brayman designed the ceramic tiles using Rhinoceros and Grasshopper, which allowed for seamless communication with BNIM. Brayman used a CNC machine to fabricate the tile dies and a hydraulic press to manufacture the tiles. Brayman dip-glazed the tiles in eight different colors. The tiles were grouted into place, on site, just before project's completed construction.

Figure 4.1.32
CoreSlab Structures in Marshall, MO manufactured the precast panels.

CoreSlab Structures in Marshall, MO, 86 mi (138 km) from the building site, manufactured the precast panels. The design team coordinated the placement of the bas reliefs and ventilation holes with the panels' reinforcing steel and pretensioning cables. Architectural polymers, a precast concrete mold fabricator, made the blockouts for the bas reliefs and ventilation holes in two separate rubber pieces. According to CoreSlab Structures (Missouri) Sales Manager Brian Goeble, CoreSlab drilled holes through the casting bed and welded bolts to the form.[15] CoreSlab set the rubber blockouts over the bolts and tightened them to the mold. With the compression force on the rubber, no caulking was needed at the joints between the rubber blockout pieces nor the rubber and the casting bed.

Figure 4.1.33
Close-up image of the panel casting bed with the rubber blockouts.
The blockouts are spaced to provide room for the reinforcing strands.

Figure 4.1.34
The blockouts come in two parts, one for the bas relief and the other for the ventilation hole.

Figure 4.1.35
A detail of the south facade. The tile surfaces have a subtle texture and are glazed in eight different colors. Photograph by Michael Robinson.

Figure 4.1.36
View of One South First from the East River. Image credit: David Sundberg/ Esto.

Figure 4.1.37
The building has views of the Williamsburg Bridge and East River.

Figure 4.1.38
Image of One S. First residential tower (left) and commercial office tower (right).

Figure 4.1.39
Detail image of the One S. First residential tower.

One South First in Brooklyn, New York, United States
By COOKFOX Architects

One South First is a mixed-use building with a three-story ground-level podium, a 22-story commercial office tower that partially supports a 42-story residential tower and totals 462,000 ft² (42,921 m²) of floor area. The building includes 150,000 ft² (13,935 m²) of leasable office space, 15,000 ft² (1393 m²) of retail space, and 332 residential units with 66 of those being affordable. One South follows the 2013 masterplan completed by SHoP Architects and commissioned by developer and construction manager, Two Trees, for the redevelopment and transformation of Brooklyn's old Domino Sugar factory. The site is prominent as it is just north of the Williamsburg Bridge and has five blocks of waterfront access along the East River. As part of New York's Uniform Land Use Review Procedure (ULURP), SHoP's masterplan informed both the urban plan of the Domino development and the massing of the development's new buildings, including the two stacked residential towers of One S. First.

According to COOKFOX Architects Partner Arno Adkins, the request for proposals (RFP) for One South included the approved zoning envelope and program that the building architect needed to follow.[16] There were two things that led COOKFOX to propose using precast concrete as the exterior building material. First was Two Trees themselves. Two Trees was the project developer, owner, operational manager, and construction manager. They have in-house architects and MEP engineers as project managers who guide the design, trade coordination, and construction process. In putting together the RFP, COOKFOX researched Two Trees' previous construction projects; they noted that the developer had used prefabricated exterior wall systems on previous projects and would be interested in systems that would lead to quick onsite erection.[17] Second, instead of following the materials indicated by SHoP's masterplan, COOKFOX took inspiration from the site's history as a sugar factory and designed a facade with materials and a play of light and shadow that would be reminiscent of sugar crystals. COOKFOX knew that a white cement-based concrete could achieve the correct color; acid washing would bring out the reflective mica in the sand-rich mixture, giving the acid-washed faces a sugar-like sparkle; and that polishing the outer flat faces would offer a different type of shine and expose the aggregate for additional visual depth.

NORTH FACADE
SPRING / FALL SOLSTICE 7 AM

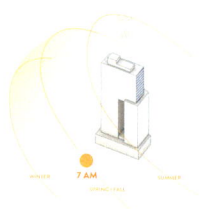

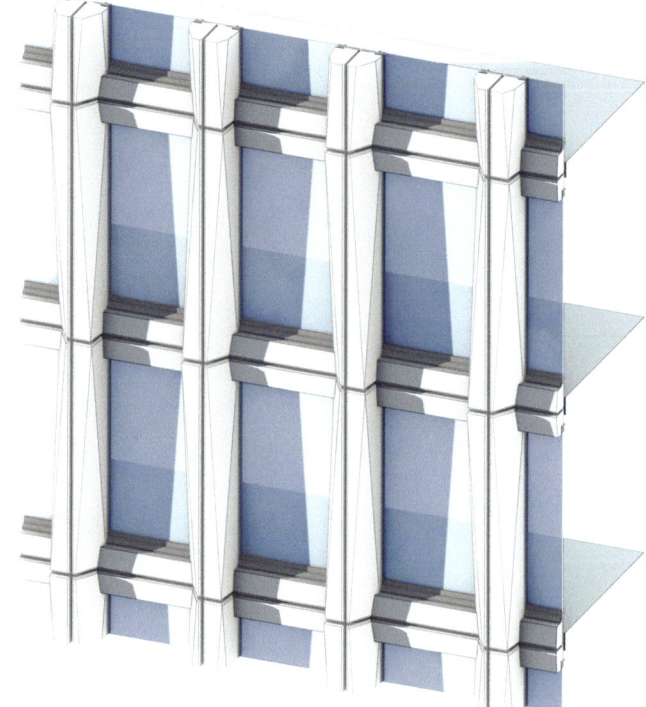

EAST FACADE
SPRING / FALL SOLSTICE 11 AM

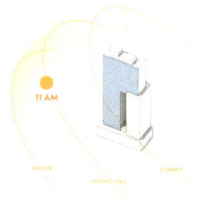

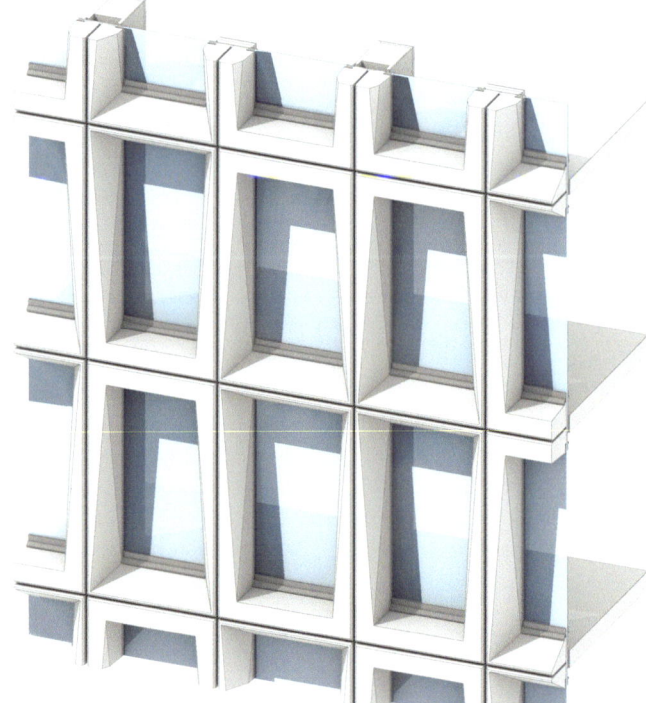

Figure 4.1.40
COOKFOX diagram of window and panel shapes based on facade orientation. Courtesy of COOKFOX
Architects. (*Continued*)

SOUTH FACADE
SUMMER SOLSTICE 12 PM (NOON)

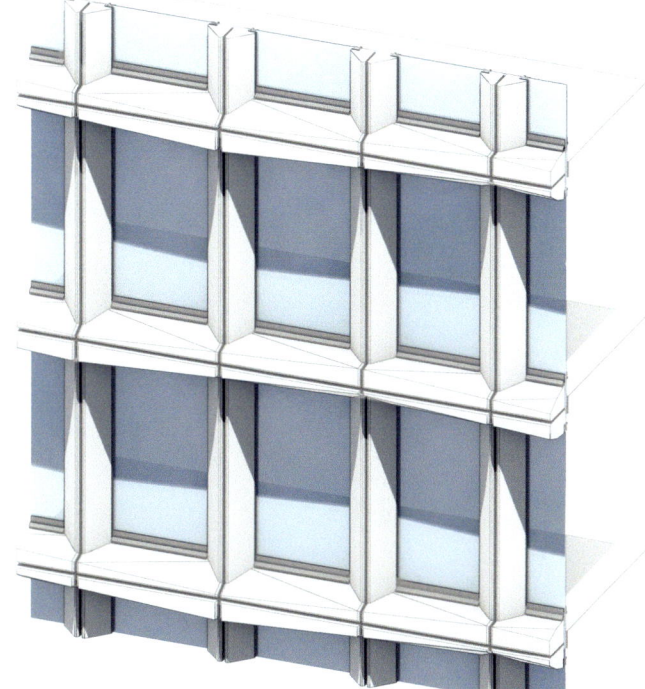

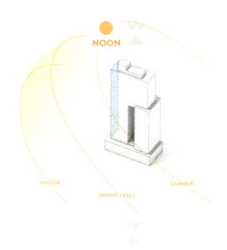

WEST FACADE
SPRING / FALL SOLSTICE 3 PM

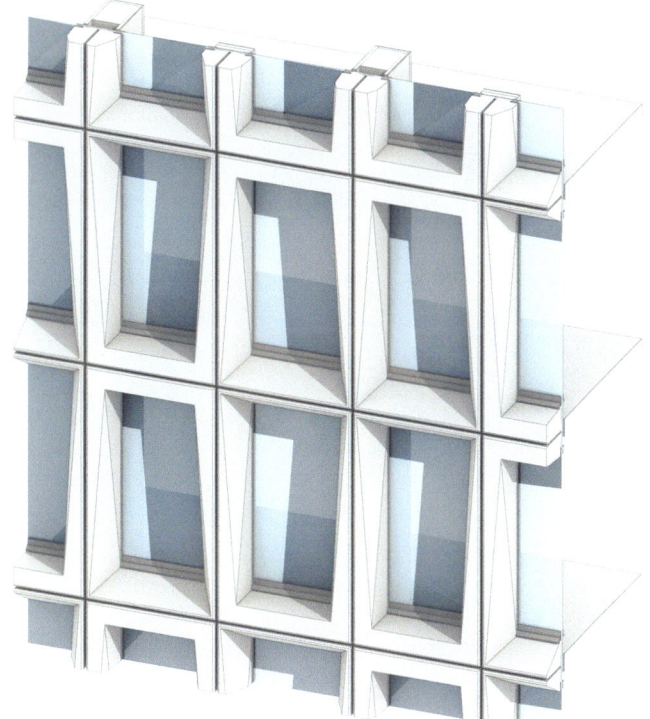

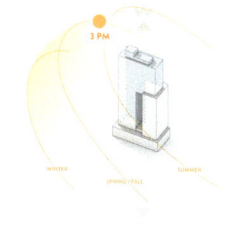

Figure 4.1.40
(*Continued*)

Figure 4.1.41
Photograph of precast molds being printed from carbon fiber–reinforced plastic.

One S. First has three distinct precast panels and window sizes, based on the building use. The ground floor of the podium has large panels that are either built up with precast pieces (i.e. separate headers, jambs, and sills) or are erected as single square donut-shaped panels. The commercial tower has large, non-operable windows and the precast panels are one to two window units wide.

The residential tower has smaller sized, sometimes operable windows, depending on its location in the apartment, and each precast panel is two to three window units wide. The panels' facets depend on their solar orientation, and COOKFOX designed them to reduce the effects of solar heat gain. The panels on the building's south face have deep horizontals to block daytime summer sun,

Figure 4.1.42
After being printed, the mold surface was milled smooth by a five-axis CNC. Courtesy of COOKFOX Architects.

the panels on the north face have deep verticals to block low west light in the summer, and the panels on the east and west faces have a waffle-like pattern with deep horizontals and verticals to block both overhead and low sun angles. The north, south, and east and west facades have a window shape that can be oriented in two different configurations across the facade, changing the windows' sills and headers within each panel.

During the design process, COOKFOX met with Gate Precast to discuss the project. It was through this discussion that the design team thought Gate would be a good partner to do design assist, as a consultant, to develop the project's precast system. After the exterior envelope bid package was issued, Gate bid on the precast and was awarded the contract. Independently, Oak Ridge National Laboratories (ORNL) has a Big Area Additive Manufacturing

Figure 4.1.43
Photograph of residential tower molds at Gate Precast. The window plugs are independent from the built-up mold, and each can be flipped 180 degrees.

(BAAM) that can print large components out of carbon fiber–reinforced acrylonitrile butadiene styrene (ABS) plastic.[18] ORNL was interested in how the technology could serve different industry sectors and reached out to the PCI, a voluntary technical and certification institute that serves the precast producers, to see if their large 3D printing technology could be of use to the precast concrete industry. As part of that study, ORNL and PCI had done some small test casts against a complex profile and were able to pull 30 casts from the mold without mold degradation.

One S. First has 1593 concrete panels and each panel requires repositionable molds, or plugs, to form its one to three window openings. During bidding, Gate had planned that all the window plugs would be built up with wood and plywood and coated with resin. However, carpenters with the necessary skill set to fabricate molds are in short supply, and Gate estimated that it would take a master carpenter 40–50 hr to fabricate each plug. A wood plug would cost $1800 to fabricate but could only be used up to 10 hr before degrading too much to be reused. Gate, a company involved with PCI, knew of the ORNL/PCI test casts and believed that the technology could be used for One South's precast panels. In contrast to the wood plugs, the 3D-printed plugs took 14–16 hr to print and mill smooth with a five-axis CNC-router and cost $9000 each; however, the 3D-printed plugs were more durable than the wood. Gate discovered that they could use the 3D-printed molds for over 200 pours before needing to be replaced. When amortized by the number of pours—$180 per pour for wood and $44.33 per pour for 3D printing—the 3D-printed plugs were much more cost effective.[19] Gate used

Figure 4.1.44
At Gate's storage yard, glazers installed windows in the precast panels prior to Gate loading the panels onto a truck for shipping.

38 3D-printed plugs and 100 wood plugs to produce the precast panels for One South.

In addition to cost, the 3D-printed plugs had other benefits over the wood plugs. First, casting concrete against the 3D-printed mold improved the mold's surface. Plywood and wood molds swell with concrete's moisture content and exothermic reaction. This causes the wood grain to become more pronounced, increasing friction during demolding. Instead, with 3D-printed molds, concrete fills the small gaps in the 3D-printed mold surface, making

Figure 4.1.45
COOKFOX detailed elevation and section. Courtesy of COOKFOX Architects.

the mold smoother and demolding easier in subsequent casts compare with earlier casts. Second, tolerances with the 3D printing were .05 in (1.27 mm), whereas wood mold tolerances are typically 1/8 in (3.2 mm). Third, the continuity of the 3D-printed corners reduced mold set-up times, as the wood mold required that the corners be checked for cracks and caulked, if needed, between pours. Ultimately, the 3D-printed molds decreased the production time for the precast panels. COOKFOX Partner Pam Campbell said, "Without the 3D printed molds it would not have been possible to create the forms that we wanted on the schedule that was necessary."[20] Fourth, Adkins contended that the quality of the panels' surface and the crispness of the corners were better with the 3D-printed molds than those cast on wood molds. The monocoque construction of the 3D-printed mold meant that the mold could

withstand greater vibration forces than wood, allowing for better concrete consolidation and fewer surface bug holes.

One South is the first building in the United States to use 3D-printed molds for precast concrete.[21] Adkins stated that COOKFOX was excited that Gate was using the 3D-printed molds. Instead of using the normal shop drawing process, Gate used the architect's Revit model to create the geometry needed to print the plugs.[22] In an interview with *PCI Journal*, Gate Senior Vice President of Engineering Steve Brock identified situations in which 3D-printed molds make sense for precast concrete: (1) the expense of steel molds cannot be justified, (2) the project requires more pours than what is achievable in wood, and (3) highly ornate profiles or shapes that are too complex to make in wood.[23]

Figure 4.1.46
Amare facade overlooking the public square that it shares with The Hague City Hall. © Ossip van Duivenbode.

Figure 4.1.47
Amare's north facade, looking back toward City Hall. © Ossip van Duivenbode.

Amare in The Hague, the Netherlands
By NOAHH | Network Oriented Architecture

Amare is the new home for the Netherlands Dance Theater (NDT), the Residentie Orkest, the Royal Conservatoire, and the Amare Foundation. The building is 581,251 ft² (54,000 m²), 131 ft (40 m) tall, and spans two city blocks. It has a mix of educational, cultural, and commercial spaces with four public performance venues—a 1500-seat or 2500-standing capacity concert hall, a 1300-seat dance theater, a 600-seat music conservatory hall, and a 200-seat studio. Amare is to the east of a public square that it shares with The Hague City Hall, a combined library and town hall, by Richard Meier. NOAHH | Network Oriented Architecture Founder Patrick Fransen referred to the building as a "super container" with the theaters acting as buildings within the building.[24] The performance spaces have their own material palette and articulation so that each is distinct and recognizable within the large Amare. Amare occupies the site of OMA's NDT, which was demolished[25] but the underground parking was retained and reused for this project.

Figure 4.1.48
Amare's bay elevation with main and secondary columns. © Ossip van Duivenbode.

Figure 4.1.49
Photograph of steel mold for main column at Westo. © NOAHH | Patrick Fransen.

Amare has no single main entrance, instead, the building has a series of entrances so that public accessibility is at the forefront of its design. To balance the acoustic isolation requirements of the performance and practice spaces with the desire to be open to the public, Fransen and his team placed Amare's structure outside of the building. This acoustically isolated the interior from the exterior and the lower public floors from the upper level performance and practice spaces. NOAHH provided space between the back of the structural elements and the building envelope, to further isolate elements and to provide access for glass cleaning. Behind the exterior structure are bird and bat houses, attached to the concrete, so that the animals could make a nest in the structure, "like in real trees."[26] Although not part of the initial design concept, Fransen recognized that Amare's facade has classical roots in Renaissance and Venetian forms.

Figure 4.1.50
Photograph of reinforcing steel being lowered into the steel mold.
© NOAHH | Patrick Fransen.

Figure 4.1.51
Close-up image of the steel mold surface as West is lowering the
reinforcing steel into the cavity. © NOAHH | Patrick Fransen.

Figure 4.1.52
Image of the main col-
umn being stored on
the ground. Note the
ribs appear on both
the back and front
faces of the column.
© NOAHH | Patrick
Fransen.

Figure 4.1.53
Looking up at the structural elements. The elements were left with their mold finish
with no post-production finishing (e.g. grit blasting and acid washing). © Jan Richard Kikkert.

Westo Prefab Betonsystemen BV, located approx-imately 125 mi (200 km) from the site, manufactured Amare's precast concrete structural exoskeleton. NOAHH worked collaboratively with the building contractor, structural engineer, and Westo for two years to design and detail the technical elements of the facade. Westo cast Amare's precast structure in closed steel molds, including the corner elements. Fransen said that the steel molds had a craftsman-ship and beauty to them, and it was almost a shame that they only poured concrete into the molds and did not use the molds themselves in the building. The existing, below-grade parking garage structural grid set the spacing for Amare's lower level columns. The upper level columns branch, creating smaller col-umns with narrower spacing that reflect the lighter loads at the upper floors. In earlier design itera-tions, NOAHH had investigated options in which the branching columns would fuse back together at the uppermost floors. Although the design team liked the option, it was too costly because of the complexity of the mold that would be required.

Figure 4.1.54
Image of the branching structure from inside the building. © NOAHH | Katja Effting.

Typically, precast concrete has only one broad finish face that is cast finish face down in the mold. The up-facing, or non-mold face, of the panel is screeded and possibly floated when surface quality is important. For Amare, the elements have two broad finish faces with articulated ribs on the front and back surfaces, and Westo used closed molds that provided a front and back finished face. Westo tilted the molds on an angle during casting to lessen air bubbles from forming on the inside face of the upper mold surface. Fransen believed that Westo chose steel for the mold material because of the precise dimensional tolerances required of the elements and the number of casts that needed to be produced. There are 50 main, lower level columns on the building with each main column branching into 3 secondary columns, totaling 150 secondary columns. Then each secondary column branches into 2 tertiary columns, totaling 300 tertiary columns. There are two different main column designs: (1) the columns along each of the four building sides, and (2) the eight corner columns with left- and right-handed versions of each. The main columns at the building corners are independent and do not connect with each other. In contrast, the corners' tertiary-level columns do fuse together at the top floor, therefore requiring a special mold for these four elements.

Figure 4.1.55
Amare during construction. The left- and right-handed main column variations can be
seen at the bottom of the image. In this photograph, the special fused tertiary column
has yet to be installed. © Sander Singor.

Figure 4.1.56
Schwabinger Tor S40, as seen from Leopoldstrasse. Photography by Michael Heinrich, München.

Figure 4.1.57
South elevation of S40. A small urban playground is in the foreground. Photography by Michael Heinrich, München.

Schwabinger Tor S40 in Munich, Germany
By Hild und K

Schwabinger Tor is a four-block dense redevelopment, located to the north of central Munich, between *Englischer Garten*, or English Garden, and *Olympiapark München*, Munich's Olympic Park. Winning a 2007 competition, 03 Architekten designed the Schwabinger Tor masterplan with nine, mixed-use buildings that total 960,000 ft² (89,000 m²) of space. The nine buildings are arranged in two rows that are offset from one another. This forms space between the buildings that have views to the west of Leopoldstrasse, a main avenue to central Munich and the former royal palace, or to a pedestrian greenway to the east. After 03 Architekten won the masterplan competition, individual design competitions were held for the buildings, which were to follow the masterplan's building height and footprint guidelines. Hild und K won the competition to design S40, a building in the southeast quadrant of Schwabinger Tor. S40 is a mixed-use building with commercial spaces on the ground floor, office spaces on the second through fourth levels, and residential units with exterior terraces on the top two floors.

Figure 4.1.58
The composite panel center pillows outward 4 ¾ in (120 mm). Photography by Michael Heinrich, München.

Figure 4.1.59
The joints between the panels are celebrated by precast concrete ornamental elements—a column capital between the windows and a colonnette at the window divider. Photography by Michael Heinrich, München.

Figure 4.1.60
A standard panel ready to be installed. The composite panels have a border of exposed concrete, giving a fabric-like appearance to the building. Photography by Sebastian Klich, Hild und K.

S40 is clad with precast brick and concrete composite rainscreen panels, which are rear ventilated with mineral wool insulation. Hild und K and Hagemeister, a German brick manufacturer, collaborated to develop custom bricks for the project. The bricks are Roman-style, thin bricks that measure 9 ½ × 4 ½ × 1 ½ in (240 × 115 × 40 mm) with a ribbed back face to increase the surface contact between the brick and the concrete. According to Hild und K Partner Matthias Haber, the office designed the precast panels so that the joints between the panels were a positive part of the building design.[27] The standard panel is an asymmetrical cross with the joints located next to integrated exposed concrete ornamental elements—a horizontal column capital and a vertical colonnette. The horizontal joints are in the middle third of the windows, while the vertical joints align with the window dividers and alternate location on every other floor.

Similar to other Hild und K projects, S40's facade has a fabric-like appearance.[28] The bricks are almost woven together, with a running bond in the panel's horizontal arms and an offset soldier bond on the panel's vertical arms. The head joint in both bond patterns has a one-third offset, as if the bricks are arranged in a bias drape. The two bonds weave together to create a herringbone-like pattern at the panel center. Like a panel in a quilt, the precast panel's center pillows outward 4 ¾ in (120 mm) from its edges. Also like a quilt, each composite panel's thin bricks are bordered by an exposed precast concrete frame, and the "stitching" between the panels is celebrated by the precast ornamental elements.

Hild und K met with precast manufacturers to develop the design, but when the project was opened to bids, prices for the precast panels came back too high. Hild und K then met with the

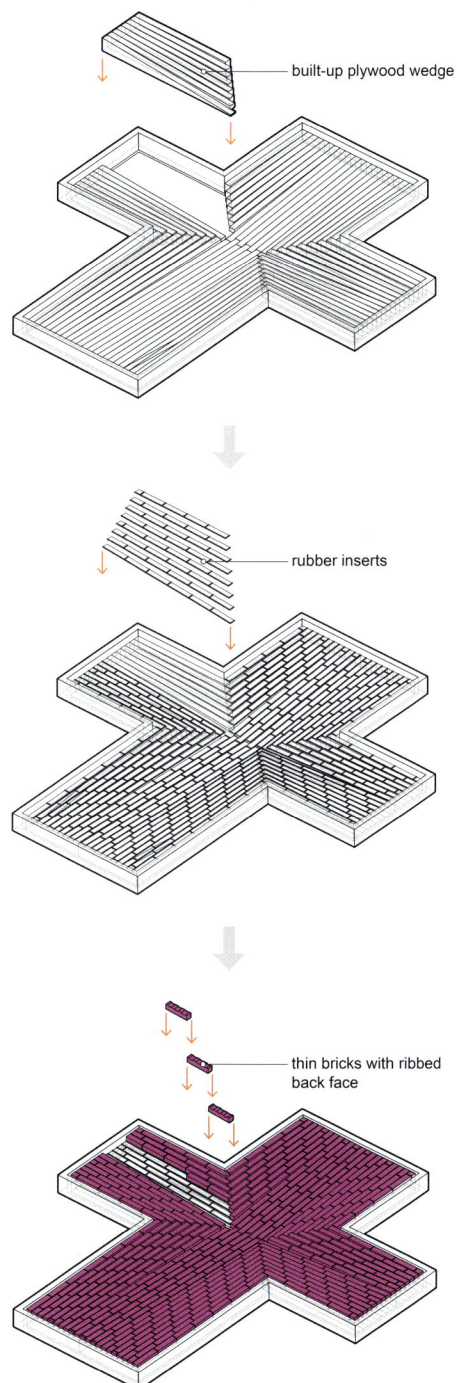

built-up plywood wedge

rubber inserts

thin bricks with ribbed
back face

Figure 4.1.61
A diagram of the mold assembly to form the panels.

Figure 4.1.62
Photographs of assembled mold with Hemmerlein in
process of placing the thin bricks into the rubber strips.
Photography by Sebastian Klich, Hild und K.

bidding precast manufacturers to discuss the project. According to Haber, after this discussion, Hemmerlein Ingenieurbau submitted a lower price and
was awarded the contract. Hild und K has worked
with Hemmerlein on other projects and is familiar
with their quality. Hild und K visited Hemmerlein's
factory to learn about their capabilities and production costs. Haber stated that the design team asked
about acid washing and sandblasting, to learn which
process was most cost efficient but still resulted in
an acceptable finish.[29] Hild und K also made small
changes to the panel's form to increase repeatability
of the molds and decrease costs.

Figure 4.1.63
At the panel edge, the rubber strips and bricks are cut to length. There is a border of exposed concrete between the brick and the panel edge. Photography by Sebastian Klich, Hild und K.

Figure 4.1.64
The molds on the ground at Hemmerlein. Photography by Sebastian Klich, Hild und K.

Hemmerlein made built-up plywood molds to form the composite panels (see Figure 4.1.61). The molds used inserted plywood wedges that provided the angled surface of the brick coursings and the stepped intersection of the brick bonds at the panel center. On the plywood wedges, Hemmerlein friction fit rubber strips that included a place for the bricks, head joints, and a single bed joint. They trimmed the rubber strips at the panel edges as needed. Hemmerlein set the thin bricks into the rubber recesses, placed the necessary reinforcement, and poured the concrete. According to Haber, the mold design came out of the post-bid discussions between the architect and the precast manufacturer. To reduce set-up time between pours, the joints between the rubber strips were not caulked. Instead, Hemmerlein used concrete with a low water content and vibrated it minimally to reduce the flow of water and cement between the rubber joints.

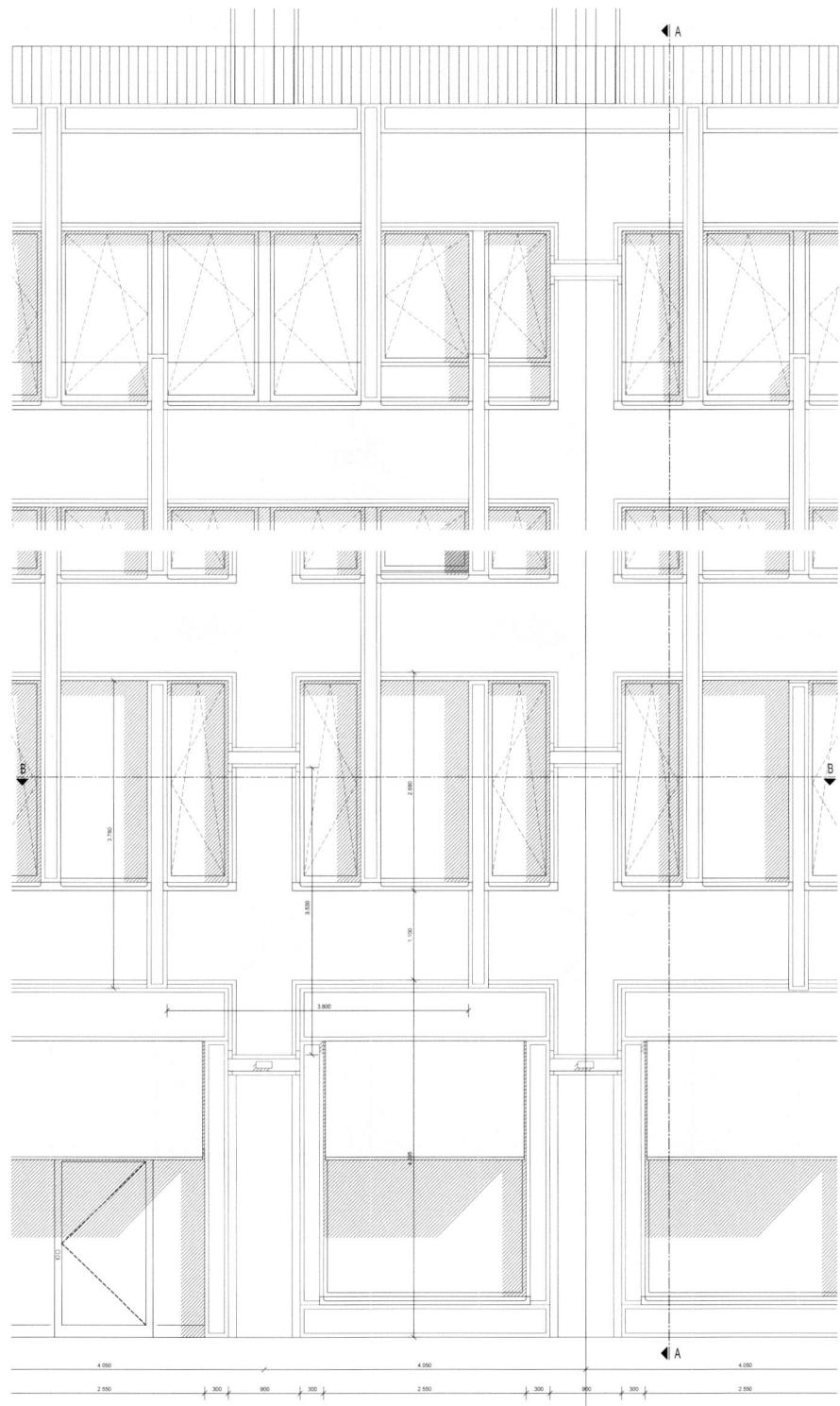

Figure 4.1.65
S40 Bay elevation and section drawing. Courtesy of Hild und K Müchen Berlin. (*Continued*)

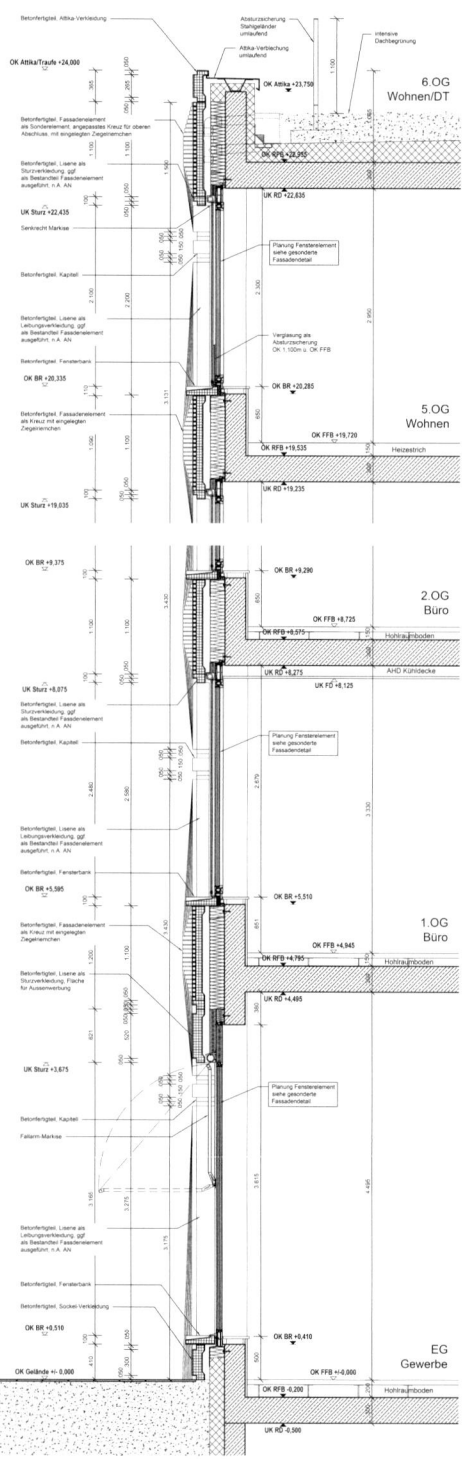

Figure 4.1.65
(*Continued*)

The Radiator House in Ichikawa, Japan
By Hiroshi Nakamura & NAP

The Radiator House is a 2400-ft^2 (223-m^2) single-family house, in a dense neighborhood in Ichikawa, Japan. The house features a warm, natural wood interior with large, full-wall windows facing a garden. To ensure privacy, the client wanted the yard to be enclosed and views to the inside of the house limited. Not wanting to restrict the flow of air for natural ventilation, Hiroshi Nakamura & NAP designed the garden privacy wall as a screen, so that air could move more freely. Like the traditional Japanese practice of *uchimizu*—in which water is sprinkled over entry doors in homes and gardens to provide cooling breezes—Hiroshi Nakamura designed the screen with a slow-drip irrigation system for evaporative cooling. Breezes come from a nearby river to the South, flow through the screen, and are cooled by the dripping water and the garden before they enter the house and exit out high second-floor windows. For Nakamura, it is important to create architecture that connects people to nature and to do this by creating "a gentle breeze for someone and to feel it on one's skin."[30]

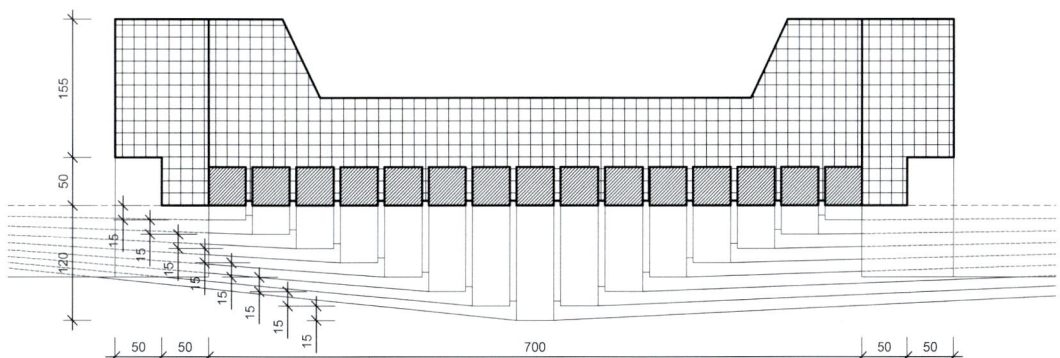

Figure 4.1.66
Plan detail of panel between windows. Courtesy of Hild und K Müchen Berlin.

Figure 4.1.67
Photograph from inside Radiator House, looking out to the garden and the screen wall. Image Nacasa & Partners Inc.

Figure 4.1.68
Photograph from outside of the wall, looking toward the house. The slow-drip
irrigation system on the wall evaporatively cools the air blowing from a nearby river.
Image Nacasa & Partners Inc.

Figure 4.1.69
The wall has ¾ in (2 cm) thick, triangular-folded fins between 1-in (2.5-cm) thick vertical
members. The non-stick, fluorocarbon polymer coating repeals water and dirt.
Image Nacasa & Partners Inc.

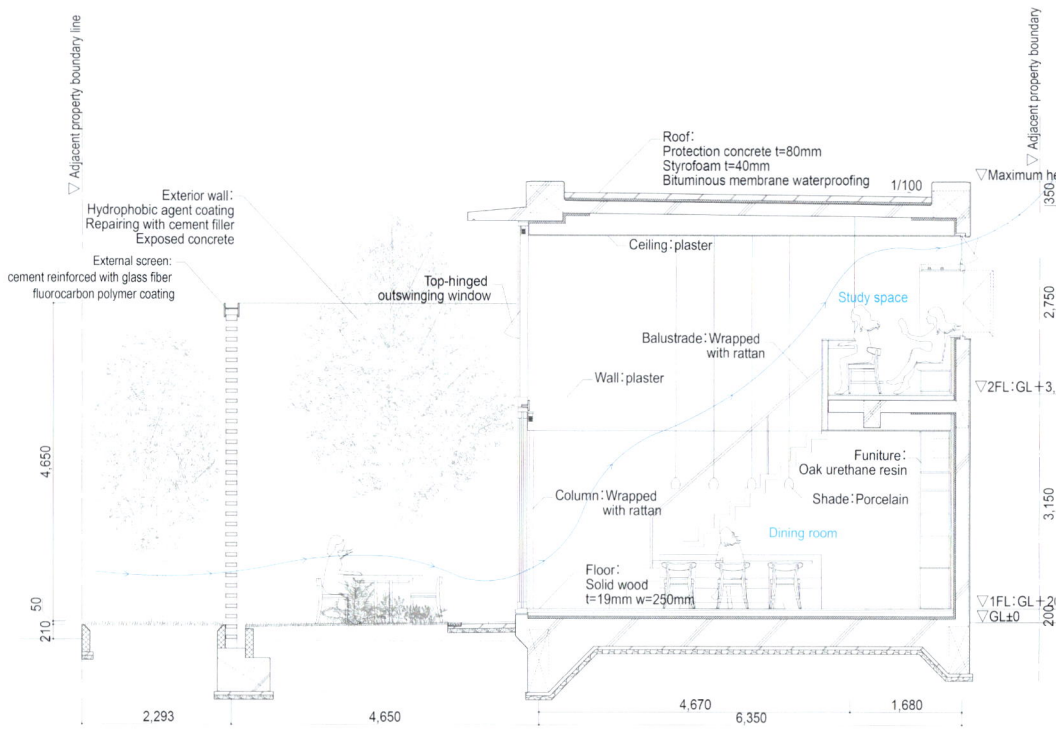

Figure 4.1.70
For passive cooling, breezes are to pass through the screen wall, across the garden, and through the house's high second-floor windows. Image Nacasa & Partners Inc.

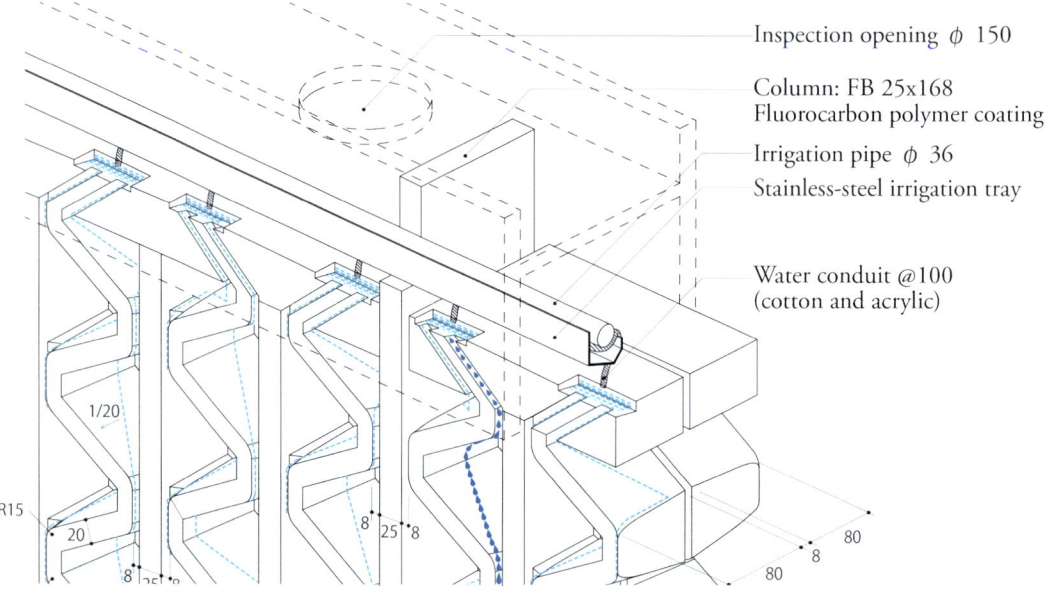

Figure 4.1.71
Isometric detail of the walls drip irrigation system and water flow. Image Nacasa & Partners Inc.

Figure 4.1.72
The GFRC was poured into an open rubber mold that was made against a CNC-milled plaster pattern.
Image Nacasa & Partners Inc.

For the screen design, Nakamura was inspired by the shape of a car engine radiator and the heat-exchanging fins of an air conditioner.[31] The screen consists of 32 wet-cast glass fiber–reinforced concrete (GFRC) panels that form the 59 ft (18 m) of screen wall length. Each panel is 16 ft 5 in (5 m) tall, 3 ft 5 in (1.05 m) wide, and 3 1/8 in (80 mm) deep. The 32 panels are arranged back-to-back, in 16 pairs, with a 5/16-in (8-mm) gap in between the pairs. Like heat-exchanging fins, the radiator screen maximizes its surface area by minimizing the thickness of its elements. The wall has ¾-in (2-cm) thick,

triangular-folded fins between 1-in thick (2.5 cm) vertical members. A slow-drip irrigation system consistently feeds water down the 5% outwardly sloped surface of the triangular fins to provide cooling. To develop the system design, the design team estimates that they built ten full-scale mockups and irrigation tests on the roof of their office building.[32]

To make the mold, liquid rubber was cast against a CNC-milled plaster pattern, or master mold. A rubber mold was more suitable than other mold materials for casting these GFRC panels because of their steep draft angles, relatively deep draw, thin cross

Figure 4.1.73
The screen panels were paired back-to-back and mechanically connected at the precast manufacturer before applying the polymer coating. Image Nacasa & Partners Inc.

sections, and large surface area. A wet, pourable mix of concrete with small aggregates and chopped glass fibers were poured into horizontal open molds. By casting finished face down, the draft angles for the mold worked with the outwardly sloped surface of the triangular fins, the mold could form the reveals that visually separate the fins from the vertical members, and it reduced the number of bug holes on the finished surface. Ultimately, the panels' unfinished face is hidden as the panels are arranged in back-to-back pairs. The GFRC manufacturer applied a non-stick, fluorocarbon polymer coating to the panels in the factory to repeal water and dirt.

Figure 4.1.74
Apertures, a six-story mixed-used building, with "sail" screens made of custom polymer concrete blocks. © Luis Gallardo/ LGM Studio.

Figure 4.1.75
The breaks between the sails allow for uninterrupted views, and the sails
are lifted at the ground floor at the building's entry points. © Luis Gallardo/
LGM Studio.

Apertures in Mexico City, Mexico
By Belzberg Architects

Apertures is a 43,900-ft^2 (4078-m^2), six-story mixed-use building. It was a collaborative project among Belzberg Architects (BA), the design architect, and Grupo Anima, the developer, architect of record, and contractor. Apertures includes three levels of below-grade parking, a ground-floor level with a restaurant and shared lobby, a floor of offices, four floors of hospitality suites, and a communal roof deck. The building exterior is a glass curtain wall covered with a series of curved, sail-like screens, made of small, custom concrete units. The sails are lifted on the ground floor to provide entry to the parking garage, restaurant, and lobby, and the spaces between the sails allow for key uninterrupted views from inside the building.

Figure 4.1.76
Image from an upper floor, through the screen. © Luis Gallardo/ LGM Studio.

Apertures is the third project of its type, as a collaboration with BA and Grupo Anima and located in Mexico City. The design team learned lessons with each project and applied their learnings to the subsequent project. Their first project, Threads (2016), used custom-fabricated aluminum fins that pass through the front curtain wall to delineate interior and exterior space. According to BA Partner Brock DeSmit, the fins were designed, engineered, and fabricated in Los Angeles (where BA is located), and the fins' lead installer traveled from LA to Mexico City to oversee the installation onto the building.[33] With the second project, Profiles (2018), the design team used Mexico's fabrication and construction resources more fully. Profiles has a perforated steel screen that covers its glass curtain wall. The screen was fabricated and installed in Mexico City. For Apertures, the design team looked toward the vernacular architecture of Mexico and wanted to use concrete masonry units (CMUs) as the building's primary feature instead of metal.

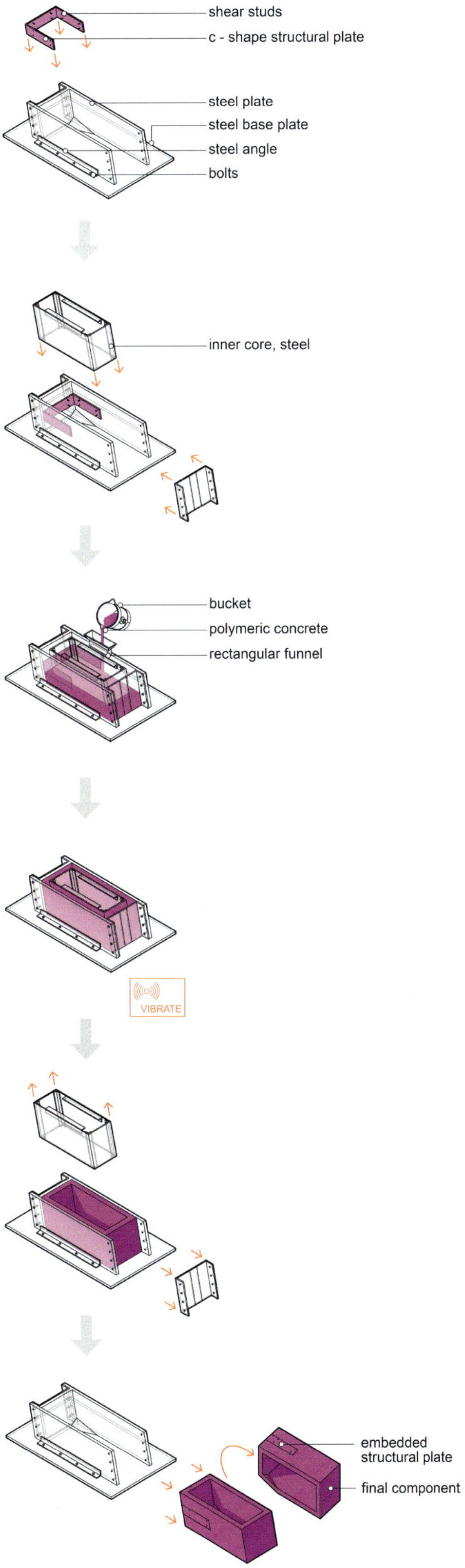

Figure 4.1.77
Diagraming the manufacturing process for the polymer concrete blocks.

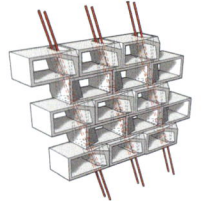 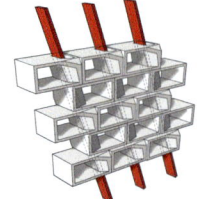 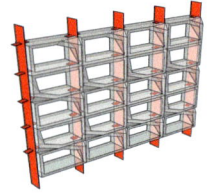 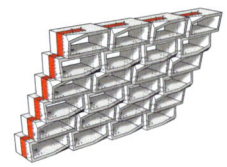

REBAR + GROUT
FAMILIAR, BUT UNSTABLE

STRUCTURAL TUBE
STRONG, BUT OPAQUE

UNITIZED PANEL
NO FORMAL VARIATION

ASSEMBLED PLATES
FEASIBLE + EXPRESSIVE

Limited volume for rebar and grout, the system is not rigid enough to withstand seismic loads.

Continuous structural steel bracing provides the necessary stiffness to support the masonry units, but their increased size obscures sightlines. Linear structural supports limit the curvature.

Vertical plate supports with clip angles preserve sightlines through the masonry units, but requires a stack bond layout in a vertical plane. This system is uniform and flat.

C-shaped structural plates mortised into the block units provide vertical and lateral support. The units are manually placed and bolted together, and can be staggered to create an overall curve in two planes while keeping sightlines clear.

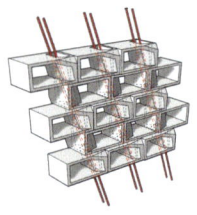 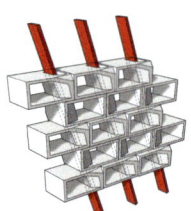 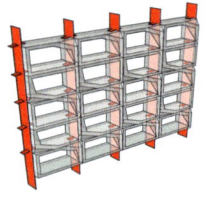 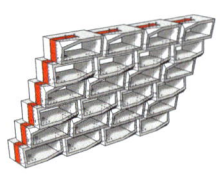

REBAR + GROUT
FAMILIAR, BUT UNSTABLE

STRUCTURAL TUBE
STRONG, BUT OPAQUE

UNITIZED PANEL
NO FORMAL VARIATION

ASSEMBLED PLATES
FEASIBLE + EXPRESSIVE

Figure 4.1.78
Structural investigations for the screen wall. The design team chose the option on the right.

Unfortunately, standard CMU is relatively weak and does not perform adequately for Mexico City's seismic requirements. In addition, according to Grupo Anima Director of Architecture and Design Hugo Balderas, the design team was not satisfied with the manufactured appearance of CMU blocks (Chapter 4.4) and, instead, investigated a wet-cast process. During this process, Grupo Anima decided

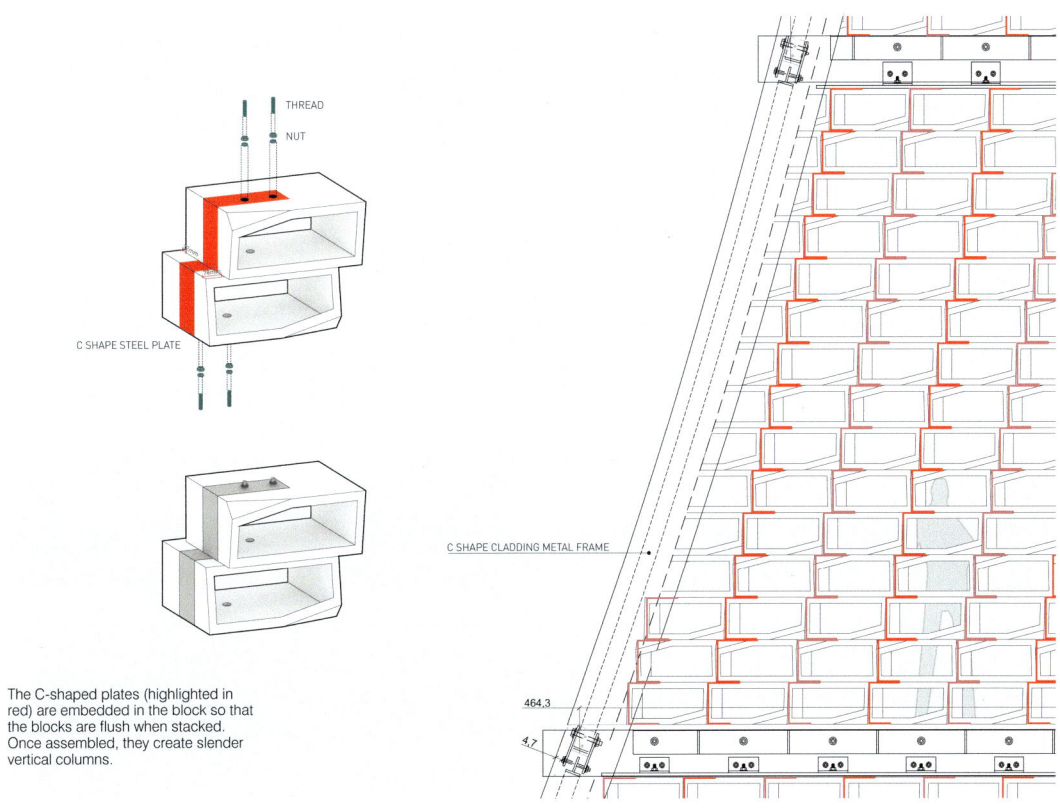

THREAD

NUT

C SHAPE STEEL PLATE

C SHAPE CLADDING METAL FRAME

464,3

4,7

The C-shaped plates (highlighted in red) are embedded in the block so that the blocks are flush when stacked. Once assembled, they create slender vertical columns.

Figure 4.1.79
Detail of assembly system of screen blocks.

to manufacture the blocks themselves, instead of subcontracting to another company, giving Grupo Anima control over the manufacturing process, production quality, and material choices. Grupo Anima investigated different types of concrete mixes and in the end chose a polymer-based concrete. The benefits of polymer concrete include a smooth surface, good material strength and resistance to water, the opportunity to adhere to other plastic-based materials, and speed of manufacturing.[34]

Figure 4.1.80
Grupo Anima mixed the polymer concrete in small batches for casting. The workers used a funnel to pour the concrete into the narrow molds.

Arup was involved in the conceptual design for the screens, investigating several different connection options for the polymer concrete blocks. In the end, the design team chose C-shaped steel plates with shear studs that would be cast into each block and bolted together in the field. The C-plates were flush with the surface of the block and hidden when installed. Grupo Anima cast holes through the blocks for the bolts, which connected through the block walls to the C-plates. The bolt heads were recessed into the block and patched in the field with polymer concrete to hide them. This C-plate system eliminated the need for internal reinforcing or exterior steel armature and maximized the screen apertures and visual porosity. After conceptual design, Neme Design Studio took over the facade's design detailing.

Figure 4.1.81
Photographs of the post-casting process: (a) a worker cleaning the mold as the polymer concrete sets, (b) workers took off the outside mold plates to demold the blocks, and (c) workers used a putty knife to knock off any flashing.

Figure 4.1.82
Grupo Anima used multiple molds for parallel productions.

The screen has left-handed and right-handed versions for its standard stretcher blocks and one block design for screen corners. Grupo Anima used built-up steel molds to cast the blocks and would disassemble the molds when demolding the blocks. The C-plate locations were inconsistent from block to block. Grupo Anima grouped the production of blocks with the same C-plate location together, because they needed to modify the molds to accommodate the new C-plate location for the next set of blocks. To reduce production time, Grupo Anima used multiples of each mold configuration for parallel productions. They did not caulk the mold joints between castings; instead, they used a putty knife to knock off any resulting flashing.

Figure 4.1.83
Seismic testing of the blocks and their connection system.

Apertures required a lot of exploration from the design team, including developing the block's shape and size; the concrete block mix design to meet the aesthetic, strength, and water shedding needs; a structural system for the screen that would vertically support the blocks and provide seismic resistance; a manufacturing process to make the custom block efficiently and effectively; and a construction process

Figure 4.1.84
Apertures under construction. A worker is installing the corner units. The holes in the blocks
for the bolts are still visible but will be patched on site.

so that blocks could be installed efficiently and safely. The team did a series of full-scale prototypes to test design, material performance, and manufacturability and mockups to test installation practices and seismic testing. When speaking to Grupo Anima's commitment to the necessary research for this project,

Balderas admitted that it was a "bold move" but thought that it was not "so uncommon in Mexico."[35] By taking on this responsibility, Balderas thought it gave Grupo Anima more control over the project budget and construction quality than if they had not manufactured the blocks themselves.

Figure 4.1.85
Ou-River Crystal Box Restaurant by AntiStatics is the first building in their
1.1 mi (1.8 km) long masterplan for Ou-River.

Harbor 1   Building 2 / 3   Building 4 / 5 with Chapel

Figure 4.1.86
The Ou-River masterplan by AntiStatics that includes a landscape design,
six new buildings, and three new harbors.

Figure 4.1.87
Ou-River Crystal Box Restaurant and its custom UHPC screen.

The Ou-River Crystal Box Restaurant in Wenzhou, China
By AntiStatics

The Ou-River Crystal Box Restaurant by AntiStatics is a five-star, fine dining restaurant that is part of AntiStatics' larger 1.1-mi (1.8-km) long masterplan for Ou-River that includes a landscape design, six new buildings, and three new harbors. The Crystal Box is at the threshold between the city and a tidal estuary which flows through the city. The restaurant stands on piers, above the Ou-River,

has approximately 11,800 ft² (1100 m²) of floor area, and with the site's comfortable, year-round climate it occupies two buildings. The buildings are an assembly of floating roof plates above a custom-manufactured concrete screen. According to AntiStatics Principal Martin Miller, the concrete screen design is reminiscent of water's movement and inspired by the tidal estuary.[36] The screen is the building's exterior design feature and is lit at night as a beacon. It also provides privacy for exterior and interior dining spaces, while allowing views from inside to outside.

Figure 4.1.88
Image from inside the restaurant, looking out the screen to the river.

Figure 4.1.89
Detail photograph of the UHPC screen units.

Reminiscent of the screens by sculptor Erwin Hauer, the Crystal Box screens are made from small cast units that are stacked and connected by a steel dowel set into mortar. The units are made from white, UHPC cast in a two-part, closed, GFRP mold. For Crystal Box, Nanjing BeiLiDa produced the screen units and has experience with UHPC and contact-molded GFRC and glass fiber–reinforced gypsum (GFRG) (Chapter 3.1). AntiStatics learned about UHPC through a previous project, MaoHaus (2017), located in Beijing. In a 2¾-in (7-cm) panel thickness the UHPC combined a wave-like form with screen apertures, forming an image of former Chinese President Mao Zedong when backlit at night.

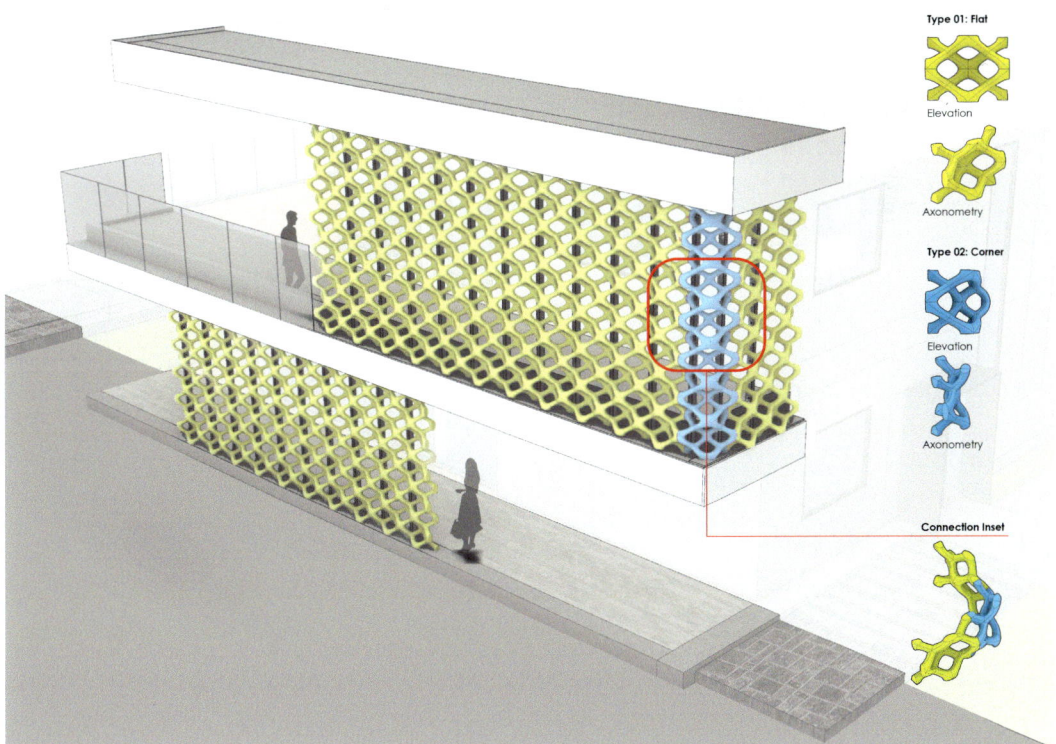

Figure 4.1.90
There are two different units: a standard shape (yellow) and a rounded corner shape (blue).

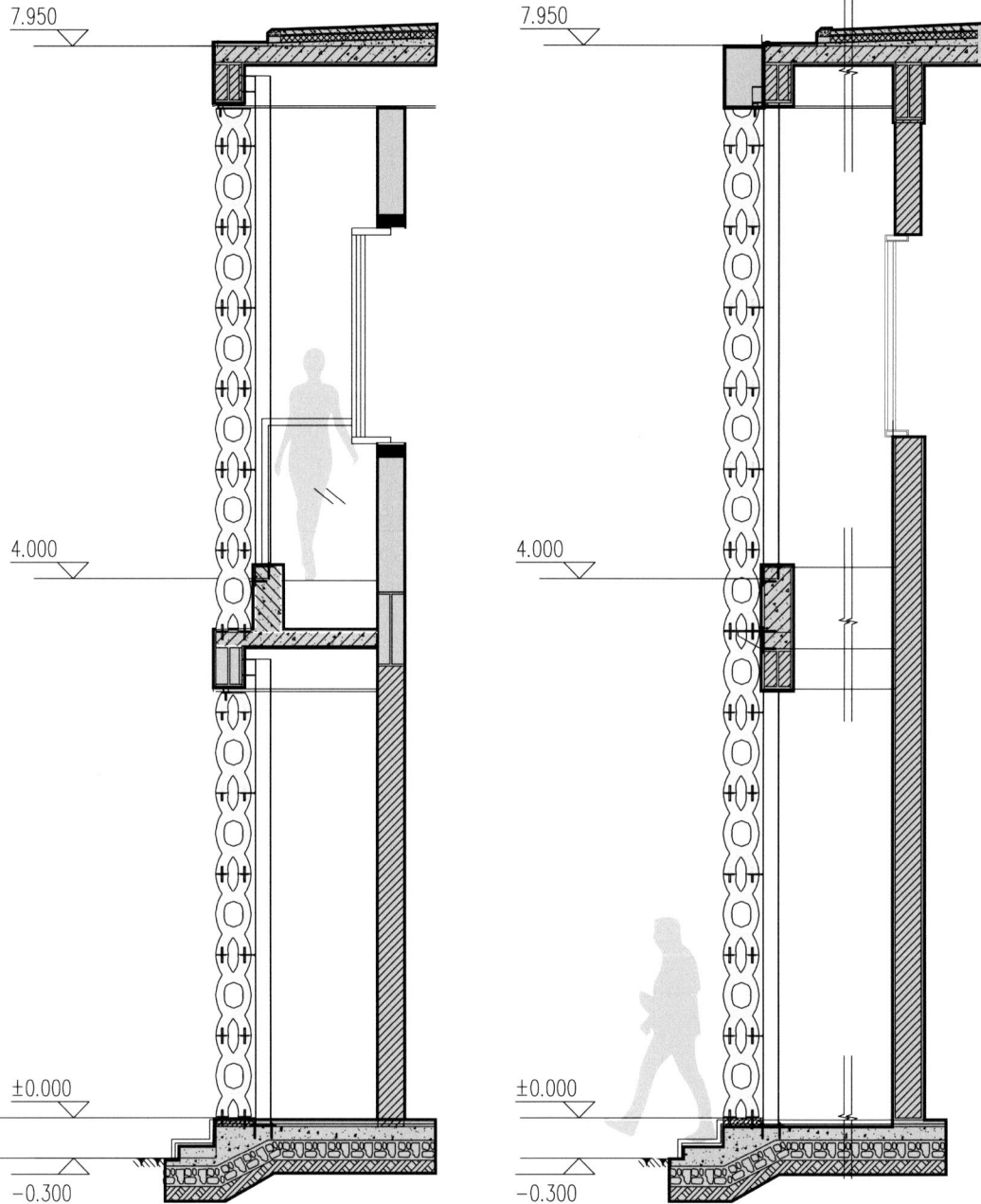

Figure 4.1.91
Section drawing at two different screen locations.

Figure 4.1.92
A standard screen unit at Nanjing BeiLiDa.

Figure 4.1.93
Although a seemingly complex shape, it has no under-
cuts. AntiStatics made a mold prototype on a three-axis
CNC mill.

Figure 4.1.94
Any bug holes that formed on the mold's vertical surfaces were patched. The units were left with their mold finish.

Miller has an undergraduate degree in fine arts and has experience as a sculptor. Miller, AntiStatics Partner Mo Zheng, and the firm are dedicated to understanding how things are made and communicating that understanding to fabricators, producers, and manufacturers. Miller stated that for AntiStatics digital tools are not just a stage set for renderings, instead the power of digital tools lies in making the design physical. AntiStatics takes on the responsibility that all details in their 3D model are resolved fully.[37] When BeiLiDa first reviewed the screen unit designs, they communicated to AntiStatics that the units could not be produced. BeiLiDa was concerned that the units had undercuts and could not be demolded without damaging the mold or the unit. AntiStatics has a three-axis CNC mill that they used to make a small prototype of the two-part mold that they sent to BeiLiDa. Since a three-axis CNC cannot make undercuts, once BeiLiDa saw the mold,

they realized that they could produce the components. AntiStatics shared the CNC-mill digital files with BeiLiDa, and it was those files that BeiLiDa used to make minor modifications. Since most of the troubleshooting was done by AntiStatics during prototyping, there was little back and forth between AntiStatics and BeiLiDa before production began.

BeiLiDa manufactured two different screen units: a standard shape and a rounded corner shape (see Figure 4.1.90). The units were cast vertically, in the same orientation as they appear in the screen, and the UHPC was poured into the top of the mold.[38] Post-production of the cast units was limited. There was minor flashing at the mold parting line that was easily knocked off during demolding. Bug holes, which often form on vertical surfaces during casting, were minimum but were patched as needed. The units were erected with their mold finish and were not grit blasted or acid washed.

Notes

1. Depending on the casting facilities, the ambient temperature and humidity, the production schedule, and the required structural performance of the concrete, the concrete may be cured in a controlled or semi-controlled environment.

2. The Precast/Prestressed Concrete Institute (PCI) recommends post-production finishing such as grit blasting or acid washing. Post-production processes remove the weak laitance layer of the concrete, roughen the surfaces to varying degrees, and expose some of the concrete's aggregate. Generally, color variations and manufacturing defects are more apparent in direct-, or mold-, finish precast components than panels that have undergone post-production finishing.

3. Polymer-based concretes are not to be confused with geopolymer-based concretes. Geopolymer-based concretes use alternative cementitious materials such as fly ash or magnesium phosphate and alkaline solution (instead of water) to set the cement within the concrete.

4. Rode, Erin. "Unusually Shaped Shade Structure Unveiled at Joshua Tree National Park." *Palm Springs Desert Sun*, 13 December 2022. https://www.desertsun.com/story/news/environment/2022/12/13/strange-shade-structure-unveiled-at-joshua-tree-national-park/69709571007/. Accessed 23 March 2023.

5. Noble, Douglas. *Personal Interview*. 28 March 2023.

6. Monsreal's tool was the basis for his Master of Building Science thesis. Monsreal, Ivan. *Precast Concrete Thin-shell Skins: A Smart Computer Tool for Generating Complex Shapes*. 2019. University of Southern California, Master of Building Science.

7. "Ribbon Cut for Award-winning Precast Concrete Project." *Civil + Structural Engineer Media*, 6 January 2023. https://csengineermag.com/ribbon-cut-for-award-winning-precast-concrete-project/. Accessed 23 March 2023.

8. Clark Pacific's Fontana plant is no longer in operation.

9. Maira, Charles and Joseph Tattoni. *Personal Interview*. 20 December 2022.

10. Maira and Tattoni.

11. Del Vento, Robert. *Personal Email*. 21 December 2022.

12. Mairiaand Tattoni.

13. Maira and Tattoni.

14. Achelpohl, Elvis. *Personal Interview*. 7 April 2023.

15. Goebel, Brian. *Personal Email*. 7 April 2023.

16. Adkins, Arno. *Personal Interview*. 26 July 2022.

17. *Ibid.*

18. The carbon fiber–reinforced plastic (CFRP) used in the 3D printing had 20% carbon fibers to improve the ABS strength and stiffness and to reduce thermal expansion.

19. Hirt, Amy Howell. "3D-printed Precast concrete Molds for Redeveloped NYC Landmark a First." *Construction Dive*, 3 April 2019. https://www.constructiondive.com/news/3d-printed-precast-concrete-molds-for-redeveloped-nyc-landmark-a-first/551903/. Accessed 15 July 2022.

20. Autodesk. "One South First Precast Construction." *Vimeo*. https://vimeo.com/344411135/c7193b7a58?embedded=true&source=vimeo_logo&owner=4474980. Accessed 15 July 2022.

21. Hirt, Amy Howell. "3D-printed Precast concrete Molds for Redeveloped NYC Landmark a First." *Construction Dive*, 3 April 2019. https://www.constructiondive.com/news/3d-printed-precast-concrete-molds-for-redeveloped-nyc-landmark-a-first/551903/. Accessed 15 July 2022.

22. Adkins.

23. Atkinson, William. "Project Spotlight: Brooklyn Redevelopment Uses 3-D Mold Printing for Window Panels." *PCI Journal*, Nov-Dec. 2018, pp. 20–21.

24. Fransen, Patrick. *Personal Interview*. 20 December 2022.

25. OMA's NDT was completed in 1987 with a limited budget of approximately 8 million dollars. Given the limited budget, the OMA building is not designed or detailed to be long-lasting. The OMA budget would be the equivalent of 19.5 million Euros at today's rates; in comparison, Amare's budget was 142 million Euros.

26. Fransen.

27. Haber, Matthias. *Personal Interview*. 8 December 2022.
28. "Brick Dress in Munich: Residential and Commercial Building Owned by Hild and K." *BauNetz*, 24 May 2018. https://www.baunetz.de/meldungen/Meldungen-Wohn-_und_Geschaeftshaus_von_Hild_und_K_5398500.html Accessed 22 November 2022.
29. Haber.
30. Nakamura, Hiroshi and Masaki Hirakawa. *Personal Interview via Email*. 4 August 2022.
31. *Ibid*.
32. *Ibid*.
33. DeSmit, Brock and Hugo Balderas. *Personal Interview*. 7 June 2022.
34. Like plastic, the polymer-based concrete units can be chemically bonded to one another, without mortar. Polymeric concrete also sets quickly; according to Balderas, the polymeric concrete units could be demolded every 10 min.
35. Balderas.
36. Miller, Martin. *Personal Interview*. 31 May 2022.
37. *Ibid*.
38. UHPC can also be pumped up from the bottom of the mold. This is best suited for closed vertical molds that are tall and narrow. In those cases, pouring the UHPC from the top will most likely result in trapped air and concrete not fully reaching all parts of the mold.

References

Achelpohl, Elvis. *Personal Interview*. 7 April 2023.
Adkins, Arno. *Personal Interview*. 26 July 2022.

Atkinson, William. "Project Spotlight: Brooklyn Redevelopment Uses 3-D Mold Printing for Window Panels." *PCI Journal*, Nov-Dec. 2018, pp. 20–21.
Autodesk. "One South First Precast Construction." *Vimeo*. https://vimeo.com/344411135/c7193b7a58?embedded=true&source=vimeo_logo&owner=4474980. Accessed 15 July 2022.
"Brick Dress in Munich: Residential and Commercial Building Owned by Hild and K." *BauNetz*, 24 May 2018. https://www.baunetz.de/meldungen/Meldungen-Wohn-_und_Geschaeftshaus_von_Hild_und_K_5398500.html Accessed 22 November 2022.
Del Vento, Robert. *Personal Email*. 21 December 2022.
DeSmit, Brock and Hugo Balderas. *Personal Interview*. 7 June 2022.
Fransen, Patrick. *Personal Interview*. 20 December 2022.
Goebel, Brian. *Personal Email*. 7 April 2023.
Haber, Matthias. *Personal Interview*. 8 December 2022.
Hirt, Amy Howell. "3D-Printed Precast Concrete Molds for Redeveloped NYC Landmark a First." *Construction Dive*, 3 April 2019. https://www.constructiondive.com/news/3d-printed-precast-concrete-molds-for-redeveloped-nyc-landmark-a-first/551903/. Accessed 15 July 2022.
Maira, Charles and Joseph Tattoni. *Personal Interview*. 20 December 2022.
Miller, Martin. *Personal Interview*. 31 May 2022.
Nakamura, Hiroshi and Masaki Hirakawa. *Personal Interview via Email*. 4 August 2022.

CHAPTER
4.2

Casting Metal

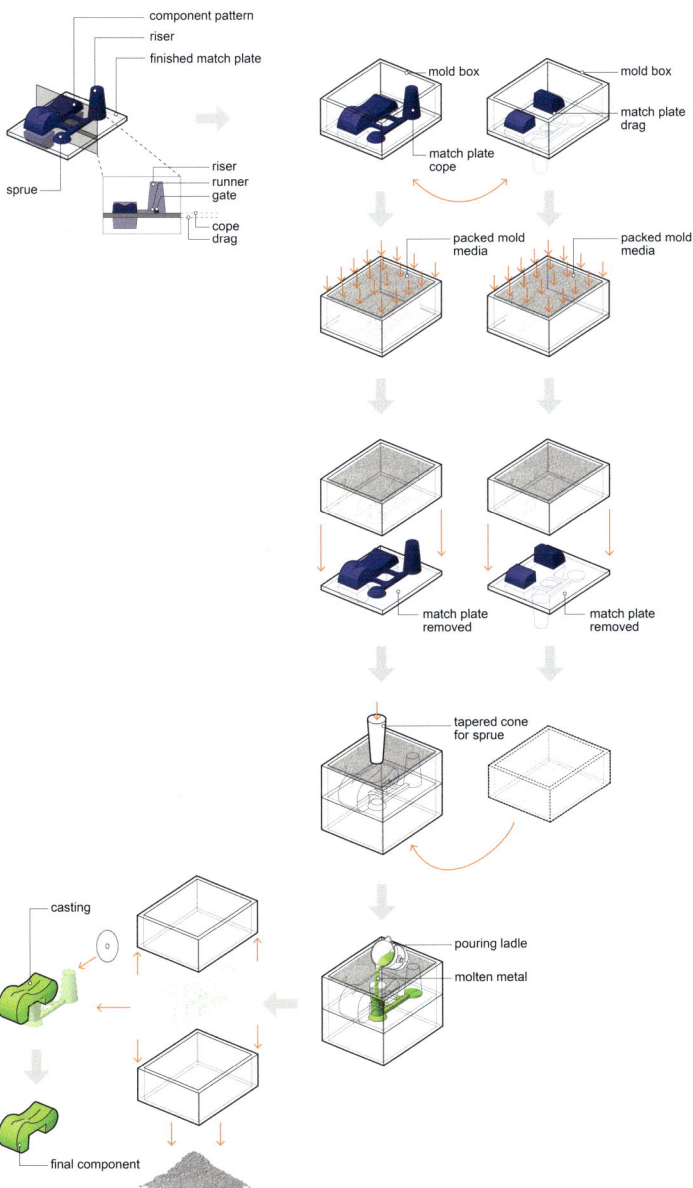

Figure 4.2.1
Casting metal process diagram.
Drawing by author. A version of this
diagram originally appeared in
Manufacturing Architecture
(Laurence King, 2018).

Casting metal is the manufacturing process that uses gravity and metal's phase change from liquid to solid to shape the metal. For architectural applications, casting metal typically uses aluminum or aluminum alloys, copper, iron or iron alloys, and bronze; however, any metal that is solid at room temperature and can be heated to a liquid state can be used for casting. Casting metal can produce solid, semi-hollow, or hollow components with complex geometries, good details, and varying cross-sectional thicknesses. Since casting metal uses the low force of gravity to move the molten metal through the mold, the mold may be made from a range of materials, including sand, plaster, and clay; therefore, capital costs with this process can be low. Components made from casting metal can vary significantly in size from small jewelry-scaled components to castings of more than 100 tons (90.7 Mg). Generally, the cycle time for metal casting is fast as metal cools quickly. Once the metal is solid enough to maintain its shape, it is demolded and continues to cool outside of the mold.

Casting metal can be done in open or closed molds and in expendable or permanent molds. Open molds are the simplest as the metal is cast directly into the mold cavity. Closed molds are more complex than open molds, and they come in two or more parts, requiring a sprue and optional gating system to get the molten metal to the mold void. Expendable molds—made from sand, plaster, or clay—are broken away from the component during demolding. Expendable molds are best suited for complex shapes that would be too difficult to demold without damage to the mold or component. Expendable molds are formed around a reusable pattern. Patterns are made from high-density foam, plastic, wood, or cast aluminum, depending on the required durability of the patterns and how many expendable molds need to be formed from the pattern. Metal casting with an expendable sand mold is cost efficient and can be used for small or large production runs. Permanent molds—made from plaster, clay, and tool steel—are best suited for simple shapes, with gentle draft angles so that demolding does not damage either the mold or the component. Molds made from plaster, clay, and steel result in a smoother surface quality than sand molds. Tool steel is the most expensive and durable of all mold material options and is best for large production runs produced over a long span of time.

In metal casting with an expendable sand mold,[1] a pattern is fabricated to form the sand into the needed mold shape. The pattern will include the mold cavity and, if needed, the gating system and riser. For repetitive casting, the pattern is often mounted onto a match plate to ensure the alignment and placement of the mold elements. An optional mold box is placed over each side of the match plate and mold sand (i.e. specialized sand with a binding agent, such as bentonite clay) is compacted against the pattern faces. The match plate is then removed from the formed sand molds and the mold halves are closed, forming the cavity for the casting. A tapered cone is placed into the top mold half, called the cope, to form the sprue. Molten metal is poured into the sprue, until the mold is full and a small puddle forms at the top. Once the metal cools enough for demolding, the mold is vibrated so that the sand breaks away from the casting. Any gating system is removed from the casting and the component is ready for post-production. Both the mold sand and the gating system are reused by the foundry.

Figure 4.2.2
The Compton, view from 30
Lodge Road. The apartment
tower is clad with custom,
sand-casted aluminum
panels, powder coated in
a bronze-gold color. Image
courtesy of Regal London.

Metal

Any metal or metal alloy that is solid at room temperature and can be heated to a molten state can be cast. Cast metals will have different mechanical properties than the same metal that has been cold worked (e.g. stamping or hydroforming) or hot (e.g. forging or rolling) worked and are generally more brittle with a lower tensile strength. All metals shrink as they cool, and each metal and metal alloy have a specific rate of shrinkage. Molds should be oversized based on the specific metal being cast.

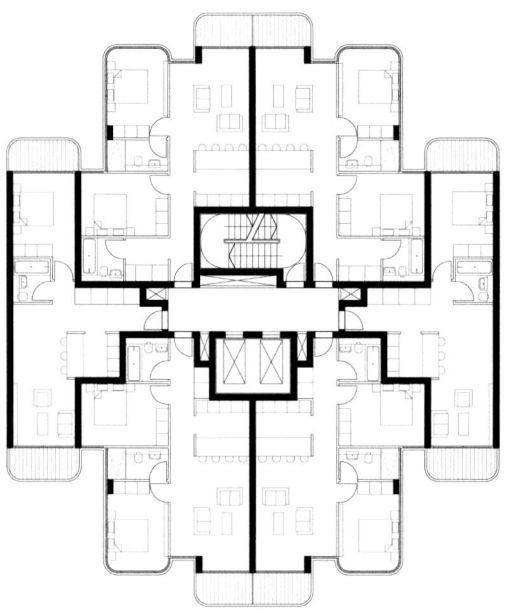

Figure 4.2.3
E8 Architecture designed the building exterior, apartment layouts, and core. Every apartment has a balcony or a private roof terrace. Courtesy of E8 Architecture.

FACING PAGE
Figure 4.2.5
Detailed image of the custom, sand-casted aluminum panels with a geometric oak-leaf motif and powder coated in a bronze-gold color. Image courtesy of Regal London.

The Compton: 30 Lodge Road, London, England
By E8 Architecture

The Compton: 30 Lodge Road is a ten-story, 70,000-ft^2 (6503-m^2), multi-family residential apartment building in the St. John's Wood district of London, near the northwest corner of Regent's Park. The building includes 49 units with a mix of one-, two-, and three-bedroom apartments with individual balconies, and penthouses with private roof terraces; a communal gym, lobby, bicycle storage, and roof terrace for all residents; and a below-grade, automated parking garage. The building's rounded corner windows allow the apartments to make the most of the views, with the best views being at oblique angles to the site.[2] The Compton was the first new building in a transitional block on Lodge Road. Prior to The Compton, the site held a vacant mail sorting office, and now, neighboring lots have buildings either under construction or planned to be under construction. Regal Homes was the project client and developer and commissioned E8 Architecture to design The Compton's exterior, interior core details, and apartment layouts and to coordinate with an interior design firm.

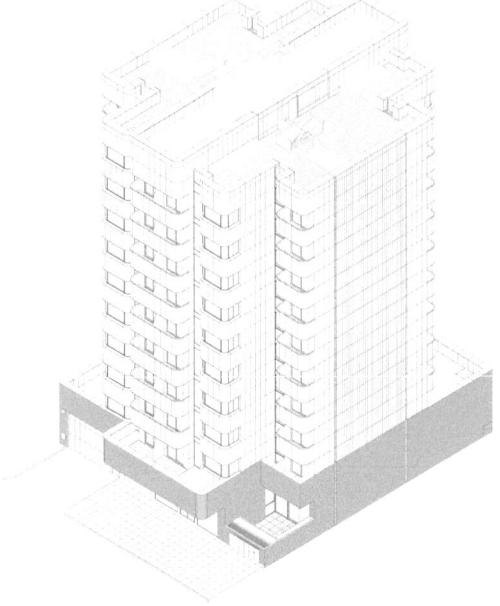

Figure 4.2.4
Following the design of neighboring apartment buildings, E8 designed The Compton on the style principles of Art Deco. Courtesy of E8 Architecture.

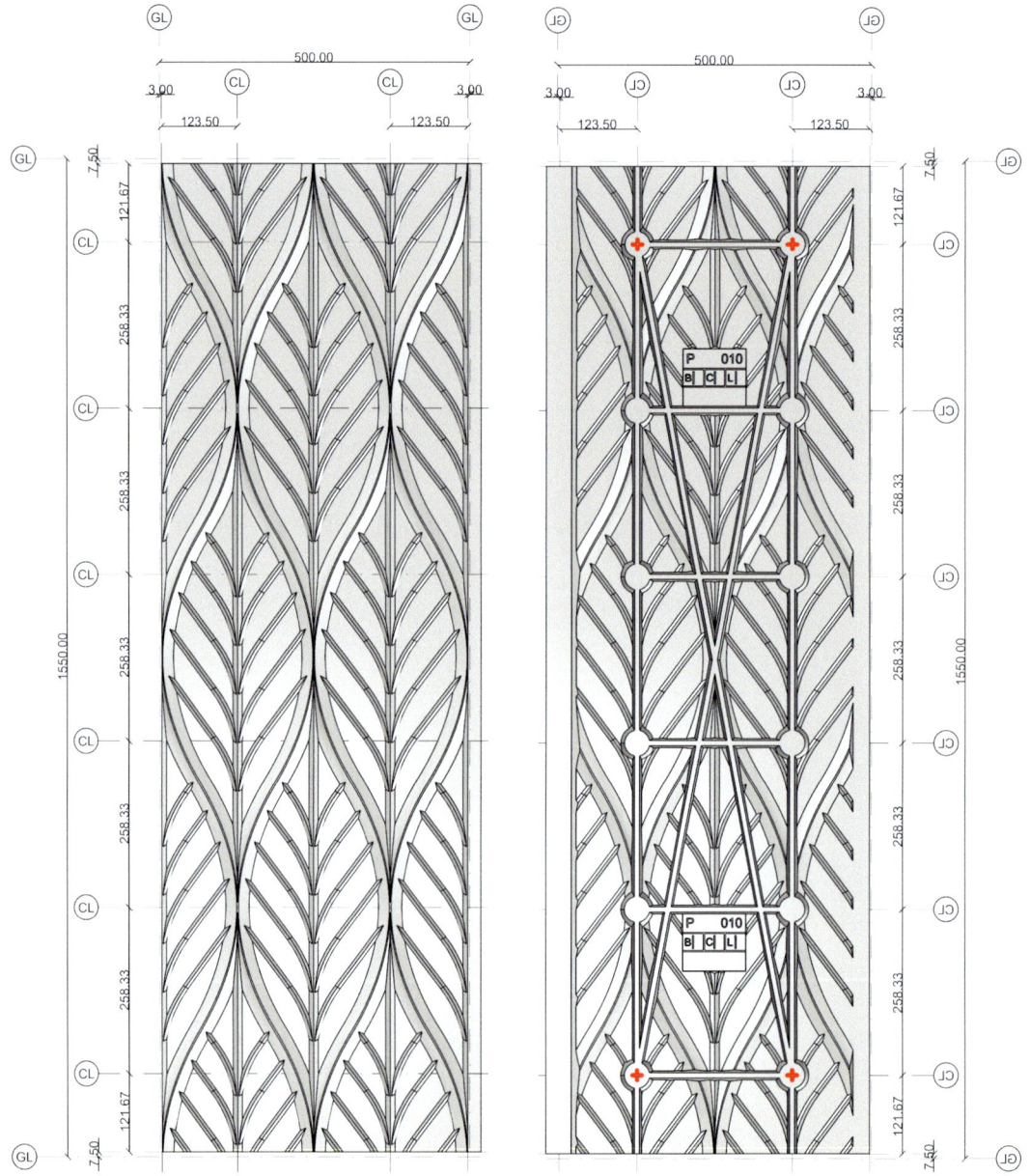

Figure 4.2.6
Front and back elevations of panel P.010. Courtesy of E8 Architecture.

E8 primarily practices in London and, according to E8 Founding Principal Simon Bowden, the firm embeds their building designs into their local district so that their buildings "feel comfort able" and reflect a sense of their place.[3] In St. John's Wood district, the established neighboring blocks are characterized by handsome, Art Deco style apartment buildings. E8 built upon the Art Deco's style principles and reinterpreted them for a modern building. At the ground level, The Compton is clad in a water-struck, long-format style brick[4] from Randers Tegl and includes four custom-molded brick shapes that form the curved exterior corners of the balconies and entry canopy. The apartment tower's upper level

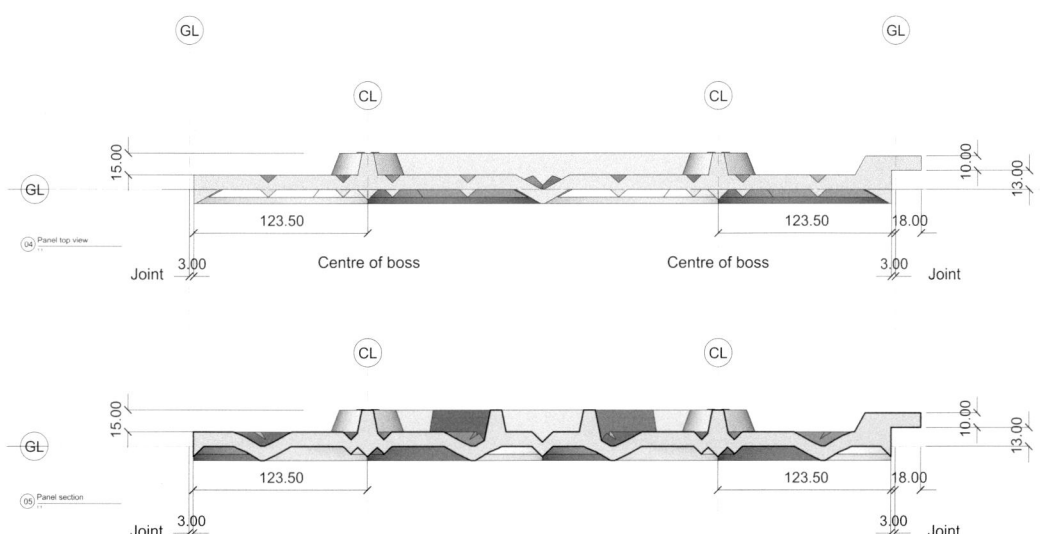

Figure 4.2.7
Plan sections of panel P.010. Note the 7/8-in (21-mm) lap joint on panel's right edge. Courtesy of E8 Architecture.

balcony rails and spandrels are custom, sand-casted aluminum panels with a repeating, geometric oakleaf motif, and powder coated in a bronze-gold color. E8 wants their buildings to read as if they are hewn from solid materials, providing a sense of solidity and quality associated with traditional buildings. The Compton uses lap joints rather than open butt joints between the cast panels so that the cladding material is visible behind the joints and shadow lines are reduced to a minimum.

Regal Homes gave E8 a facade budget for The Compton that Bowden said was the equivalent of a facade made of cast, or reconstituted, stone (Chapter 4.5). With this budget in mind, E8 explored alternative materials. In addition to the cast metal, the design team explored architectural precast concrete (Chapter 4.1), glass fiber-reinforced concrete (GFRC) (Chapters 3.1 and 4.1), and clay (Chapter 4.7). E8 tested each of the options against the design brief, the budget, and talking

with representatives from those industries. Bowden stated that the cast metal was the best choice as it met the project cost, quality, and schedule.

Regal had prior experiences using stone, brick, or metal panels to clad their buildings, but not custom cast aluminum. E8 is indebted to Regal for working with them to explore alternatives.[5] E8 structured their design process so that if at any point the design team realized that the custom cast metal would not be possible, then they would revert to a more familiar material such as cast stone. Bowden acknowledged that this project had a lot of unknowns: the foundry had not previously cast metal to this scale, this was the first time E8 designed with cast aluminum, and the facade contractor had not done anything similar before. Bowden credits E8 Founding Partner Mark Fleming with coordinating the process and the project constituents and developing the details to make the material work.

Figure 4.2.8
Image of transferring aluminum to crucible at the foundry. Photograph by E8 Architecture of AATi Foundry.

For this project, E8 wanted to work directly with manufacturers, and they organized workshops and visits with several foundries. "They would cast something, and then we would sketch, and then we would have prototypes that we had printed."[6] Bowden estimates that the firm reviewed almost 200 pattern options through 3D-printed prototypes and digitally rendered visuals. Since the site is close to Regent's Park, E8 started with a leaf for the pattern; however, the final objective was to design a pattern with depth and texture that would capture and reflect the light. E8's workflow included parametrically designing the patterns in Rhinoceros and then bringing them into Revit to see how the patterns would read on the building. The pattern was sized so that it book-matched across the seams from one panel to the next and coursed correctly across the building.

Figure 4.2.9
Image at AATi, casting the aluminum into closed molds. Photograph by E8 Architecture of AATi Foundry.

Figure 4.2.10
Image at AATi, demolding a curved panel. Photograph by E8 Architecture of AATi Foundry.

Figure 4.2.11
Image at AATi, with the sand mold, broken away from the casting. Photograph by E8 Architecture of AATi Foundry.

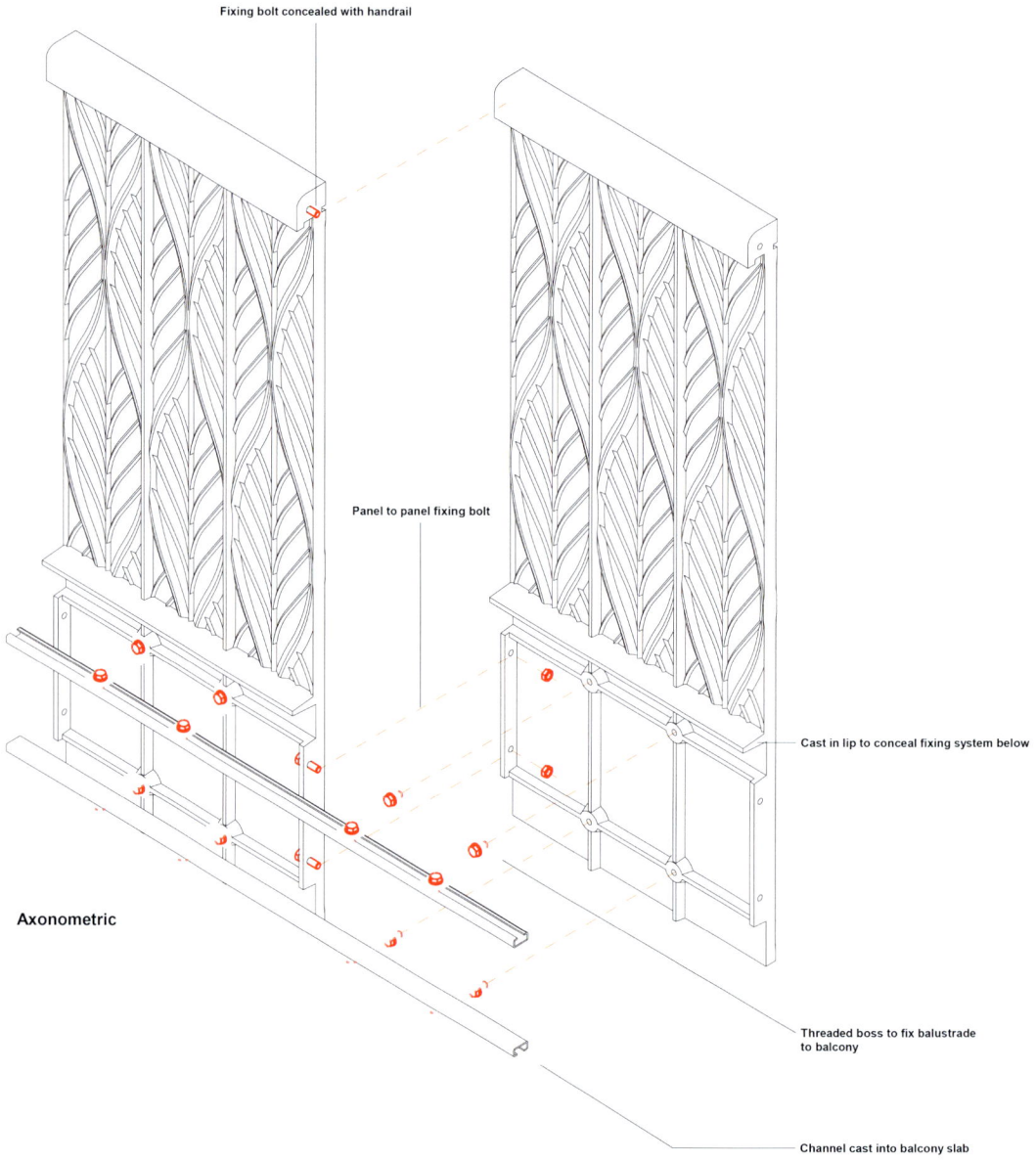

Figure 4.2.12
Axonometric of balcony rail panel with integrated bosses for mechanically
connecting to substructure channels. Courtesy of E8 Architecture.

Through their research, E8 found three foundries in the United Kingdom that could cast the aluminum panels. The design team selected AATi Architectural Limited, formally Antislip Antiwear Treads International Limited, located approximately 50 mi (80 km) northeast from Lodge Road. Fleming noted that E8 worked within the confines of AATi's manufacturing process by tailoring the panels' design and dimensions to AATi's manufacturing equipment. This increased the number of panels that AATi could produce each day and kept costs within the budget. AATi cast the panels in closed, two-sided molds. The balcony rails had the oak-leaf motif for both the interior and exterior faces, and the spandrel panels had a supporting rib and boss structure to reduce material cost and weight, while providing the necessary strength and material for mechanical connections.

Figure 4.2.13
Image of panel fully demolded before the sprue and risers are removed. Photograph by E8 Architecture of AATi Foundry.

Figure 4.2.14
Checking the panel's radius against a template for quality control. Photograph by E8 Architecture of AATi Foundry.

The Compton has 1800 cast aluminum panels in 52 different configurations. AATi formed the sand molds on 12 different "parent" patterns. The parent patterns were CNC milled out of resin and took a long time to produce. When needed to form the different configurations, they cut the parent patterns down to create smaller "child" patterns.[7] Front and back patterns would be ganged together so that the

Figure 4.2.15
A detailed image of the cast panel surface and leaf motif. Photograph by E8 Architecture of AATi Foundry.

Figure 4.2.16
An image of the apartment tower. Image courtesy of Regal London.

different pattern configurations would be used to form different sand molds. After casting, the panels were powder coated by another company and then stored at AATi for just-in-time delivery to the site. E8 did the cast panel shop drawings, and their drawings were used to CNC mill the patterns. By being responsible for the shop drawings, Fleming noted that E8 had control over the design, but both Fleming and Bowden acknowledged it also meant that the firm carried the risk if anything went wrong.

For The Compton, Bowden said that E8 did a lot of testing of the cast panels' mounting system. "We spent a year working up prototypes, running analysis, and testing. Our final prototype was rigorously tested at a research facility to accurately measure the resilience and strength of the structure—to ensure it surpassed building standards."[8] E8 used their experiences on The Compton to use custom cast metal for subsequent projects, including balustrades for Marylebone Square and 204 Great

Portland Street. With the firm's design process, Bowden believes their clients are getting something more bespoke and with more provenance than their clients would have gotten otherwise.

Notes

1. Since casting metal in a closed, expendable sand mold is the most common and the most complex of the different mold types, I am using that as the process diagram for metal casting.
2. E8 Architecture. "30 Lodge Road." https://www.e8architecture.co.uk/home#/thirty-lodge-road/. Accessed 15 November 2022.
3. Bowden, Simon and Mark Fleming. *Personal Interview*. 28 November 2022.
4. The Compton's bricks are longer format than Roman-style bricks and measure approximately 18 ½ in long, 1 ½ in tall, and 4 ¼ in deep (468 × 38 × 108 mm, respectively).
5. Bowden.
6. *Ibid.*
7. Fleming, Mark. *Personal Email*. 5 December 2022.
8. Sengupta, Debanjali, Ed. "Regal London Delivers One of the World's First Cast Aluminium Clad Buildings." *End User News: Industrialized Buildings*. AlCircle, 1 May 2019. https://www.alcircle.com/news/regal-london-delivers-one-of-worlds-first-cast-aluminium-clad-buildings-45795. Accessed 15 November 2022.

References

Bowden, Simon and Mark Fleming. *Personal Interview*. 28 November 2022.

E8 Architecture. "30 Lodge Road." https://www.e8architecture.co.uk/home#/thirty-lodge-road/. Accessed 15 November 2022.

Fleming, Mark. *Personal Email*. 5 December 2022.

Sengupta, Debanjali, Ed. "Regal London Delivers One of the World's First Cast Aluminium Clad Buildings." *End User News: Industrialized Buildings*. AlCircle, 1 May 2019. https://www.alcircle.com/news/regal-london-delivers-one-of-worlds-first-cast-aluminium-clad-buildings-45795. Accessed 15 November 2022.

Casting Glass

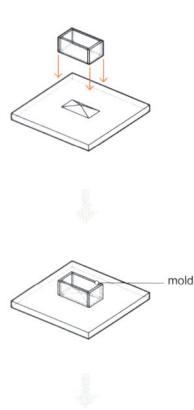

mold

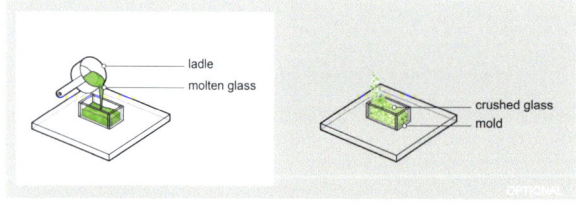

ladle
molten glass

crushed glass
mold

OPTIONAL

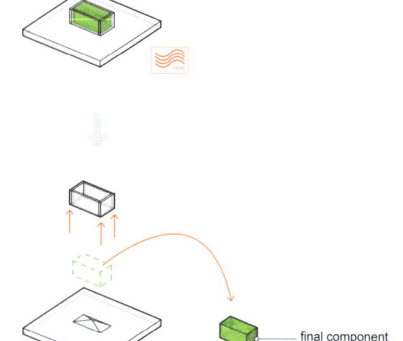

final component

Figure 4.3.1
Casting glass process diagram. Drawing
by author. A version of this diagram
originally appeared in *Manufacturing
Architecture* (Laurence King, 2018).

DOI: 10.4324/9781003299196-22

Casting glass is the manufacturing process that heats glass until it is sufficiently fluid and then uses gravity to deform the heated glass into the shape of the mold. This process includes *hot casting*, which uses heated, liquid glass that is poured into a mold outside of a kiln to make transparent castings, and *kiln casting*, which uses small glass pieces placed into a mold and they are heated inside of a kiln to make translucent or opaque castings.[1] Casting glass can make shapes that are simple or complex, and sizes can range from small beads to telescope lenses that are over 25 ft (7.5 m) in diameter. Generally, casting glass is not a highly mechanized process and can be done by artists, craftspersons, or foundries. Since molten glass is not very fluid and casting glass uses gravity for deformation, very little pressure is placed on the mold and the mold can be made of a range of inexpensive materials. Generally, casting glass is time intensive and often multiple molds are used for parallel productions to reduce the overall production schedule.

With hot casting glass, a multipart mold can be used with little concern because molten glass is too thick to seep into mold joints. Depending on the mold material, the mold is typically heated to reduce thermal shock and possible cracking when the molten glass meets it. Once the molten glass is poured into the mold, heat is applied to both the mold and the glass so that the glass cools slowly. The glass is further annealed in a kiln under controlled temperature conditions. Annealing may be done with the glass still in the mold, or the glass may be removed from the mold and placed in the annealer by itself.

Molds for glass casting may be made of a wide range of materials, including water-soaked wood, sand, refractory plaster, cast iron, or machined steel. Molds for casting glass may be open or closed. Open molds work for both hot and kiln casting, whereas closed molds are best with hot casting.[2] Sand molds are almost always sacrificial, as they are destroyed during the demolding process; however, the sand can be reused to form another mold. Refractory plaster molds may be sacrificial or reusable, depending on the manufacturer's needs; however, unlike sand, refractory plaster is not reusable. The mold material will affect the cast glass surface. If a glossy finish is required, then the casting should be removed from the mold and flame finished before it is annealed.

Glass

Any glass that can be heated to its thick, fluid-like state can be hot cast. Different glass types have different degrees of workability with lead glass being the easiest to hot cast and borosilicate glass being one of the most difficult. Generally, melted glass has the consistency of thick honey and takes a while to fill a mold during the casting process. Since cast glass components are solid, annealing times for cast glass components are longer than those for slumped (Chapter 1.1) or pressed glass (Chapter 4.7) components. This increases cycle times and production schedules.[3]

Figure 4.3.2
Crystal Houses facade on Pieter Cornelisz (PC) Hooftstraat. The buildings have custom
cast glass bricks in three different sizes.

FACING PAGE
Figure 4.3.3
The window sills, jambs, architraves, and dividers are
custom cast glass. Interior glass brick buttresses stabilize the glass brick facade, as seen through the store
windows and entrance.

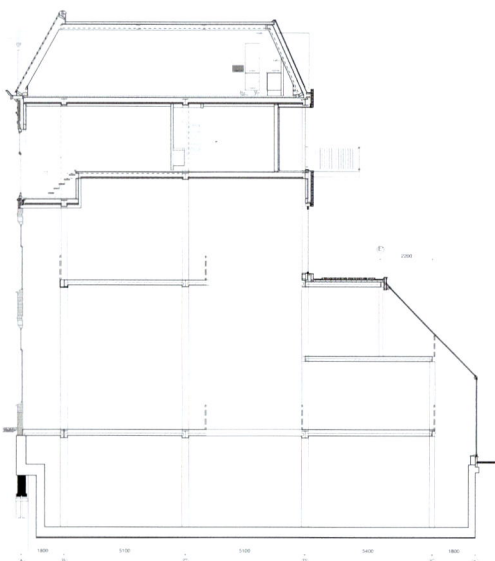

Figure 4.3.4
Crystal Houses section through the commercial level mezzanine. The brick cavity wall construction starts above the commercial levels.

Crystal Houses in Amsterdam, the Netherlands
By MVRDV

Crystal Houses is a mixed-use building that includes 6,674 ft² (620 m²) of retail with 2,368 ft² (220 m²) of housing above. It is located on Pieter Cornelisz (PC)

Hooftstraat, a street in Amsterdam well known for luxury brand stores. Originally on the site had been three-story, traditional brick Amsterdam houses with partial attics and commercial storefronts added later to the ground floor. The original buildings could not meet the needs of the developer-client, so they were demolished, and Crystal Houses built in their place. MVRDV designed Crystal Houses with a taller facade than the original structures so that a mezzanine could be added to the commercial space; also added were an addition on the back, a basement, and a full attic story.

According to MVRDV Associate Design Director Gijs Rikken, the client wanted the building to be an

Figure 4.3.5
Delft University experts on adhesives and glass structures helped develop the facade.

Figure 4.3.6
Poesia Glass Studio hot cast the bricks in an open, two-part mold.

art piece for the commercial flagship store.[4] MVRDV conceived of the building with glass bricks on the commercial level that transitioned to clay bricks on the upper levels as a way to meet the client's desire to be distinct, the need to provide transparency at the lower levels for the storefront, and the requirements of Amsterdam's Committee for Aesthetics and Monuments (CWM). The result is a loadbearing glass brick facade constructed from custom cast glass bricks bonded together with a transparent adhesive. In addition to the glass brick, the buildings' two lowest levels have custom cast glass windowsills, jambs, architraves, and window pane dividers.

Knowing that the design required expertise, MVRDV and structural engineers ABT reached out to experts on adhesives and glass structures at Delft University to help develop the facade.[5] It was through this partnership that the project team decided on the manufacturing method for the bricks, the adhesive types, brick tolerances, glass strength, wall assembly strength, and the method of bricklaying. The bricks needed to be solid, as hollow bricks would not be strong enough to carry the loads and would have required a supporting backup structure.[6] Internal glass brick buttresses provided lateral support to the front facade. The team investigated using pressed glass bricks rather than cast glass bricks; however, that was rejected due to cost concerns.

Poesia Glass Studio, located just west of Venice, Italy, cast the project's glass bricks and window surrounds. The over 6,500 bricks were hot cast in

an open, two-part steel mold with a four-sided ring and a base plate (see Figure 4.3.6). There are three different brick sizes, and all the bricks have a less than 1/8-in (3-mm) radial fillet on the brick edges and corners. The window surrounds were cast in two-part graphite molds that can be less expensive and provide sharper details than steel molds.[7] After casting, once the glass cooled enough to maintain its shape, it was removed from the mold and placed in the annealing oven. Depending on the size of the cast glass component, annealing took 8–38 hr.

The adhesive used between the bricks set brick tolerances and joint sizes. The adhesive was fluid, which required a flat glass surface to reduce the flow of the adhesive across the surface. The cast glass bricks had a 1/100-in (0.25-mm) tolerance across its bed joint surface. Small tolerances are difficult to achieve with this manufacturing process for two reasons. First, casting thick components often results in sink marks due to the glass shrinking during cooling. Second, casting the components in open molds results in a face that is not in contact with the mold and is less accurate than the mold faces. For Crystal Houses, a CNC mill ground the brick sailor faces until they were within tolerance and then the bricks were polished.[8] The brick adhesive was only placed in the bed joints and a clear sealant was used in the head joints to keep out weather and insects. A more viscous adhesive was used for the window surrounds that could accommodate looser tolerances and wider joints.

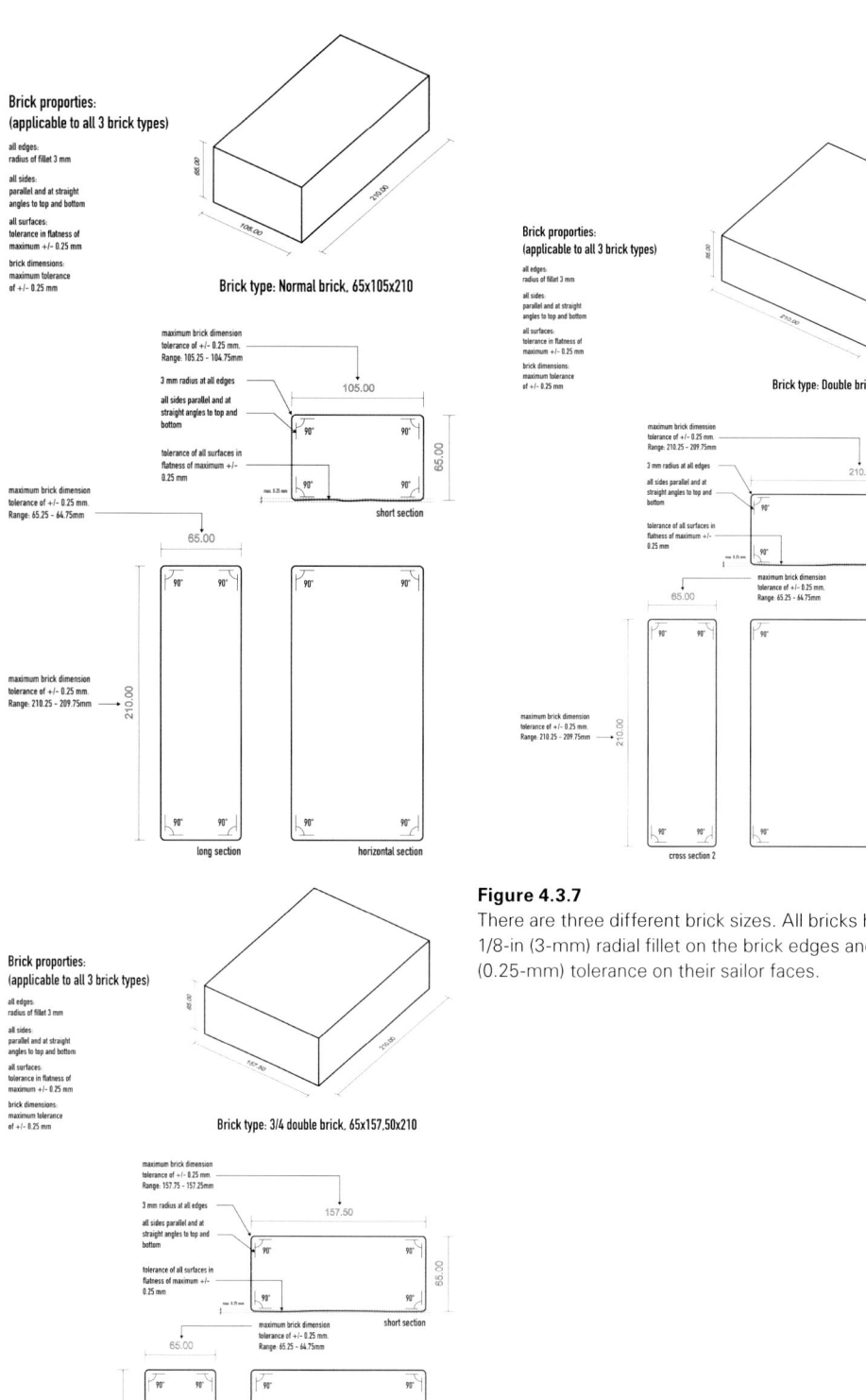

Brick proporties:
(applicable to all 3 brick types)

all edges:
radius of fillet 3 mm

all sides:
parallel and at straight
angles to top and bottom

all surfaces:
tolerance in flatness of
maximum +/- 0.25 mm

brick dimensions:
maximum tolerance
of +/- 0.25 mm

Brick type: Normal brick, 65x105x210

maximum brick dimension
tolerance of +/- 0.25 mm.
Range: 105.25 - 104.75mm

3 mm radius at all edges

all sides parallel and at
straight angles to top and
bottom

tolerance of all surfaces in
flatness of maximum +/-
0.25 mm

maximum brick dimension
tolerance of +/- 0.25 mm.
Range: 65.25 - 64.75mm

short section

maximum brick dimension
tolerance of +/- 0.25 mm.
Range: 210.25 - 209.75mm

long section

horizontal section

Brick proporties:
(applicable to all 3 brick types)

all edges:
radius of fillet 3 mm

all sides:
parallel and at straight
angles to top and bottom

all surfaces:
tolerance in flatness of
maximum +/- 0.25 mm

brick dimensions:
maximum tolerance
of +/- 0.25 mm

Brick type: Double brick, 65x210x210

maximum brick dimension
tolerance of +/- 0.25 mm.
Range: 210.25 - 209.75mm

3 mm radius at all edges

all sides parallel and at
straight angles to top and
bottom

tolerance of all surfaces in
flatness of maximum +/-
0.25 mm

cross section 1

maximum brick dimension
tolerance of +/- 0.25 mm.
Range: 65.25 - 64.75mm

cross section 2

maximum brick dimension
tolerance of +/- 0.25 mm.
Range: 210.25 - 209.75mm

horizontal section

Figure 4.3.7
There are three different brick sizes. All bricks have less than
1/8-in (3-mm) radial fillet on the brick edges and a 1/100-in
(0.25-mm) tolerance on their sailor faces.

Brick proporties:
(applicable to all 3 brick types)

all edges:
radius of fillet 3 mm

all sides:
parallel and at straight
angles to top and bottom

all surfaces:
tolerance in flatness of
maximum +/- 0.25 mm

brick dimensions:
maximum tolerance
of +/- 0.25 mm

Brick type: 3/4 double brick, 65x157,50x210

maximum brick dimension
tolerance of +/- 0.25 mm
Range: 157.75 - 157.25mm

3 mm radius at all edges

all sides parallel and at
straight angles to top and
bottom

tolerance of all surfaces in
flatness of maximum +/-
0.25 mm

maximum brick dimension
tolerance of +/- 0.25 mm.
Range: 65.25 - 64.75mm

short section

maximum brick dimension
tolerance of +/- 0.25 mm.
Range: 210.25 - 209.75mm

long section

horizontal section

FACING PAGE
Figure 4.3.8
In the transition from glass to clay bricks,
thin clay brick faces are adhered to the glass
bricks and an acrylic-based mortar simulates
a traditional cement-based mortar.

Although Crystal Houses facade appears to transition from glass bricks to clay bricks, this is an illusion. Due to differences in assembly materials (e.g. adhesive versus mortar) and thermal expansion rates, the glass and clay bricks could not be laid in the same bonds. Instead, the glass bricks fully cover the facade until the second story. Above this, the clay bricks sit on a continuous shelf angle and fully cover the remainder of the facade up to the cornice. The transitional zone has thin clay brick faces adhered to the front and back stretcher faces of glass bricks. The stretcher faces were ground so that the clay brick face is coplanar with its neighboring glass brick faces. An acrylic-based mortar is used around the thin clay brick to simulate a traditional cement-based mortar.

Rikken acknowledges that the thermal performance of the solid glass brick wall "is terrible"; however, within Dutch regulations, commercial shops and storefronts have fewer performance requirements as they may open their doors to invite shoppers inside.[9] Crystal Houses have ground source heat pumps to reduce the energy loads, and the upper floors have cavity wall construction for the clay bricks. MVRDV developed a replacement plan so that if any of the glass bricks got damaged, they could be removed safely and replaced. Poesia also cast extras of all their components so that the part could be replaced quickly without Poesia restarting their manufacturing process.

Notes

1. In kiln casting, air bubbles get trapped between the melted glass pieces. The smaller the glass pieces used, the more air bubbles trapped and the more opaque the casting.

2. This can also include a variant to hot casting, called dribble casting. Dribble casting is done in a kiln. Dribble casting uses small glass pieces held in a crucible with a hole in its bottom. The crucible is placed onto the mold and as the glass pieces melt, they dribble through the crucible hole into the mold. The resulting glass quality of dribble casting is like hot casting.

3. It is not uncommon for castings to require 48 hours or more of annealing time.

4. Rikken, Gijs. *Personal Interview*. 1 July 2022.

5. Frederic Veer and Rob Nijsse, respectively.

6. Oikonomopoulou, F., et al. "The Construction of the Crystal Houses Façade: Challenges and Innovations." *Glass Structures & Engineering*, vol. 3, no. 1, 2018, pp. 87–108

7. Oikonomopoulou, F., et al. "Rethinking the Cast Glass Mold." Challenging Glass 7 Conference on Architectural and Structural Applications of Glass. Belis, Bos & Louter (Eds.), Ghent University, September 2020.

8. The brick's header and stretcher faces were left with their mold-finish dimensions. The 3-mm radial fillet was added at the edges of the ground faces for appearance consistency.

9. Rikken.

References

Oikonomopoulou, F., et al. "The Construction of the Crystal Houses Façade: Challenges and Innovations." *Glass Structures & Engineering*, vol. 3, no. 1, 2018, pp. 87–108.

Oikonomopoulou, F., et al. "Rethinking the Cast Glass Mold." *Challenging Glass 7 Conference on Architectural and Structural Applications of Glass*. Belis, Bos & Louter (Eds.), Ghent University, September 2020.

Rikken, Gijs. *Personal Interview*. 1 July 2022.

Vibration Press Casting

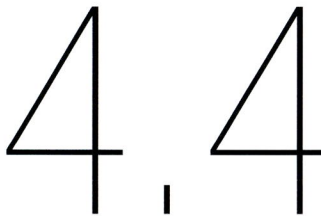

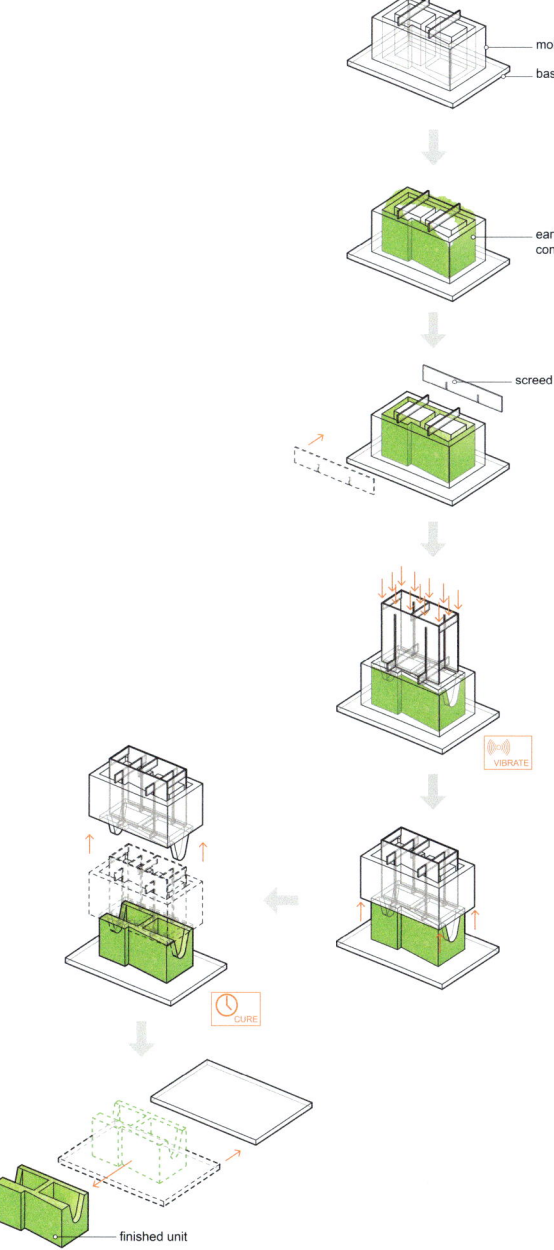

mold

base plate

earth moist concrete

screed

VIBRATE

CURE

finished unit

Figure 4.4.1
Vibration press casting diagram.Drawing
by author. A version of this diagram originally
appeared in *Manufacturing Architecture*
(Laurence King, 2018).

DOI: 10.4324/9781003299196-23

Vibration press casting is the manufacturing process that uses vibration and top pressure to distribute a dry concrete mixture throughout a mold. Vibration press casting is used to manufacture concrete masonry units (CMU). CMU shapes can be hollow or solid and side walls can be sloped. Unfortunately, side-wall holes and undercuts are difficult to produce and only a few manufacturers have the equipment to do so. Fine surface details are almost impossible to achieve. The ability for a manufacturer to produce certain shapes will be limited by their equipment. Mechanized manufacturers use large block-pressing machines, conveyors to move blocks from one station to another, steam rooms for curing, and robots for packaging. Other manufacturers may use hand-operated presses, move CMUs by hand to transfer them from one station to another, cure the CMUs in open air, and pack pallets by hand. CMU sizes are limited by the manufacturer's equipment.[1] Capital costs for this process can vary greatly, depending on the mechanization of the manufacturer. Typically, multiple CMUs are produced with each vibration press cycle and therefore production times can be short.[2]

In vibration press casting, a four-sided mold, open at the top and bottom, is lowered onto a base plate. The mold is then filled from the top with the dry concrete mixture and any excess is screeded off. The stripping shoe is lowered from the top, pressing the concrete into the mold. The mold vibrates to ensure that the concrete mixture fills the mold cavity. The four-sided mold is raised as the stripping shoe stays in place, stripping the CMU from the mold and leaving it on the base plate. Both are removed from the press and the CMU is cured on the plate. Once the CMU is sufficiently cured, it is removed from the plate and the plate is returned to be reused for another cycle.

The tooling for vibration press casting includes the mold, stripping shoe, and core pulls if forming side-wall holes; the tooling size is dictated by the press. For standard mechanized presses in the United States, the mold size will be limited by the *mold box*, a 16-in wide by 24-in long by 8-in high (410-, 610-, 200-mm, respectively) steel box that includes multiple mold cavities, the vibration mechanism, and fits into the press. The dry concrete mixture places a lot of friction on the mold surface, causing it to wear quickly. For high production runs, molds will be built up from several tool steel plates, allowing individual plates to be removed and replaced as needed. Stripping shoes experience less wear than the mold walls and might be made of a lower quality steel to reduce tooling costs. For mid-production runs or for hand presses, molds may be made from painted steel with no removable parts. For prototypes or small production runs, molds can be made of fiberglass.

Earth-Moist Concrete

The concrete used in vibration press casting is a mix of cement (Portland or white), course and fine aggregates, water, and optional additives (e.g. color and fly ash). A minimum amount of water is included so that the earth-moist concrete can fill the mold yet hold its shape during demolding without slumping. Since the amount of water is minimal, additional moisture may be added during the curing process to help the concrete gain its full strength. The dry concrete mixture often results in a rough surface with small pits. If a smooth surface is desired, then the blocks can be ground or polished.

Figure 4.4.2
Phra Pradeang House in the Phra Pradeang district of Bangkok, in an old neighborhood where
the streets are narrow, and the existing houses are small and built close to one another.
Photography by Soopakorn Srisakul, courtesy of all(zone).

Figure 4.4.3
The house's client desired privacy so all(zone) designed screen made with custom CMU
for the upper floors. Photography by Soopakorn Srisakul, courtesy of all(zone).

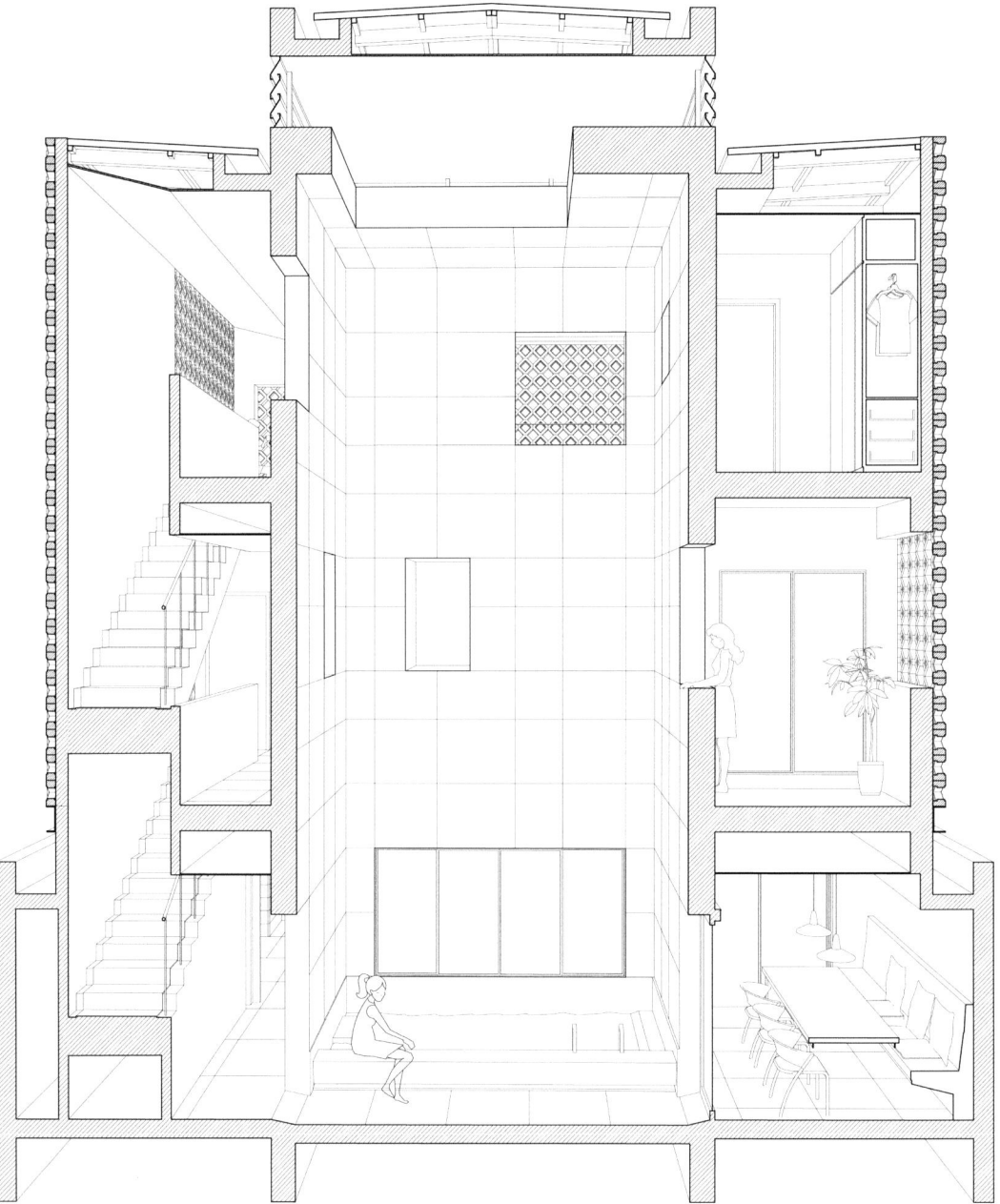

Figure 4.4.4
Section perspective through the three-story central atrium. Drawings courtesy of all(zone).

Phra Pradeang House, in Bangkok, Thailand
By all(zone)

Phra Pradeang House is three stories tall, has 4176 ft² (388 m²), and located in the Phra Pradeang district of Bangkok. The neighborhood is old, the streets are narrow, and the existing houses are small and built close to one another. The client chose the site because it is in the neighborhood where she grew up and is close to her relatives, but she did not like the lack of privacy. According to all(zone) Design

Figure 4.4.5
Interior photograph of the house and screen. Photography by Soopakorn Srisakul, courtesy of all(zone).

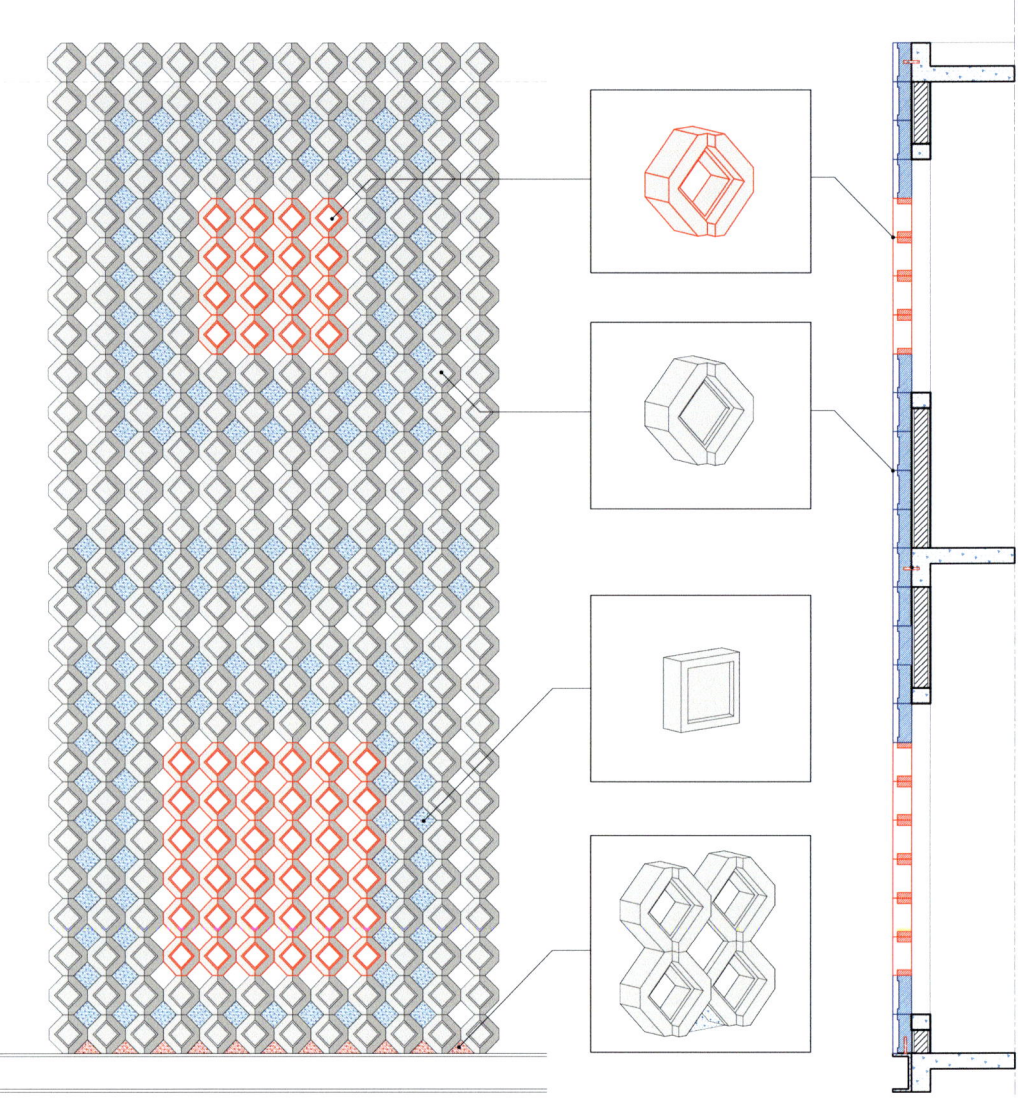

Figure 4.4.6
Bay elevation and section of screen with three different CMU units. Drawings courtesy of all(zone).

Director Rachaporn Choochuey, the inception for the house's design came from the client's desire to achieve privacy despite the project's location.[3] The house has a three-story, central atrium with a swimming pool at the ground level and a skylight and clearstory windows at the top for passive ventilation. The house has a solid perimeter wall on the first floor and a screen made with custom concrete blocks on the upper two floors. By wrapping the house in a concrete block screen, it provided privacy, while allowing for natural ventilation. The client did not want a standard concrete block to be used as she felt that a horizontal and vertical grid would look like a cage.[4] After many iterations with the client, the design team selected a lattice-like pattern on a 45-degree angle. Choochuey appreciates CMU and refers to it as a primitive material. In Thailand, Choochuey finds that CMUs are easy to make and have the potential to be used in a more architectural manner than they are typically used. All(zone) has

Figure 4.4.7
About 20 block prototypes were made and laid by masons to test their stability during layup. Photography by Soopakorn Srisakul, courtesy of all(zone).

designed different custom concrete blocks for their office headquarters, the Sadao Immigration Detention Center, and for a 500-unit, affordable housing project in Samut Prakan, Thailand.[5]

The manufacturer of the custom blocks, Tangrungjareon Engineering, is small and has facilities in Samut Songkram and Songkhla, Thailand. Tangrungjareon made three different concrete blocks for Phra Pradeang House—an open octagon, a closed octagon, and a closed square that is used as an infill piece between the closed octagons when privacy is needed. In addition to manufacturing the CMU blocks, Tangrungjareon also fabricated the equipment and molds that make the blocks. All(zone) found the manufacturer through the internet and Choochuey recalled researching "hundreds" of manufacturers that would be willing to collaborate with them for the custom concrete block. She found that the owner of Tangrungjareon was open to working with them to develop the mold. All(zone) and Tangrungjareon have collaborated on all(zone)'s other custom concrete block projects. For this house, all(zone) and the manufacturer worked back and forth on the concrete block's design with the manufacturer providing the design parameters—such as size limits, minimum wall thicknesses, and angles—so that the concrete blocks could be efficiently produced.

The manufacturer produced approximately 20 prototypes of the open octagon block on a hand press, using coarse sand and Portland cement. These prototypes were laid by masons to test their stability during layup. Since there is more material on half of the block's front face, and the octagon bed joint is limited to a small area, approximately 2 3/8 in by 3 ¾ in (60 mm by 95 mm), the blocks tend to tip forward. Through these tests, the team discovered that the projecting faces needed to be oriented vertically rather than horizontally[6] and that they needed to stack on top of one another rather than alternate back and forth from one coursing to the next. In addition, the masons switched from a Portland cement-based mortar to an epoxy mortar as it sets faster and is stronger. For production, the CMU blocks were made on a mechanical press, using a mixture of coarse sand, stone dust, and Portland cement.

The project's structural engineer required that the screen wall be arranged in vertical three-meter-wide, independent sections across the facade. Due to differences in construction tolerances for the site-cast concrete structure and the concrete block, the vertical joints between the sections are larger than designed and are different dimensions on the house's front and the back than its two sides. The exterior corners of the screen are joined on site by the epoxy mortar, cast against a custom mold fitted between the screen elements.

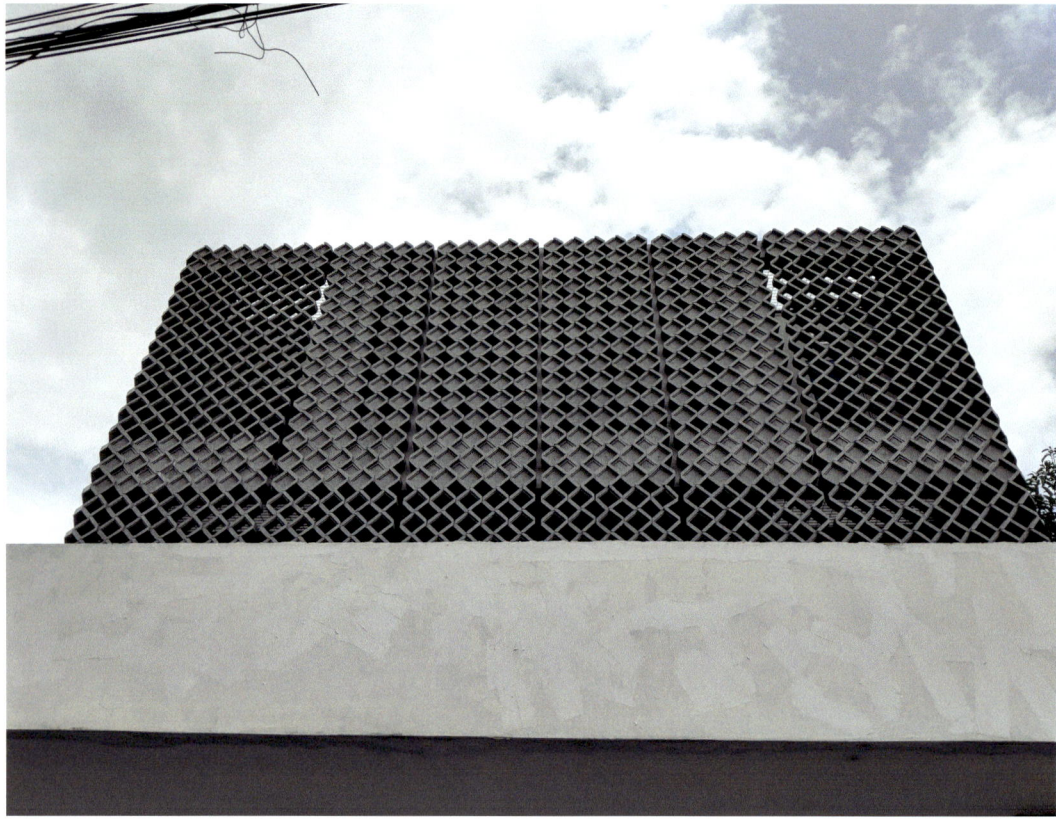

Figure 4.4.8
The structural engineer required that the screen wall be arranged in vertical three-meter-wide, independent sections across the facade. Photography by Soopakorn Srisakul, courtesy of all(zone).

Notes

1. A standard CMU press with a mechanized manufacturer is limited to a mold box size of 16 in wide by 24 in long by 8 in high (410, 610, and 200 mm, respectively). Some mechanized manufacturers may have a press that makes larger landscape units. These mold boxes will be larger, producing units up to 36 in (91 cm) but limited to only 4 in (10 cm) high.

2. For a mechanized press, a cycle can take as little as 10 sec to complete. Depending on the size of the CMU and the equipment's mold size limits, one cycle can produce 1–12 masonry units.

3. Choochuey, Rachaporn. *Personal Interview*. 11 July 2022.
4. *Ibid.*
5. Kitmungsa, Mai. *Personal Email*. 27 October 2022.
6. Choochuey noted all(zone) preferred the vertical orientation to the horizontal orientation, because the horizonal projection would be a shelf for water to sit.

References

Choochuey, Rachaporn. *Personal Interview*. 11 July 2022.
Kitmungsa, Mai. *Personal Email*. 27 October 2022.

CHAPTER
4.5

Vibration Tamping

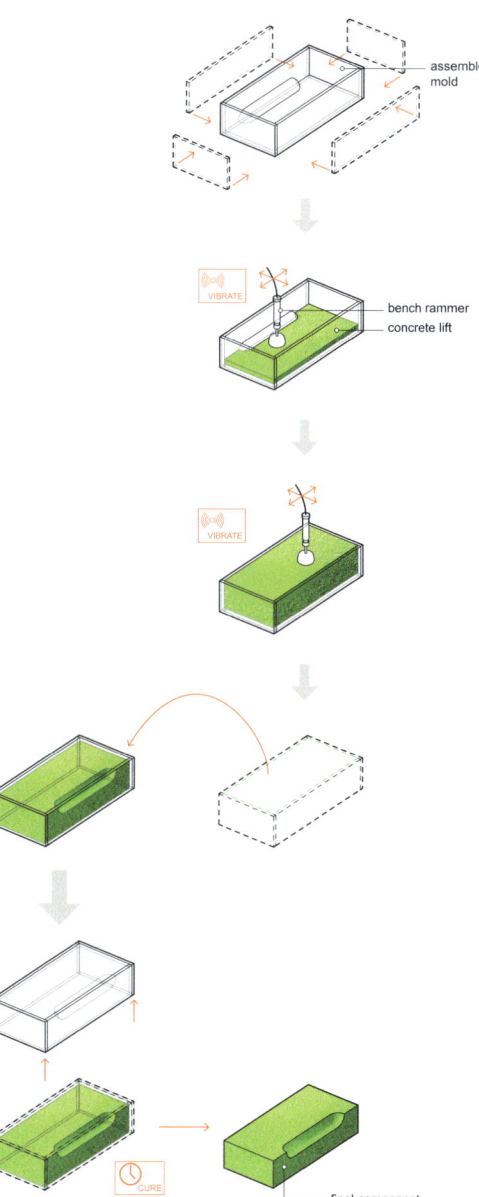

assembled mold

bench rammer
concrete lift

VIBRATE

VIBRATE

CURE

final component

Figure 4.5.1
Vibration tamping process diagram. Drawing by author.
A version of this diagram originally appeared in
Manufacturing Architecture (Laurence King, 2018).

DOI: 10.4324/9781003299196-24

Vibration tamping is the manufacturing process that uses a vibrating tamper to compress layers, or lifts, of a dry, stiff concrete mixture into a mold. This is the manufacturing method to make architectural cast stone. (The term *vibration tamping* describes the manufacturing process, but most architects and cast stone manufacturers will use the terms *making* or *manufacturing cast stone* when referring to the process.) The shapes made from vibration tamping are generally solid and can be fairly complex with moderate surface details. Sizes can range from brick size to large decorative cast stone components such as cornices.[1] Vibration tamping is a labor-intensive process that has not been mechanized and can easily accommodate custom designs. Like vibration press casting (Chapter 4.4), the concrete mixture is dry when tamped into the mold and therefore can be demolded once the mold is filled. Cycle times depend on the dimensions of the component; generally, cycle times are longer in vibration tamping than vibration press casting but shorter than wet-casting concrete (Chapter 4.1).[2]

In vibration tamping, molds are typically built-up from smaller parts, and since the concrete mix is dry, joints between mold pieces do not need to be caulked between casts. The dry concrete mixture is placed into the mold in approximately 2 in (50 mm) lifts. The worker then moves a pneumatic rammer across the full surface of the lift, tamping it to approximately half of its original depth. Then another lift is added, and vibration tamped. This is repeated until the component is the desired thickness. The component is then demolded and set aside to cure so that the cycle can begin again. Because the concrete mixture is dry, cast stone components are often cured with steam to ensure that there is enough water for hydration.

Molds for vibration tamping are typically open, and the opening is on the component's broad side so that workers can add lifts and access all the mold with the pneumatic rammer. Molds can be made from plywood, medium-density fiberboard (MDF), or particle boards with a melamine surface. The surface quality of a cast stone component is smoother than vibration-pressed concrete masonry units (Chapter 4.4) and is more akin to saw-cut or honed-finished sedimentary rock, like limestone or sandstone. The concrete mix for vibration tamping is dry and the components will not pick up the mold surface like wet-cast concrete does (Chapter 4.1).

This process may also be known as *vibroramming*, *vibrant dry tamping (VDT)*, and *manufacturing cast stone*.

Earth-Moist Concrete

The concrete used in vibration tamping is a mix of cement (Portland or white), sand and small finely crushed stone aggregates, inorganic pigments, waterproof admixture, and just enough water to have the tamped mixture hold its shape but not slump when demolded. The waterproof admixture helps to make the cast stone less water absorptive than sedimentary stones. Since little water is in the concrete mix, additional moisture, such as steam, may be added during the curing process to help the concrete reach its full strength.

Figure 4.5.2
Photograph of Lakeside Residence's private street facade and its custom cast stone cladding.
Image credit: Robert Tsai.

Figure 4.5.3
Photograph of the lakeside facade with its floor-to-ceiling windows for Lakeside Residence. Image credit: Robert Tsai.

Lakeside Residence in Rowlett, Texas, United States
By FAR + DANG

Lakeside Residence is on a small land outcropping into Lake Ray Hubbarb, northeast of Dallas, Texas. The house has more than 180-degree expansive views of the lake to the north and east but is relatively close to its neighbors on its street side. For privacy, glazing on the street facade is limited to a single large, screened window or to a band of high clerestory windows, while the lakeside facade is almost all floor-to-ceiling glazing. Far + Dang designed the house in three wings that knit together, and the central wing contains the house's taller, shared functions. Far + Dang Partner Rizwan Faruqui stated that the design team wanted to give the house a lightness and openness, despite the solidness of its street facade.[3] The side wings are clad with a dark aluminum composite panel that almost matches the clerestory glazing, while the central wing is visually grounded and clad with custom cast stone panels.

Figure 4.5.4
The central wing is visually grounded and clad with custom cast stone panels. Image credit: Robert Tsai.

Figure 4.5.5
Photograph of the CNC-milled MDF mold at Advanced Architectural Stone. Image credit: Far + Dang.

Inspired by Dallas's Perot Museum of Nature and Science by Morphosis and in consultation with the client, Far + Dang were interested initially in pursuing precast architectural concrete (Chapter 4.1) for the central wing. Through early investigations, the design team realized that the scale of the project was not compatible with large precast producers; the team reached out to Advanced Architectural Stone (AAS) a local precast supplier to see if the project was feasible for them. AAS is in Fort Worth, Texas, located only 60 mi away (97 km) from Lakeside, and was the producer for the cast stone interior of Louisiana Museum & Sports Hall of Fame by Trahan Architects. AAS suggested to Far + Dang that they consider cast stone instead of precast. Faruqui recalled that for the scale of the application, AAS said that cast stone offered greater flexibility in shaping within budget than precast concrete, while appearing similar.

Figure 4.5.6
Photograph of workers vibration tamping cast stone.

Figure 4.5.7
Photograph of color and texture of cast stone units at AAS. Image credit: Far + Dang.

Far + Dang designed the cast stone panels in Rhinoceros and Grasshopper, and they wanted to give the central wing some movement and fluidity to distinguish it from the more muted sides. The design team created the overall pattern then developed the pattern to be logical with the cast stone units. Far + Dang developed the design to work within the parameters that AAS set for the cast stone. AAS used 44 CNC-milled medium density fiberboard (MDF) molds in 25 different designs to make 92 cast stone units. AAS was able to use Far + Dang's digital files to fabricate the molds. AAS produced prototypes for Far + Dang's and the owners' reviews before production started.

The cast stone units are approximately 2 ft by 5 ft (61 cm by 152 cm) and 4 in (101 mm) thick elements with projections 2 ¾ in (70 mm) or less. The cast stone units are set with 3/8 in (9.5 mm) thick mortar joints and are in a stacked bond pattern. The sidewall keys provided a location for the lateral anchors

Figure 4.5.8
Photograph at night with racking down lights. The projections are just under 2 3/4 in. Image credit: Robert Tsai.

and a lock for the mortar bed. Since the side-wall anchors formed undercuts in the mold, AAS needed to remove the mold sides before demolding the cast stone. Far + Dang used AAS' standard light gray color that best approximated the color of precast concrete. Far + Dang had investigated using off-the-shelf panels made from fiber cement manufacturers but realized that they could get the cast stone, custom, locally made, for half the price. Faruqui was impressed with the cast stone and AAS, "as we were somehow able to do it all in what I would consider to be an exceptional and reasonable budget."

Notes

1. Large cast stone components can be reinforced with conventional reinforcing steel or welded-wire fabric as needed.

2. Wet-casting concrete typically requires that the concrete be left in the mold for up to 24 hr before demolding. Wet-casting concrete also requires that all open joints be caulked prior to the concrete being cast so that the wet cement does not seep into the joints, which would damage the mold and the component when demolding.

3. Faruqui, Rizwan. *Personal Interview*. 12 April 2023.

Reference

Faruqui, Rizwan. *Personal Interview*. 12 April 2023.

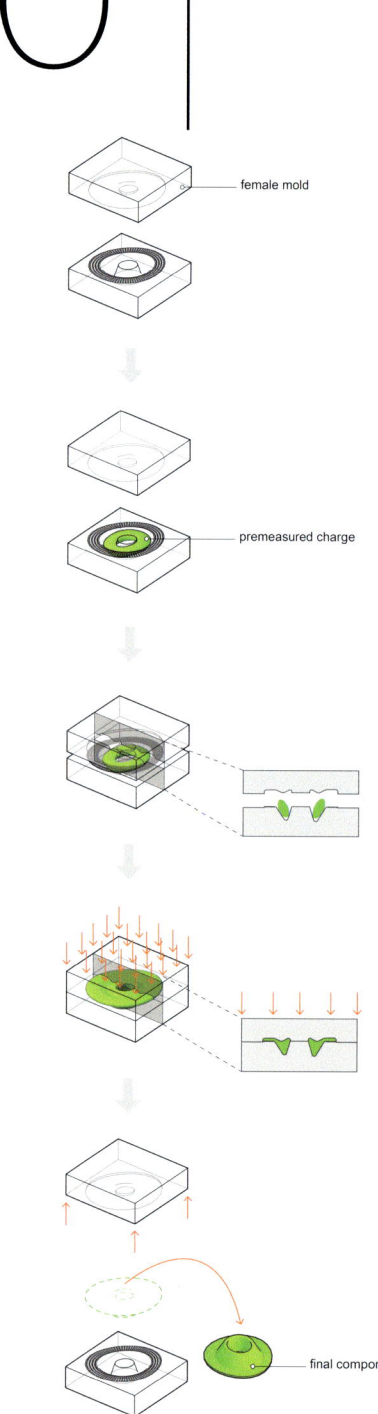

CHAPTER 4.6

Pressing

female mold

premeasured charge

final component

Figure 4.6.1
Pressing process diagram. Drawing
by author. A version of this diagram
originally appeared in *Manufacturing
Architecture* (Laurence King, 2018).

DOI: 10.4324/9781003299196-25

Pressing is the manufacturing process that presses a soft medium into a mold or a die[1] so that it fully fills the tooling cavity. Once the medium is stiff enough to retain its shape, it is removed from the tooling. For architectural applications, pressing typically uses clay, stiff mud, or glass, but the manufacturing process can also use plastic or fiber-reinforced plastic.[2] Pressing produces components with good surface quality and can impart intricate details on the component's surface. If the component is small, then it can be pressed into a solid shape; if it is large, then it should be pressed into a semi-hollow shape with interior wall structures, as the walls help reduce cracking. Clay can be pressed by hand or by mechanical or hydraulic press, while glass is pressed by mechanical or hydraulic press only. Tooling can be made from a wide range of materials depending on the pressing medium and the production run; therefore, capital costs can vary. Cycle times for this process depend on the medium and the tooling material, if pressing is done by hand or mechanical or hydraulic press, the intricacy of the component design, and if the manufacturer has automated its processes.

In pressing, a pre-measured, softened medium is placed into the mold or die cavity and a force pushes the medium to fill the cavity. The force may be provided by a hand—such as in hand-pressing clay—or by a male plunger[3] or a second-shaped die half. Once the medium fills the mold, it stays in the mold until it is stiff enough that it can be demolded. If pressing clay, then the clay needs to lose enough water before demolding. Depending on the wetness of the clay, the thickness of the component, the mold material, and the manufacturing conditions, dehydration times range from several seconds to several

hours. If pressing stiff mud, then demolding can be done immediately after forming. If pressing glass, then the glass needs to cool enough for demolding. Once the component is demolded, the component further solidifies away from the mold so that the mold can be reused for the next cycle. The component is then ready for optional post-production.

Tools for pressing may be open molds or closed dies. Manufacturers use open molds for hand-pressing clay, and they are typically for complex component designs, with limited production runs, and may come in multiple parts to facilitate demolding. Closed dies are used for mechanically or hydraulically pressing clay, stiff mud, or glass and may include a plunger or another die half, depending on the component's design. With closed dies for clay and glass pressing, the female die half may come in multiple parts to facilitate demolding. For pressing clay, tools are typically made from plaster as plaster draws water out from the clay, thus stiffening the clay and shortening cycle times. However, manufacturers may use wood, resin, or steel tooling to press clay. For pressing stiff mud, tools are typically steel to withstand the forces needed to shape the rigid material. For pressing glass, tools are typically made from cast iron (Chapter 4.2) or machined tool steel; however, refractory plaster can be used for small production runs.

Clay

Clay is used for hand and mechanical or hydraulic pressing and has enough water content that it is moist to touch. Clay for hand-pressing has a higher water content than mechanical- or hydraulic-pressed clay and is often left in the mold longer to stiffen enough before demolding. The clay must be free of stone, debris, and organic materials and may contain small particles of grog as filler. Generally, clay is more malleable than stiff mud and forms components that are more complex with thinner walls.

Stiff Mud

Stiff mud is the material used for forming bricks. Stiff mud includes clay and has larger particles (e.g. sand and feldspar) but less water than clay. Unlike clay, stiff mud does not feel moist when touched and when hand squeezed does not leave a wet residue behind. The particles in stiff mud adhere to each other only under pressure. Generally, stiff mud is pressed at higher pressures than clay, requires thicker walls, and results in a rougher surface texture.

Glass

Any glass that can be heated to its thick, fluid-like state can be pressed. Different types of glass have different amounts of workability, with lead glass being the easiest to work and borosilicate glass being one of the most difficult. Generally, melted glass has the consistency of thick honey, and a multipart female mold can be used with little-to-no flashing resulting at the mold seams. Annealing times for pressed glass components will be longer than those for slumped glass (Chapter 1.1) but depending on the glass wall thickness can be shorter than those for cast glass (Chapter 4.3).

Figure 4.6.2
The Fitzroy by Roman and Williams is clad in custom-hand-pressed, green-glazed terracotta.
Photograph by Jimi Billingsley, courtesy of Boston Valley Terra Cotta.

Figure 4.6.3
The mottled green glaze varied from one terracotta element to the next. Photograph by Michael Young, courtesy of Roman Williams.

The Fitzroy in New York, New York, United States
By Roman and Williams

The Fitzroy is a 65,300 ft² (6067 m²), ten-story tall residential building, with for-sale units that range from two to five bedrooms. Building amenities include an attended lobby, wine cellars, a children's playroom, bicycle storage, climate-controlled storage, fitness center, and rooftop deck with kitchen. The building is located on West 24th Street and is visible from the High Line; it is just over a block from Hudson River Park and minutes from Chelsea Piers. The Fitzroy is clad in green glazed terracotta, with copper-clad windows and doors. Roman and Williams Founder Stephen Alesch said that the firm chose terracotta for this project for two reasons.[4] First, Alesch and firm partner Robin Standefer have a love of terracotta. They have a large Gladding McBean sample in their office, and they admire the historic terracotta buildings near their office in Tribeca. Second, Roman and Williams had met previously with The Fitzroy developer and contractor, JDS Development Group, to collaborate on the renovation of an Art Deco Ralph Walker building. Although Roman and Williams were not awarded that commission, a few years later, JDS approached them to design a new building in the spirit of Ralph Walker. Alesch stated that JDS knew how much they loved terracotta and the Art Deco style, "so it was a dream call." The firm brought its Gladding McBean sample to the first meeting with JDS, and the design for The Fitzroy started from there.

Figure 4.6.4
Roman and Williams design drawings over three-dimensional renderings to develop the terracotta profiles. Drawings by Stephen Alesch. Courtesy Roman and Williams.

Figure 4.6.5
Detail photograph of pilaster detail. Courtesy Roman and Williams.

Initially, Roman and Williams proposed a facade that used terracotta as the primary cladding material, but they thought it would be value engineered to

Figure 4.6.6
Image at Boston Valley Terra Cotta of hand-pressing clay into a plaster mold.

a smaller amount, used only for building highlights, like windowsills. In the end, JDS fully supported using terracotta and the front and rear facades of The Fitzroy are clad in it. During early design, the design team was going to work with another terracotta manufacturer. Alesch recalled going through the first manufacturer's catalog to specify the terracotta elements, and the team had design meetings with them. As they completed drawings, JDS chose to work with Boston Valley Terra Cotta, instead.

Figure 4.6.7
Image of The Fitzroy under construction. Masons hand-laid the terracotta elements with a 3/8 in (9.5 mm) joint. Photograph by Stephen Penta, courtesy of Boston Valley Terra Cotta.

Figure 4.6.8
Image of The Fitzroy under construction. Masons used a dark gray mortar, and the terracotta elements were anchored to the facade. Heavy terracotta elements were supported by steel angles when needed. Photograph by Stephen Penta, courtesy of Boston Valley Terra Cotta.

Figure 4.6.9
Museum of Art, Architecture, and Technology (MAAT) faces south to the Tagus River. MAAT's roof is accessible from the riverwalk and by a pedestrian footbridge that connects to the historic neighborhood to the north. The newly repurposed Central Tejo Power Station is seen on the left. Photograph by Fernando Guerra/ FG + SG.

Boston Valley manufactured the terracotta elements for The Fitzroy, using all four of their manufacturing processes: extrusion, hydraulic press, hand press, and slip casting. Some elements are stock profiles while other custom pieces were used for specialty items like the curved caps and buttresses.[5] There are 5600 terracotta elements and 500 unique shapes. Boston Valley hand presses rather than hydraulic pressing terracotta elements when there is a small production run or when an element has a lot of non-linear detail.[6] Roman and Williams wanted a custom green glaze with a depth of color that was based on a Japanese tile that they had in their office. The glaze was designed to be mottled with variation from one terracotta element to the next. The first terracotta manufacturer sent color samples that Roman and Williams felt were too simplistic, flat, and even. When Boston Valley became involved with the project and submitted initial glaze samples, Alesch thought that their submissions were perfect. Alesch stated that the original manufacturer had given us perfection, but what we were looking for was a more hand-crafted look.

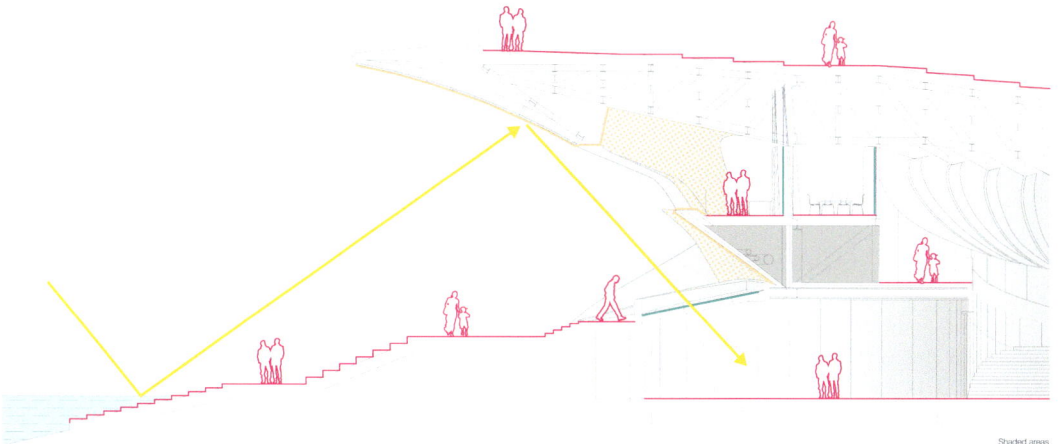

Figure 4.6.10
MAAT with Ponte 25 de Abril Bridge seen beyond. Photograph by Fernando Guerra/ FG + SG.

Shaded areas

Figure 4.6.11
The profile and plan of the south elevation reflects light into the lower level gallery while providing shade to the public space and entrance. Courtesy of AL_A.

Figure 4.6.12
The white glazed glossy ceramic tile reflects the southern light. Photography by Hufton + Crow.

Museum of Art, Architecture, and Technology (MAAT) in Lisbon, Portugal
By AL_A

The Museum of Art, Architecture, and Technology (MAAT) is part of Energias de Portugal (EDP) Foundation's masterplan for a Tagus River art campus that includes a newly repurposed Central Tejo Power Station. MAAT is 79,653 ft² (7400 m²) with four exhibit spaces, a restaurant, and a museum shop. According to AL_A Director Maximiliano Arrocet, EDP Foundation did not have a set brief for the project, and instead they used the project to give back to the city of Lisbon.[7] MAAT is low-slung and rises just 41 ft (12.4 m) from the river boardwalk to maintain views of the river and the Ponte 25 de Abril bridge from the historic Belém neighborhood to the north. Since the river was separated from Lisbon by railroad tracks and a highway, the riverfront has been difficult to access for people. AL_A incorporated a public at-grade pedestrian footbridge that connects Belém to MAAT's occupiable green roof and spans over the road and railroad tracks. The roof is an outdoor room of the museum; during the day, there are 180-degree views of the river to the south and Lisbon to the north, and at night, visitors can watch movies. The building is 623 ft (190 m) long, with ramps on both ends to access the roof. MAAT's lowest level gallery is sunk 20 ft (6 m) down from the boardwalk. To get light to the lower level, AL_A wanted sunlight to reflect off the Tagus and bounce off MAAT's facade into the lower level gallery (see Figure 4.6.11). AL_A's studies of the solar path dictated the angles of the south facade in section and its curve in plan to allow for this reflected light while providing shade during the summer to the building's southern public space and entrance.

Figure 4.6.13
When the sun is low, the building reflects the colors of the water and sky. Photography by Hufton + Crow.

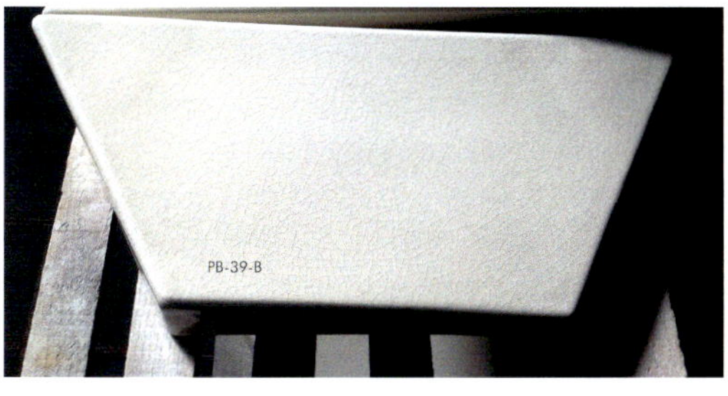

Figure 4.6.14
Photograph of the white crackle-glazed glossy finish.

Figure 4.6.15
Prototypes of the three tiles. The tile's three-dimensional shape hides any joint size inconstancies in the shadow between the tiles.

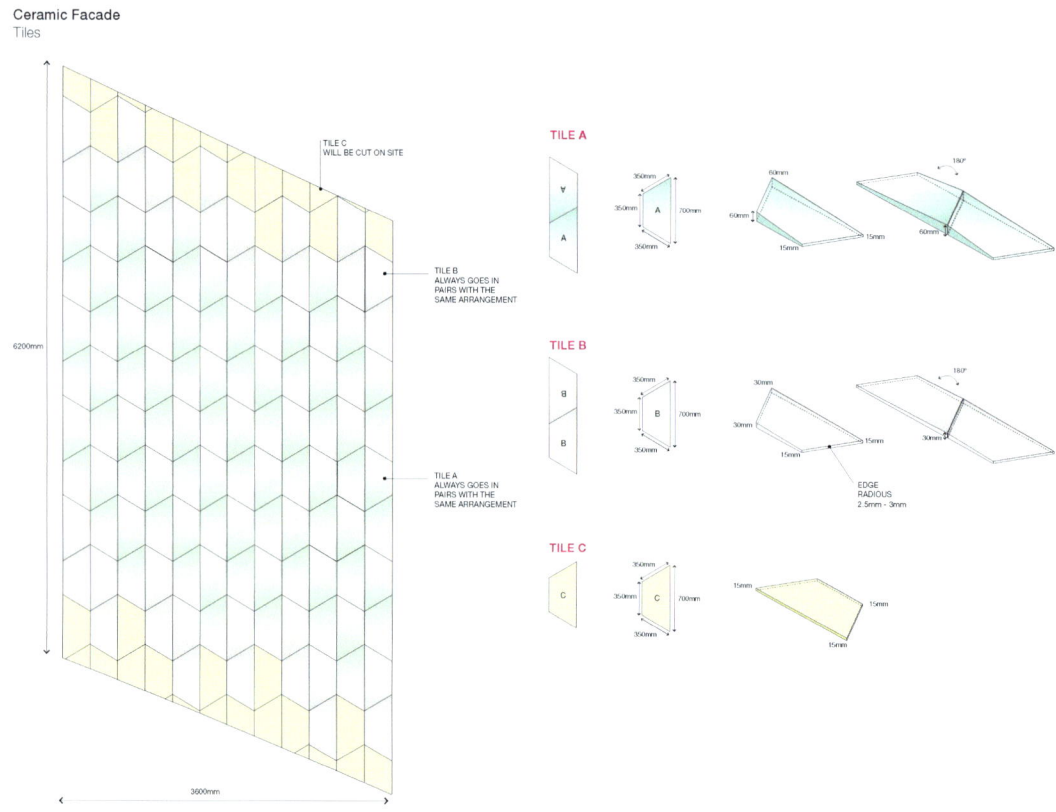

Ceramic Facade
Tiles

Figure 4.6.16
The three tiles and where they generally appear on the facade. Tile A (green) is PR-60.
Tile B (gray) is PR-45. Tile C (yellow) is the flat extruded tile. Courtesy of AL_A.

Arrocet noted that the light at the site is fantastic. AL_A chose to use a white crackle-glazed glossy ceramic tile to reflect that light, and when the sun is low, the water, sky, and building all reflect the colors of sunrises and sunsets.[8] The shape of the tiles developed from two desires. First, AL_A wanted a tile shape that captured and reflected light in a similar way to the surface of the Tagus.[9] Second, the tiles were going to be mounted on a rainscreen system. The design team knew that MAAT's changing section angles dictated a ruled surface for the rainscreen tracks; however, the MAAT's curved plan resulted in the spacing between the ruled lines being larger at the top or bottom, depending on the location along the facade. This meant that the gaps between the ceramic tiles would be inconsistent in dimension from the bottom of the facade to the top. Any deviations could be compounded by the rainscreen installer and more visible than AL_A wanted. To hide the inconsistency, the design team chose to use a three-dimensional tile that would hide any inconsistent gaps in the shadow between the tiles.

Figure 4.6.18
The three-dimensional tiles were formed on a hydraulic press (seen in background).

Figure 4.6.17
The flat extruded tile. Cumella shipped some extra-long tile that the installer cut to fit onsite to minimize small pieces at the building edges.

AL_A had collaborated with Ceràmica Cumella for custom terracotta elements for the Victoria and Albert Museum entrance renovation, and they approached Cumella for collaborating with them for MAAT. Initially, AL_A proposed two different ceramic tiles for MAAT—a three-dimensional tile and a flat tile. The three-dimensional tile would hide the joints, but the flat tile was needed so that when cut for the top and bottom of the building, the flat tiles would form a consistent and predictable edge. Through discussion with Antoni Cumella, the design team introduced an additional third tile for MAAT that provides a transition from the deeper standard tile to the flat tile.[10] Cumella manufactured the two three-dimensional tiles on a hydraulic press. The standard tile is PR-60 and has a 60-mm (2 3/8 in) maximum tile depth and a 30-mm (1 1/8 in) minimum depth. The transitional tile is PR-45 with a 45-mm (1 ¾ in) maximum relief and a 30-mm (1 1/8 in) minimum depth. The flat tiles are extruded and trimmed to shape with a die. Any flat tiles at the building edges were cut onsite to the needed length. All the tiles are a half hexagon, measuring 12 in (30 cm) on the short face and 23 ½ in (60 cm) on the long face.

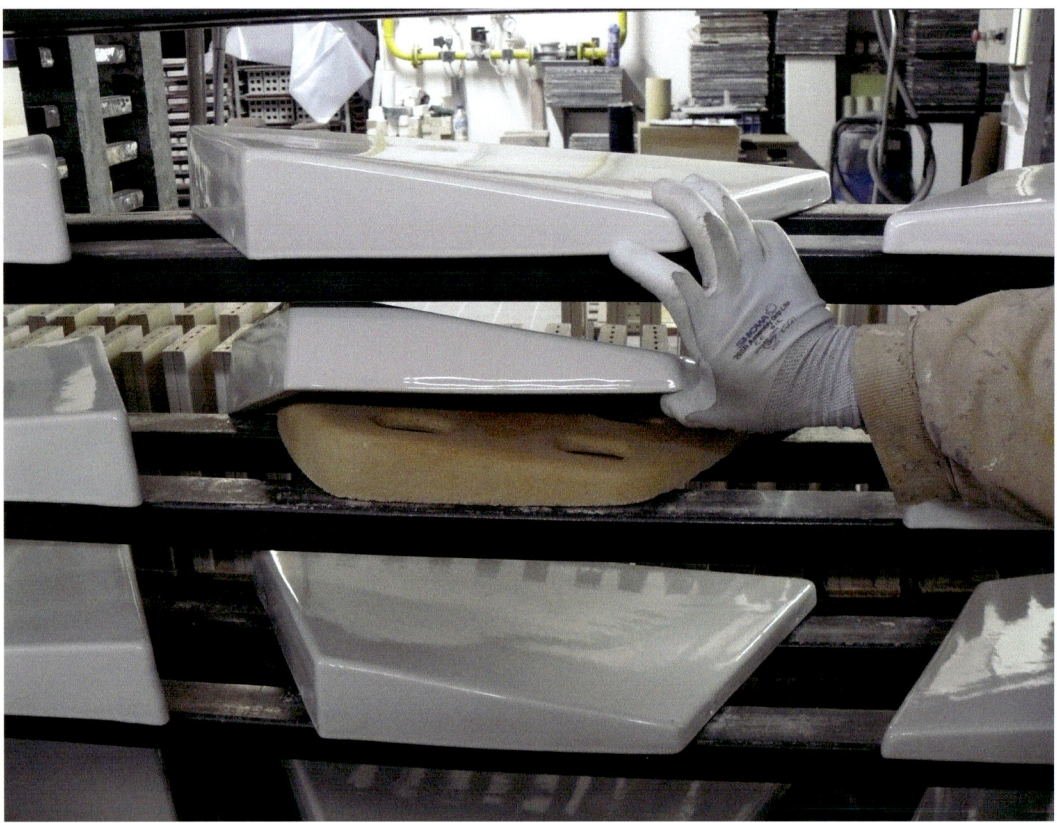

Figure 4.6.19
Cumella made approximately 200 reusable refractory molds to support the tile face during firing.

Cumella manufactured 14,751 ceramic tiles for MAAT. With the geometry of the pressed tile surface, the face of the three-dimensional tile twists. To produce the tile geometry on the press, Cumella manufactured the tiles with the hollow side down. The dimensions of the tile were such that the face of the tile would sag without support. Cumella tested putting internal supporting walls on the backside of the tile, but the face would sag between the walls. To limit sagging, Cumella used a CNC-milled polystyrene mold placed underneath the tiles during drying and a refractory mold to place underneath the tiles during firing. Before firing, Cumella placed the tiles in a jig and drilled out four holes in its side walls that fit over the rainscreen clips. After firing, to achieve the crackled glaze, Cumella tossed cold water on the tile.

The changing geometry of MAAT required that the tiles function as a ventilated rainscreen and as an overhead soffit. The project's regulating agencies needed rigorous testing of the facade elements. The outcomes of one test required that a glass fiber-reinforced plastic (GFRP) backing be placed on the inside face of the tile, so that if a tile broke, then its pieces would stay in place and not fall. Cumella installed the GFRP backing on the tiles.

Figure 4.6.20
A mockup of the tiles mounted to the rainscreen system. Plastic sleeves were added to the aluminum clips to protect the ceramic tiles from the aluminum supports' thermal expansion.

Figure 4.6.21
The tile and rainscreen installer, Disset, used string to locate the clips that attached the vertical rainscreen tracks to the substructure. Each tile was installed individually on site. On the right of the bucket lift, you can see where the site-cut flat tiles will be installed. Courtesy of AL_A.

FACING PAGE
Figure 4.6.23
Kohan Ceram headquarters has a custom-pressed brick.
© Deed Studio.

Kohan Ceram Headquarters in Tehran, Iran
By Hooba Design Group

Kohan Ceram[11] Brick Manufacturing Company has been making bricks since 1376 and its current factory is in an industrial and agricultural area, approximately 35 mi (60 km) southeast of central Tehran. Hooba Design Group designed the new mixed-use headquarters for Kohan Ceram, its surrounding landscape, and its custom-pressed bricks. The six-story, 11,302 ft² (1050 m²) headquarters include a basement, ground-level parking, a reception area and product showroom, office space, and an apartment unit that occupies the top two floors with access to the roof terrace. The headquarters is in a mixed-use neighborhood with surrounding buildings of similar height; it is across the street from a small neighborhood park, and next to a limited-access expressway.

Figure 4.6.22
Kohan Ceram headquarters is in a mixed-use neighborhood with surrounding buildings of similar height.
It is across the street from a small neighborhood park, and next to a limited-access expressway.
© Parham Taghioff.

Figure 4.6.24
Hooba Design Group designed two bricks for the project—a brick with round core
holes and a solid brick with round niches. Courtesy of Hooba Design Group.

Figure 4.6.25
Detail image of Spectacle Brick. When used for the building's interior and exterior finishes, the two brick wythes are separated by insulation. Courtesy of Hooba Design Group.

Figure 4.6.26
The back or inside face of Spectacle Brick has reliefs to allow for air between the brick and the insulation. Courtesy of Hooba Design Group.

Building the headquarters of a brick manufacturing company was an opportunity for Hooba Founder and Managing Director Hooman Balazadeh to develop custom bricks for the project. Some of the inspiration for the brick design came from Balazadeh's childhood.[12] Balazadeh grew up in Tabriz, a city in Iran known for its historic bazaar that occupies approximately 1730 acres (7 km²) with 27 sub-bazaars grouped by products.[13] In addition, Balazadeh's father made eyeglasses, or spectacles, and so Balazadeh found himself interested in spectacles. Balazadeh's inspiration also came from balancing Iran's history of building in brick with Tehran modern buildings combining many different materials together. Balazadeh's conceptual question for the headquarters seemed to be how to reconcile the need to use modern materials while acknowledging the positive influence that brick had in creating Tehran's identity. The final inspiration came from Balazadeh's desire to combine architectural dualities such as solidity and transparency, impenetrable and penetrable, and homogenous and heterogeneous

Figure 4.6.27
Initial investigations by Hooba into Spectacle Brick included a glass-casting factory near the Kohan Ceram factory casting molten glass directly into brick core holes. This method was not used because the brick cracked. Courtesy of Hooba Design Group.

that could be combined to create a new material. Balazadeh then researched combining glass and brick into a single custom brick unit, called the *Spectacle Brick*.

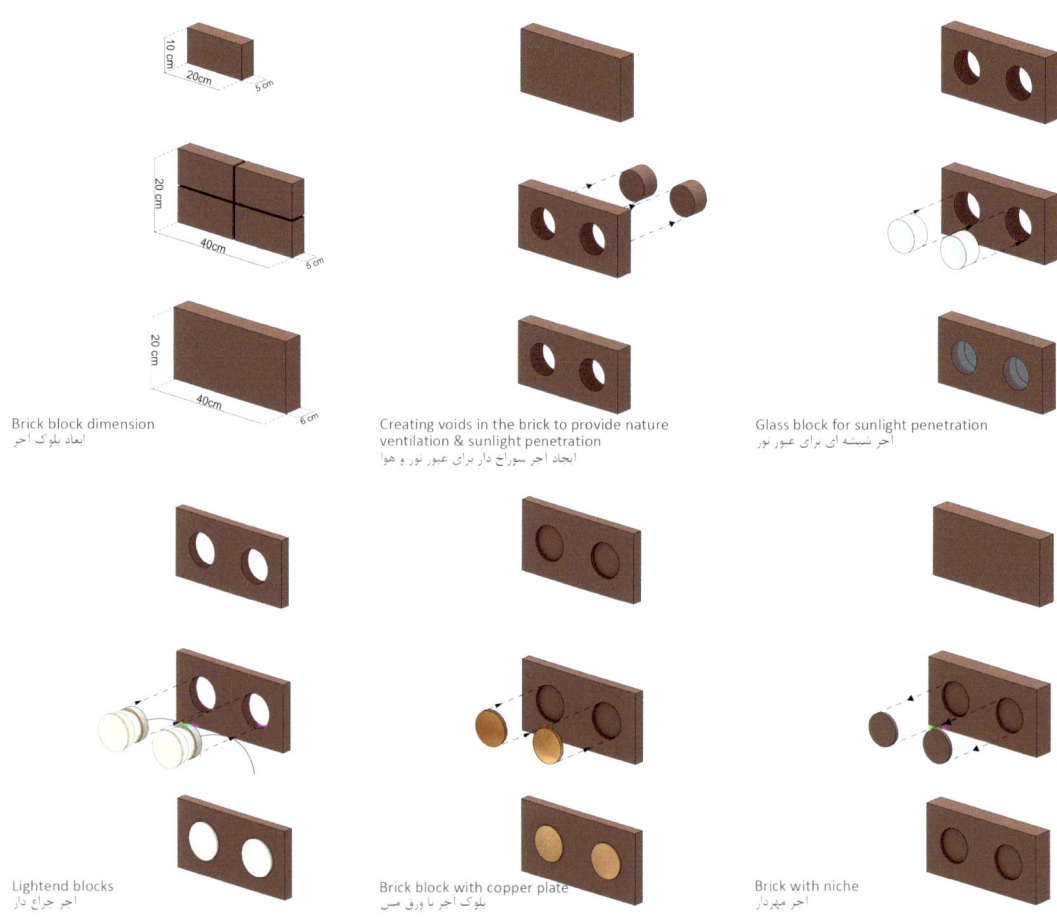

Brick block dimension
ابعاد بلوک آجر

Creating voids in the brick to provide nature
ventilation & sunlight penetration
ایجاد آجر سوراخ دار برای عبور نور و هوا

Glass block for sunlight penetration
آجر شیشه ای برای عبور نور

Lightend blocks
آجر چراغ دار

Brick block with copper plate
بلوک آجر با ورق مس

Brick with niche
آجر مهردار

Figure 4.6.28
Two different pressed brick shapes made five different Spectacle Bricks. Courtesy of Hooba Design Group.

Initial investigations by Hooba into Spectacle Brick included a glass-casting factory near the Kohan Ceram factory casting molten glass directly into different shaped brick core holes. Unfortunately, these tests demonstrated that the brick would crack during this process, so instead the design team elected to adhere premade glass lenses into the brick frames. There are approximately 12,000 custom bricks in this building, using two different shapes to form five different spectacle bricks. The first brick has circular holes that go all the way through the bricks; those bricks may be left open for venting, glass lenses may be added for natural lighting, or small lights may be added

Figure 4.6.29
The bricks are attached to metal straps that are welded to a metal framework. The bricks in this building are recyclable and the metal framework addresses the seismic concerns for the exterior walls. Courtesy of Hooba Design Group.

for artificial lighting. The second brick has circular, partially inset niches that are either left blank, or circular copper blanks are adhered to the back face of the niche. The overall dimensions of both bricks are 15 3/4 in long, 7 7/8 in high, and approximately 2 3/8 in deep (40 cm, 20 cm, and 6 cm, respectively).[14] Kohan Ceram manufactured the two brick shapes in an automated assembly line by pressing stiff mud into multi-cavity steel molds that can produce multiple bricks in each cycle. Spectacle Brick is part of Kohan Ceram's catalog of products and is available for purchase.[15]

The headquarter's brick walls are two wythes separated by foam insulation when bricks are used for both the interior and exterior building finishes.[16] The bricks are supported by a metal framework that includes horizontal square tubes in the bed joints welded to metal straps attached to the top of each brick and vertical tubes embedded at the head joints. The holes for the dowels were created after the bricks were pressed by placing the bricks into

Figure 4.6.30
The assembly has a high-strength mortar in the brick bed joint. Courtesy of Hooba Design Group.

a jig and drilling out the stiff mud. The bricks are in a stacked bond pattern with deep-raked bed joints and closed, mortarless head joints. Special, high-strength mortar was used in the bed joints to bond the metal and brick together. The bricks in this building are recyclable[17] and the wall assembly addresses the seismic concerns for the exterior walls.[18]

Figure 4.6.31
The headquarter's brick showroom. Spectacle Brick is part of Kohan Ceram's catalog of products and is available for purchase. © Parham Taghioff.

FACING PAGE
Figure 4.6.33
During the day, natural light hits the upper panel lenses.
© Hélène Binet

Figure 4.6.32
Le Prisme is meant to be seen from afar. At night, exterior projectors mounted at the top and bottom of the interstitial space project lighting programs through the custom-pressed glass lenses. © Brisac Gonzalez.

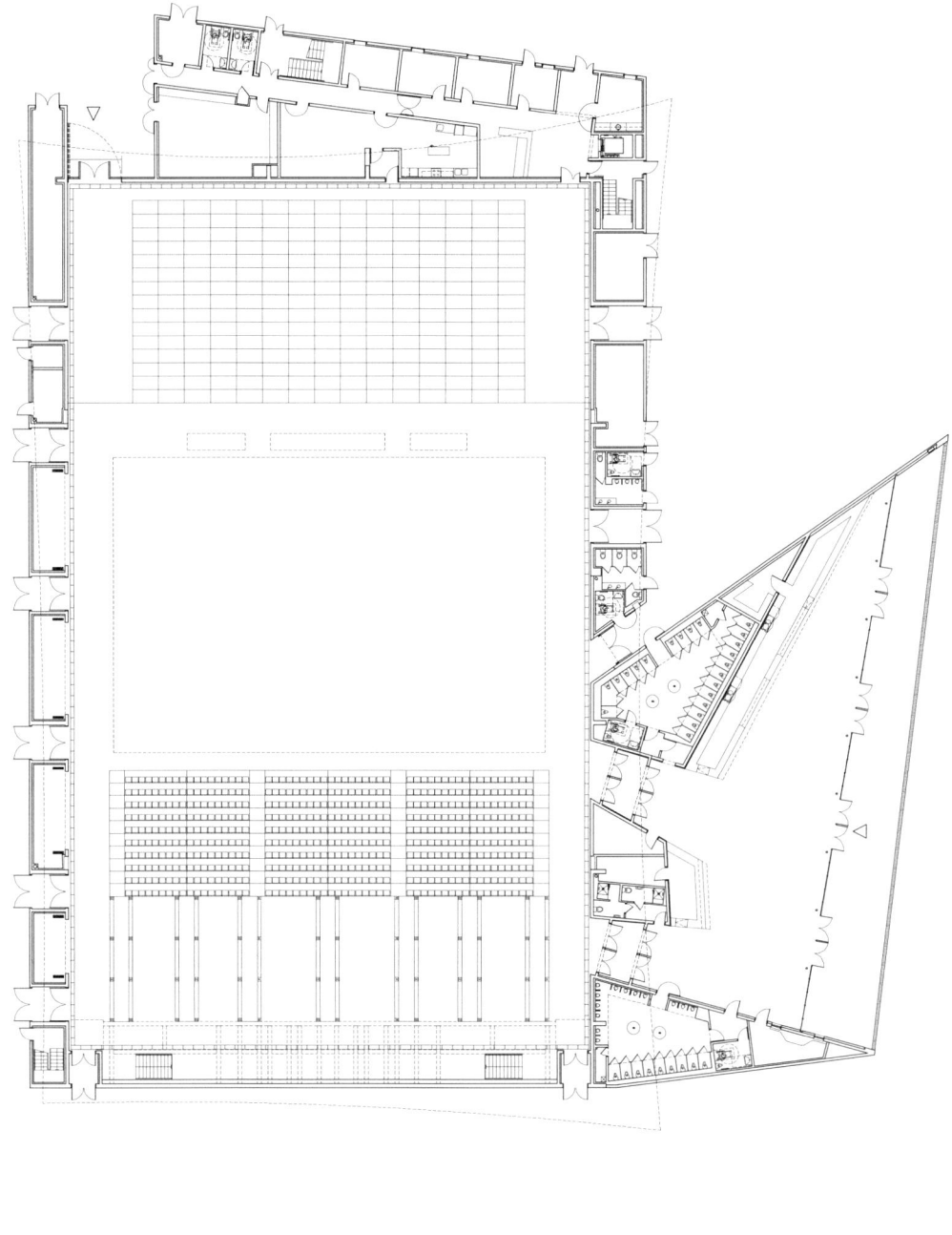

Figure 4.6.34
The building program includes the central space, surrounded by support spaces such as an entrance foyer, changing rooms, offices, catering facilities, and restrooms. © Brisac Gonzalez.

Le Prisme in Aurillac, France
By Brisac Gonzalez

Le Prisme is a 56,671 ft^2 (5265 m^2) multipurpose event hall that accommodates 4500 people and is for concerts and other performances, sports events, or trade shows. The building program includes the central space—a black box with retractable seating and a demountable stage—surrounded by support spaces such as an entrance foyer, changing rooms, offices, catering facilities, and restrooms. The building is surrounded by a parking lot and is next to an industrial neighborhood in Aurillac; it is bordered by a regional vehicle artery to the west and train tracks to the east. The building is visible from all four sides and is meant to be seen from afar and traveling at a speed.[19] According to Brisac Gonzalez Founding Partner Cecile Brisac, because the project essentially was a large opaque box, measuring approximately 230 ft long, 130 ft wide and 50 ft tall (70 m, 40 m, and 15 m, respectively), the design team wanted to give the building life and lightness.[20] Toward that end, Brisac Gonzalez designed the building with a lower base and an upper, gently curving screen that peels away from the building surface.

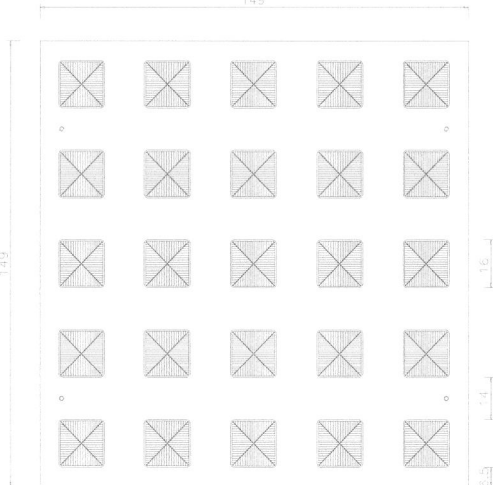

Figure 4.6.35
Elevation of the GFRC panel with embedded pressed glass lenses. The panels' measurements appear in cm.
© Brisac Gonzalez.

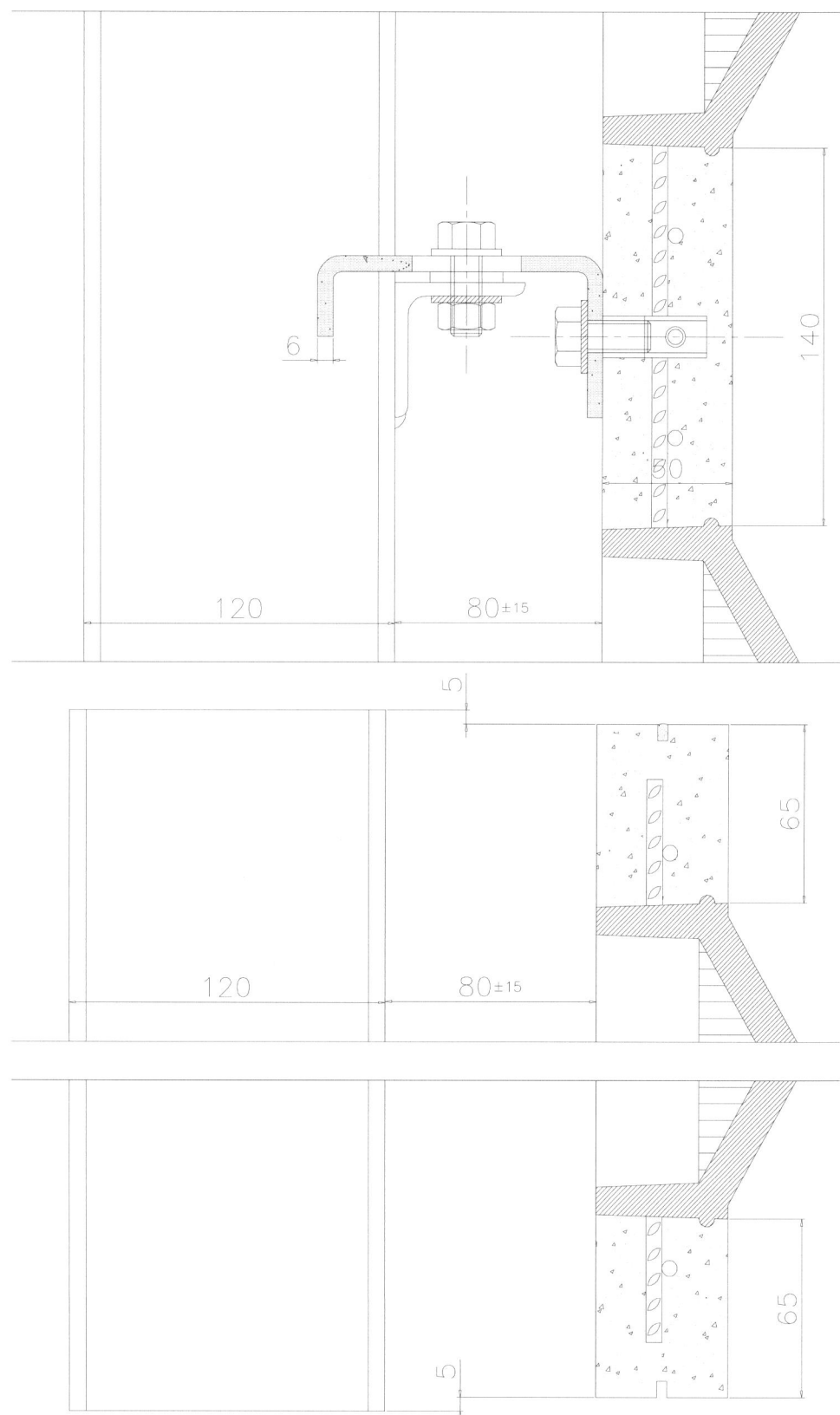

Figure 4.6.36
Section details through the GFRC panels. The panel interior reinforcing bars are made of stainless steel.
© Brisac Gonzalez.

The screen is made with custom-pressed glass lenses and wet-cast glass fiber-reinforced concrete (GFRC) (Chapter 4.1). Each glass lens measures 6 ¼ in × 6 ¼ in (16 × 16 cm) in plan and is approximately 3 ½ in (9 cm) tall. Its exposed, outside face is a smooth pyramid, while its inside surface is facetted to make it

Figure 4.6.37
Saverbat and the design team are evaluating prototypes of the lenses. © Brisac Gonzalez.

a Fresnel lens.[21] The lens is open to the back and has a half-round collar on the outside to mechanically bond the glass to the surrounding GFRC panel. The lenses were set into a rubber mold with stainless steel reinforcing bars. The GFRC panels are 2 in (5 cm) thick and their inside face is coplanar to the legs of the glass lenses (see Figure 4.6.35). The GFRC panels are supported by a steel substructure that provides a technical space between the panels and the building's central volume. During the day, natural light hits the upper panel lenses, and at night, exterior projectors mounted at the top and bottom of the interstitial space project lighting programs through the lenses. The lenses became the distinguishing feature of the building, and it is their pyramidal shape that gave the building its name, *Le Prisme*.

Figure 4.6.38
The glass lenses being pressed at the subcontractor. © Brisac Gonzalez.

Figure 4.6.39
The glass lenses coming out of the annealer where they cooled under controlled conditions. © Brisac Gonzalez.

There are 25,000 custom glass lenses in Le Prisme. During design, Brisac Gonzalez contacted the two glass block manufacturers in France—Saverbat and La Rochère. Brisac recalled that La Rochère would not produce a custom press glass for a production run of less than 100,000 units, but Saverbat could. Brisac said that Saverbat were amazing and that they were confident that they could do the job with the given production run. Saverbat has experience reproducing historic pressed glass slabs and blocks for building restorations and can produce small production runs of custom units. Saverbat produced prototypes of the lenses to evaluate the balance among the Fresnel facets, the transmission of light, and the lens transparency. For production, Saverbat subcontracted the lens pressing to a manufacturer in Portugal with whom they had worked previously. It took the manufacturer nine months to produce all of the lenses. Brisac felt working with Saverbat was a "true collaboration," and she enjoyed working together and problem-solving the design. Brisac kept in touch with the owner of Saverbat for several years after the project was completed.

Notes

1. Both the terms *mold* and *die* may be used to describe the tooling for pressing. The term *die* is typically reserved for two shaped tooling halves that are forced together. This includes dies that are used for metal stamping, hydroforming, and mechanically or hydraulically pressing clay.

2. Manufacturers use the term *compression molding* when describing pressing plastic or fiber-reinforced plastic throughout the mold.

3. A male plunger is a generic shape that is typically used when the interior of the press component is not exposed. Sometimes, one plunger can be used for multiple pressed component shapes.

4. Alesch, Stephen. *Personal Interview*. 14 April 2023.

5. Marani, Matthew. "The Fitzroy Harkens Back to Old New York with Art Deco-inspired Terra-Cotta Blocks." *Facades+*, 20 November 2020. https://facadesplus.com/the-fitzroy-harkens-back-to-old-new-york-with-art-deco-inspired-terra-cotta-blocks/. Accessed 30 March 2023.

6. Boston Valley Terra Cotta. "NYC's The Fitzroy Shines in a Video From JDS Development Group." *News*, 12 January 2018. https://bostonvalley.com/fitzroy-nyc-video-bvtc-plant-tour/. Accessed 30 March 2023.

7. Arrocet, Maximiliano. *Personal Interview*. 28 March 2023.

8. Wright, Herbert. "New Wave: Lisbon's MAAT Has Opened a Curvaceous New Waterfront Building by Amanda Levete's Practice AL_A. It Transforms the Riverside, Reframes the Idea of Exhibition Space, and Reconnects This Edgeland with the City behind It. We Review the Building, Explore Its Urban and Architectural Context, and Catch Amanda Levete for an Interview in Her Showcase Structure." *Blueprint (London, England)*, vol. /Dec., Nov. 2016, pp. 46–62.

9. Arrocet.

10. Costa Arespacochaga, Ines. "Architecture and Matter: The Ceramic Piece by Toni Cumella." *Archivo Digital UPM*, June 2020. https://oa.upm.es/65284/. Downloaded 21 February 2023.

11. Ceram is translated from Persian language and alphabet and is also translated as Saram or Seram.

12. Balazadeh, Hooman. *Personal Interview*. 15 August 2022.

13. The Tabriz Historic Bazaar is located on the old Silk Road and has been in operation since the 13th century. It is a UNESCO World Heritage site.

14. A measurement of 1 1/8 in (3 cm) thin bricks were also produced and attached to adjustable, folding metal screens to shade upper level exterior spaces and large glazed openings.

15. *Kohan Ceram Manufacturer of Refractory Brick 2019 Catalogue.*
16. There is no airspace or moisture barrier inside the wall. Instead, silica is added to keep any moisture from condensing on the glass lenses and lights.
17. Hooba Design Group. "Project Information." Kohan Ceram. http://www.hoobadesign.com/show.aspx?pid=10079. Accessed 15 September 2022.
18. Balazadeh.
19. Woodman, Ellis. "Rock Chic." *Building Design*, 18 April 2008, pp. 10–15.
20. Brisac, Cecile. *Personal Interview.* 12 December 2022.
21. Used in lighthouses, a Fresnel lens is a compact lens that can intensify light as it passed through the glass without adding thickness.

References

Alesch, Stephen. *Personal Interview.* 14 April 2023.

Arrocet, Maximiliano. *Personal Interview.* 28 March 2023.

Balazadeh, Hooman. *Personal Interview.* 15 August 2022.

Boston Valley Terra Cotta. "NYC's The Fitzroy Shines in a Video From JDS Development Group." *News*, 12 January 2018. https://bostonvalley.com/fitzroy-nyc-video-bvtc-plant-tour/. Accessed 30 March 2023.

Brisac, Cecile. *Personal Interview.* 12 December 2022.

Costa Arespacochaga, Ines. "Architecture and Matter: The Ceramic Piece by Toni Cumella." *Archivo Digital UPM*, June 2020. https://oa.upm.es/65284/. Downloaded 21 February 2023.

Hooba Design Group. "Project Information." *Kohan Ceram.* http://www.hoobadesign.com/show.aspx?pid=10079. Accessed 15 September 2022.

Kohan Ceram Manufacturer of Refractory Brick 2019 *Catalogue.*

Marani, Matthew. "The Fitzroy Harkens Back to Old New York with Art Deco-Inspired Terra-Cotta Blocks." *Facades+*, 20 November 2020. https://facadesplus.com/the-fitzroy-harkens-back-to-old-new-york-with-art-deco-inspired-terra-cotta-blocks/. Accessed 30 March 2023.

Woodman, Ellis. "Rock Chic." *Building Design*, 18 April 2008, pp. 10–15.

Wright, Herbert. "New Wave: Lisbon's MAAT Has Opened a Curvaceous New Waterfront Building by Amanda Levete's Practice AL_A. It Transforms the Riverside, Reframes the Idea of Exhibition Space, and Reconnects This Edgeland with the City behind It. We Review the Building, Explore Its Urban and Architectural Context, and Catch Amanda Levete for an Interview in Her Showcase Structure." *Blueprint (London, England)*, vol./Dec., Nov. 2016, pp. 46–62.

CHAPTER
4.7

Injection Molding

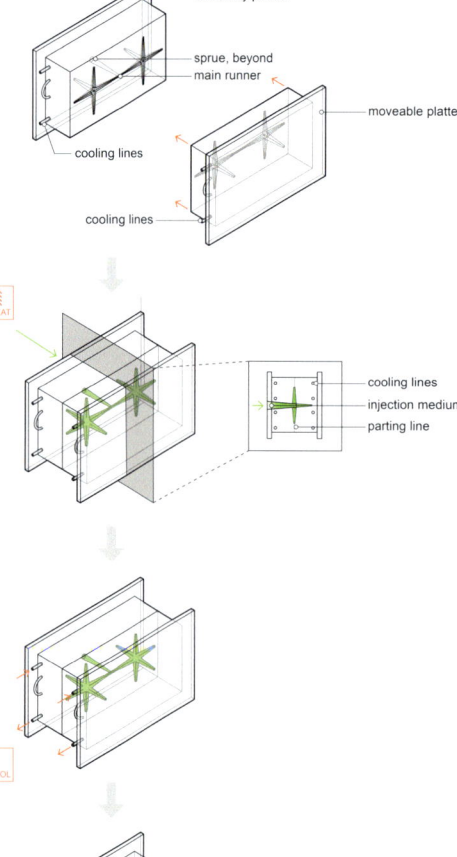

stationary platten

sprue, beyond
main runner

moveable platten

cooling lines

cooling lines

HEAT

cooling lines
injection medium
parting line

COOL

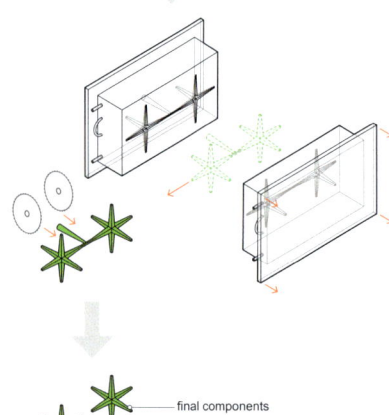

final components

Figure 4.7.1
Injection molding process diagram.

DOI: 10.4324/9781003299196-26

Injection molding is the manufacturing process that uses high pressure to push a liquid medium into a mold to form a desired shape. Injection molding typically uses thermoplastic but can include thermoset plastic, fiber-reinforced plastics (FRPs), and metal powders in a polymeric matrix. If injection molding liquid metal or metal alloys, then the process is called *die casting*.[1] Injection molding produces a wide range of shapes, including open thin walled, semi-hollow, or solid shapes. Since high pressures are needed to inject the medium into the mold, generally large sizes are limited, and shopping cart baskets are close to the upper size limit for this process. Molding pressures also require that molds be made of high-grade tool steel, and molds typically have embedded cooling lines. If a high production run is required or the component shape is complex, then the mold may use ejector pins, retractable cores, or other moving parts, which increase mold costs. Generally, lead times are long for injection molding due to mold complexity, but cycle times are short, making this process best suited for components with large production runs.

In injection molding, the mold, with its embedded cooling lines, is placed between the injection molding machine platens. The hydraulic press closes the mold, and the machine injects the liquid medium into the mold. Cooling liquid is sent into the embedded lines to cool the medium so that it can be demolded. The mold opens and the component is removed. If a high-production run is required, then moveable ejector pins will push the component out of the mold, making demolding automated. Then the mold closes, and the cycle begins again. The gating system is removed from the component.

Molds for injection molding are closed, made with solid tool steel, and come in two or more parts. Plastic shrinks as it cools, with the shrinkage rates depending on the specific type of plastic (e.g. acrylic, polypropylene, and polyethylene) used; therefore, injection molds are designed for a particular plastic. Molds can have multiple, moving parts such as hydraulic slides, collapsible cores, or spinning cores to form components with undercuts. Molds can be designed so that a single gating system serves multiple component cavities. This means that multiple components can be molded during the same cycle, decreasing the number of cycles required and the overall production time.

Thermoplastic

Almost all thermoplastics can be injection molded with the common plastics being acrylonitrile butadiene styrene (ABS), polyethylene (PE), polycarbonate (PC), nylon, polystyrene (PS), and polypropylene (PP). Additives, added to the plastic to improve the material's resistance to fire and UV-degradation, may impact its moldability.

Figure 4.7.2
Aggregate Pavilion 2018 is made from two sizes of injection-molded granular material. The smaller units can be seen at the bottom of the pavilion, and the larger units at the top. © Institute for Computational Design and Construction (ICD), University of Stuttgart.

Figure 4.7.3
The main space in the pavilion's interior. © Institute for Computational Design and Construction (ICD), University of Stuttgart.

Figure 4.7.4
The ICD used a custom-built robot suspended on tension cables to pour inflated
balls and the pavilion's hexapods and decapods. © Institute for Computational
Design and Construction (ICD), University of Stuttgart.

Aggregate Pavilion 2018 in Stuttgart, Germany
By University of Stuttgart Institute for Computational Design and Construction (ICD)

Aggregate Pavilion 2018 was constructed of non-bonded, granular materials that were poured in small amounts by a robot to form the walls and the ceiling of the structure.[2] The pavilion was approximately 16 ft wide by 23 ft long by 10 ft tall (5 m by 7 m by 3 m, respectively) and its interior included two interconnecting vaults, one for the entry and the other for the main space. The floor of the pavilion was made from gravel and the walls were made of a mix of 70,000 hexapods that were custom injection-molded for this pavilion and

50,000 decapods that were also custom injection molded but were reused from an earlier University of Stuttgart Institute for Computational Design and Construction (ICD) research project. The ICD used a custom-built robot suspended on tension cables to construct the pavilion, pouring the granular material where it was needed. The containers that stored the hexapods and decapods prior to pouring were used as formwork for the pavilion's exterior. To support the walls and ceiling during construction 750 off-the-shelf inflated balls were used on the pavilion's interior to form the vaulted space. After construction was completed, the balls flowed out of the pavilion, leaving the vaulted interior spaces; both the balls and the containers were removed prior to being completed.

Figure 4.7.5
The 750 off-the-shelf inflated balls flowed out of the pavilion, leaving vaulted interior spaces. A few inflated balls were stuck in the hexapods and removed.
© Institute for Computational Design and Construction (ICD), University of Stuttgart.

The value of granular materials is if the granules have enough friction to interlock, then they can form a self-supporting structure while maintaining their individual integrity. This allows the granular material to be reused and reconfigured into another structure. The ICD's research in granular material for architectural application started in 2009 when now Humboldt University of Berlin Matters of Activity Professor Karola Dierichs joined the University of Stuttgart as a doctoral student under the direction of Professor and ICD Director Achim Menges.[3] Prior to joining the doctorate program at University of Stuttgart, Dierichs worked with sand, another granular material, while earning her master's degree at the Architectural Association Emergent Technologies and Design (EmTech). Dierichs worked on several ICD constructions using granular material, including the 2015 Aggregate Pavilion, the 2017 Aggregate

Figure 4.7.6
A precursor study by Dierichs of using granular material to form and arch. © Institute for Computational Design and Construction (ICD), University of Stuttgart.

Figure 4.7.7
The approximately 12 in (300 mm) hexapod custom injection molded for this pavilion. © Institute for Computational Design and Construction (ICD), University of Stuttgart.

Wall, and smaller studied prototypes such as columns and an arch. The different granule shapes and sizes affect how the granular constructions compact under loading, stability of the granular material, what loads it could take, and the density of the granular material in the structure. Dierichs' primary research contribution is systematizing granular materials and taking it to the large scale.[4]

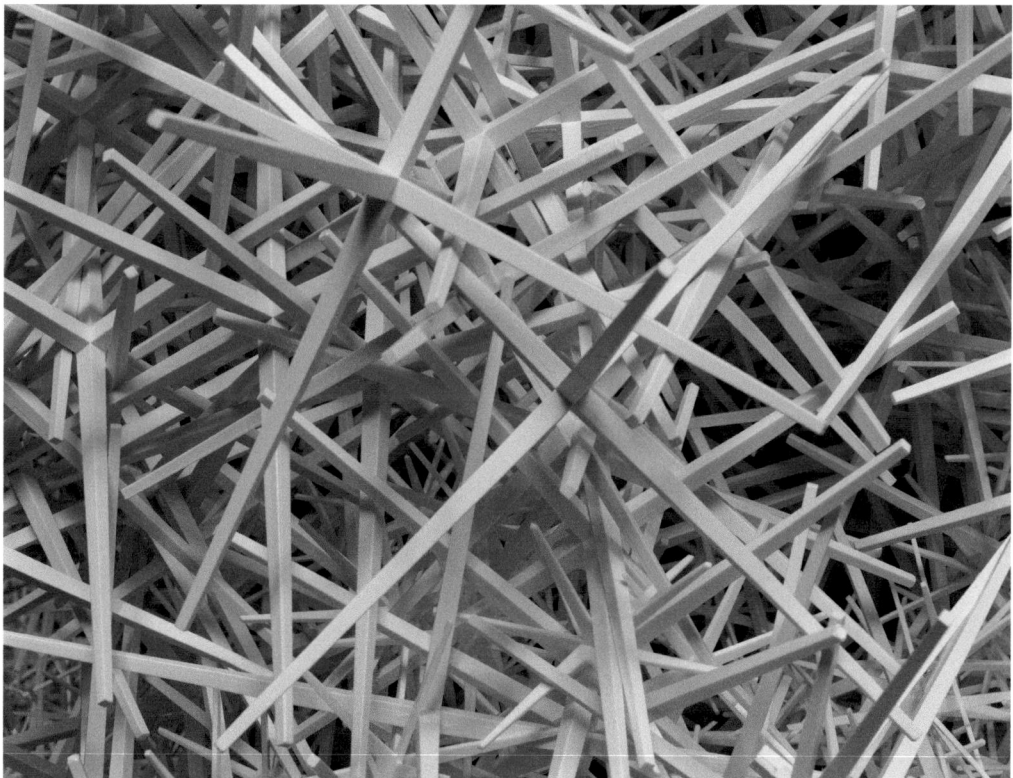

Figure 4.7.8
The shape of the hexapod allows it to interlock with others to form a stable pavilion. © Institute for Computational Design and Construction (ICD), University of Stuttgart.

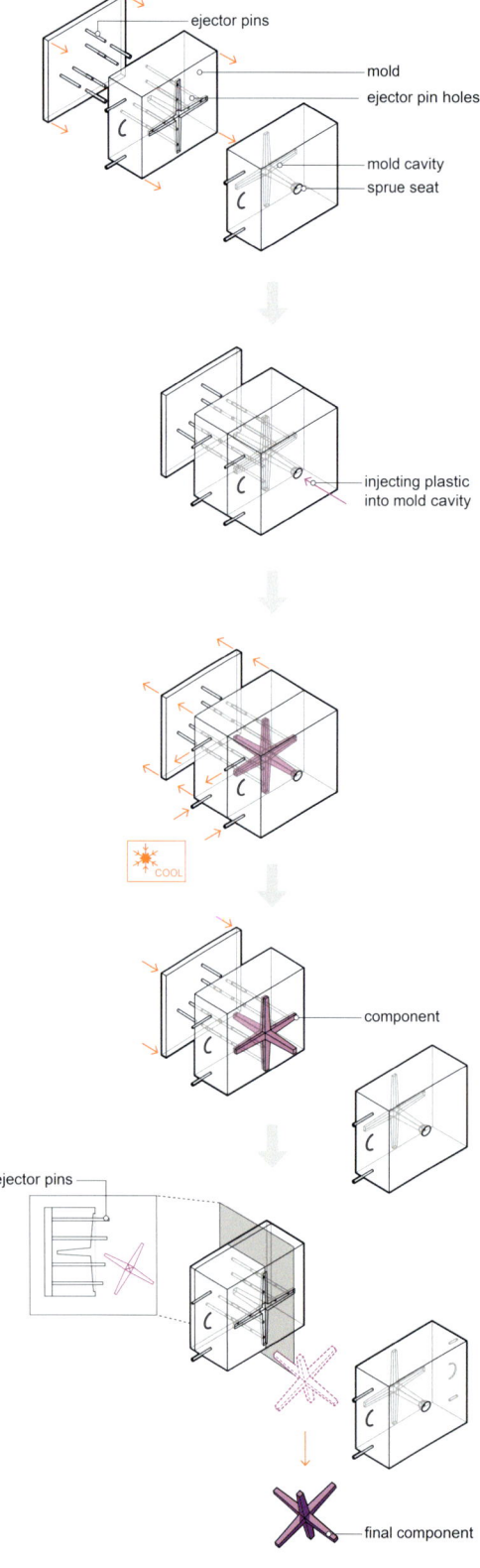

ejector pins
mold
ejector pin holes
mold cavity
sprue seat
injecting plastic into mold cavity
COOL
component
ejector pins
final component

Figure 4.7.9
An arm of the hexapod served as the sprue, and no gating system was required. Injector pins made cycle times short. © Institute for Computational Design and Construction (ICD), University of Stuttgart.

The 2018 Pavilion uses approximately 12 in (300 mm) hexapods at the top of the structure for lightness and volume and also uses smaller decapods on the outside of the base.[5] The decopods were recycled from a former study and were reused to add volume as the hexapods were compressing more than anticipated during construction. Wilhelm Weber GmbH, an injection mold fabricator, made the hexapod's custom, two-part mold and subcontracted their manufacturing to another company. The injection molder used plastic recycled from the local car manufacturing industry. Collaboration among ICD, Wilhelm, and the injection molder was straightforward, but according to Dierichs, there was some delay during fabrication as the hexapod's long arms required the shape to be in the mold for longer than hoped, slowing down cycle times.[6] To automate the injection molding process, Wilhelm designed the molds to use ejector pins to push the hexapods out of the mold after they

Figure 4.7.10
Half of the injection mold. The mold's smooth surface reduced friction, making the granules compact more than anticipated. © Institute for Computational Design and Construction (ICD), University of Stuttgart.

had solidified. There was no gating system in the production of the hexapods as the sprue formed one of the arms; therefore, no post-production was needed. The surface of the hexapods was smooth, to facilitate demolding; however, its smooth surface allowed the granules to slide past one another and compact more than early ICD studies with 3D-printed granule prototypes.

Notes

1. Die casting and injection molding are essentially the same process in which a liquid medium—metal or plastic, respectively—is forced under high pressure into a closed mold. The difference in terms is only related to the difference in medium. Die casting is for metals and metal alloys, while injection molding is for plastic.
2. The pavilion has since been disassembled and its granular parts stored to be used on another project.
3. Dierichs, Karola. *Personal Interview*. 15 July 2022.
4. *Ibid*.
5. Dimensions are the diameter of an invisible boundary circle in which the star-like shapes of the pods would fit.
6. The hexapod shapes were restricted by performance requirements, and therefore, changes to the hexapods design to shorten cycle times could not be accommodated.

Reference

Dierichs, Karola. *Personal Interview*. 15 July 2022.

Acknowledgments

Thank you to those who helped me make this book possible.

Thank you to my family: my husband, Sean Tobin, who just smiled and nodded when I told him my next book idea and how that one would be easier, and my daughter, Adele Tobin-Gulling, who just asked, "Is this one done, yet?"

Thank you to my editors for agreeing to this project and answering my questions along the way.

Thank you to the NC State students, Brianna Creviston, Nicole Cuneo-Williams, and Daniel Garrett, who worked with me to correspond with architecture firms, gather case study documents, gain copyright permissions, and draw graphics and diagrams.

Thank you to NC State College of Design and School of Architecture administration, particularly SoA Head David Hill, for giving me time and support to work on this.

And thank you to the architecture firms for providing information about each project and taking the time to answer my questions. This would not have been possible without your participation.

Glossary

Acid washing   A post-production finishing process for cast concrete, in which the manufacturer sprays diluted acid over the cured concrete to remove its cement surface, or laitance layer. Generally, the acid wash does not expose the aggregate and gives the concrete a finish similar to cast stone.

American Society of Testing and Materials (ASTM)   An organization that sets international, voluntary technical standards for a range of materials, products, systems, and services. www.astm.org

Anisotropy   The property of a material to behave differently along different axes. Most notable anisotropy is wood's greater compression strength for loads applied parallel rather than perpendicular to its grain.

Annealing   *Glass*: Annealing glass is a post-forming process, necessary to relieve internal stresses that form during shaping. When annealing glass, it is heated and then cooled slowly. Annealed glass is more durable, stronger, and less prone to cracking when exposed to temperature changes or mechanical stress than glass that has not been annealed. *Metal*: Annealing metals may be done before, during, or after forming the metal workpiece. The metal is heated and then some metals are cooled slowly, while others are quenched in water. Annealed metal is more ductile and easier to form than metal that has not been annealed.

Architectural precast concrete   Precast concrete in which its primary objective is to provide an architecturally acceptable surface. Architectural precast concrete may or may not be load bearing or provide lateral stiffness.

Axisymmetric   A symmetry along a rotated axis.

Band clamp   An adjustable clamp that encircles the item or items being clamped.

Batch kilns   Kilns that operate in cycles. They turn on and off between loading and unloading.

Bending metal   Use of a break press to create a bend in a localized area of a metal blank or workpiece.

Bendy plywood   Laminated wood-veneer sheet products that are made to bend.

Bent plywood   Plywood with simple, or one-directional, curves.

Billet   In aluminum extruding, the billet is a cylinder of solid aluminum that is loaded into the extruder and forced through a die to form the extrusion.

Bladder   A flexible membrane that can apply even pressure.

Blank   A sheet material, not yet in its final shape. In metal sheet forming (e.g. stamping, spinning, or hydroforming), a blank is a precut sheet of metal that has not been formed. In bending plywood, a blank is the bent plywood form that has not yet been trimmed.

Blank holder   In metal forming, it holds the blank in place during forming.

Blanking   Cutting a piece of sheet metal from a larger sheet of metal. The metal that is cut away is the piece kept, while the remainder of the sheet is scrap.

Blowpipe   Hollow tubes for blowing-molding glass. Typically there is a mouthpiece on one end for blowing and opening at the other end for the workpiece.

Bosses   A solid circular feature added to the wall of a component. Bosses give a self-taping screw a place where it can be drilled without damaging the component.

Bowing   Out of plane curvature, along the components length.

British Standards (BS)   Set by the British Standards Institution (BSI), voluntary, technical standards for a range of materials, products, systems, and services. www.bsigroup.com

Bug holes   Small surface pockets formed by trapped air bubbles on the surface of precast concrete.

Bush hammering   Fracturing the surface of concrete (or masonry) with a mechanically- or pneumatically driven hammer so that the interior aggregate is exposed.

Calendaring   A manufacturing process in which a material is passed through a series of rollers to form into long, thin sheets. Calendaring is used to manufacture rolls of paper, textiles, rubber, or plastic.

Capital costs   Fixed, one-time expenses associated with production. It can include the cost of land, buildings, equipment and machinery, and tooling.

Casting   *Noun*: The object resulting from the casting process. This term particularly refers to cast metal components. *Verb*: Pouring a liquid medium into a mold.

Catenary curve   The draped shape that a material forms naturally, due to gravity and self-weight.

Caul   Used often in wood working to provide even pressure between clamps.

Charge   *Explosive forming*: The explosive material used for each explosion or cycle. *Glass*: The starting batch of molten glass.

Chopped strand mats (CSM)   Fiber mats made from short, chopped fibers held together by a resin-soluble binder.

Clay   Natural, fine-grained earth that can be plastically shaped when wet.

Clay slip   Clay that is suspended in water.

Closed molds   Molds that have limited openings to access their inner cavity. Generally, closed molds produce components with all faces finished.

Computer-aided manufacturing (CAM)   Manufacturing that relies on CNC equipment.

Computer numeric controlled (CNC)   Computer-controlled electronics and motors that operate a machine's production in a precise and reproducible manner.

Coefficient of thermal expansion   The fractional change in material size per degree temperature change.

Cold working   Working metal at room temperature.

Complex curve   A shape in which two or more curves intersect one another.

Compound die   A metal stamping die that can perform more than one operation in a single stroke.

Continuous filament mat (CFM)   A non-woven mat that has random, swirled, and indefinitely long fibers held together by a resin-soluble binder.

Continuous kilns   Kilns that are kept continuously at an elevated temperature.

Continuous manufacturing   Manufacturing processes that operate continuously, rather than in cycles.

Contract manufacturers   Manufacturers that manufacture items to fill a contracted-order and do not produce their own final products.

Cope   The upper half of a closed mold.

Crazing Fine cracks that appear on a surface.

Creel In pultrusion or filament winding, a rack or collection of fiber strands.

Cullet Recycled glass.

Cycle times The amount of time it takes for a full cycle to be completed.

Cyclical manufacturing Manufacturing processes that operate in repeatable cycles. Components are removed from the tooling before the cycle begins again.

Deep draw Stretching the sheet material far out of plane, generally over a small surface area.

Demold Removing the component from a mold.

Die A tool that impresses a shape onto a material.

Die angle In extrusion, the die angle is on the die's upstream side, which funnels the medium toward the die opening. The steeper the die angle, the more pushing force needed to move the medium through the die.

Die clearance In metal stamping, the distance between two die halves. The die clearance affects the wall thickness of the component.

Die swell In plastic extrusion, the property of plastic to change shape when exiting the die. The die swell is specific to the type of plastic being extruded.

Dip mold In glass blowing, a partial or open mold used to shape glass. The molten glass is dipped into the mold, inflated, and then removed from the mold.

Direct extrusion A type of metal extrusion in which the billet is pushed against a die.

Do-it-yourself (DIY) The term for doing work without the direct help of professionals or experts. Skills within this group can vary greatly.

Draft angle The angle needed to remove the component from its mold without damaging either the component or mold.

Drag The bottom half of a closed mold.

Drape forming In thermoforming, this is forming the component over a male mold.

Draw angle See draft angle.

Draw bead In metal stamping, a rib-like projection along die edges that lock the blank in place and control its flow rate into the die.

Draw radius The radius needed on the exterior corners of the die cavity entrance to keep the medium from tearing upon entry.

Draw depth The depth that sheet material can deform out of its original plane.

Draw ratio The ratio of the surface area of a deformed workpiece to the surface area of the original blank.

Ductility The ability of a material to stretch plastically, without damage.

Earth-moist For clay or concrete, so little water that when the earth-moist medium is squeezed in one's hand, no water residue is left behind; however, there is just enough moisture that the medium can hold its shape during demolding.

Ejector pins Movable pins as part of a tool that help with demolding.

Elastic behavior A material's behavior if stressed below its yield point. When the stress or load is removed, the deformed materials return to their original shape.

Electronic discharge machine (EDM) A CNC machine that uses electronic discharge to cut or shape metal.

Envelope molds In casting concrete, a five-sided mold that is left intact when demolding.

Expendable mold A mold that is temporary and not reusable from one cycle to the next.

Extrusion ratio In metal extrusion, the ratio of a billet's cross-sectional area to the cross-sectional area of the extruded profile.

Extrudate The material as it comes out of an extrusion die.

Faience Glazed clay.

Feedstock Loose medium in plastic extrusion (e.g. pellets or flakes) that is fed into a hopper.

Female molds Molds with an interior cavity, in which the component is formed.

Ferrous alloys Metal alloys that contain iron.

Fiber fabric matt Fibers that are woven together to create a fabric.

Five-axis CNC A CNC mill with five directions of movement, typically able to produce undercuts.

Flame finished See flame polished.

Flame polished The natural finish for glass when it is exposed to high temperatures. Flame polished glass with be smooth, shiny, and—if the glass is transparent—clear.

Flange In metal stamping, the part of the blank that does not enter the die cavity. Typically removed during post-production.

Flashing Excess medium that squeezes out between mold and die parts. Typically removed during post-production.

Flask In sand casting, the box that holds the compressed sand mold.

Follower block See following block.

Formability The ability of a metal to undergo plastic deformation without tearing.

Formliner Placed on the inside of a concrete mold and generally is used to affect the surface of architectural precast. Formliners are typically made of rubber or thermoformed plastic.

Gage Sheet material thickness.

Gating system The network of channels that molten media flows into a closed-mold cavity.

Glory hole A reheating chamber for working glass, typically in a glass studio setting.

Grain orientation The direction of the crystalline structure within a metal alloy. Affects the direction that the metal can be bent.

Green clay Clay that just has been removed from the mold. Green clay is stiff enough to hold the shape and can easily be carved with hand tools and little pressure.

Grog Fired clay scraps that are ground into small pieces.

Hybrid composite A fiber-reinforced composite with two or more different fibers.

Hot working Working metal at a high temperature. The temperature is low enough that the metal is solid but high enough that the metal can be plastically deformed.

Indirect extrusion A type of metal extrusion in which the die is pushed against the billet.

Injection molding A manufacturing process that uses pressure to force a medium (e.g. plastic) through a gating system into a closed mold.

In-line (processes) A process that is integrated into the production or assembly line.

Isotropy The property of a material to behave the same along different axes. The opposite of anisotropy.

Jig A device that holds the workpiece intermittently so that it can be worked on. In bending, this is usually done with a series of clamps, positioned along a track.

Kirksite   A zinc alloy that can be cast at a relatively low melting temperature.

Laitance   A layer of fine sand and hydrated cement that appears on the surface of the concrete. It is weaker than the interior of the concrete and is prone to cracking.

Lampworking   Using a concentrated flame to heat glass to a working temperature.

Layup   *FRP*: The method of applying fibers and resin onto a mold. *Plywood*: A stack of ply-wood veneers and wet adhesive.

Leather hard   Clay that has dried to a similar stiff-ness of leather, if the leather were the same thickness of the clay. Leather-hard clay can be handled without damage and trimmed or gouged with pressure and hand tools.

Leadership in Energy and Environmental Design (LEED)   A third-party certification program to measure the environmental impact of buildings.

Lehr   A temperature-controlled tunnel or continu-ous kiln to anneal glass.

Lift   For casting concrete, a layer of poured concrete.

Male molds   A positive mold on which the compo-nent is formed.

Master   A mold used to make a pattern that is then used to make the mold used in manufacturing.

Match plate   Two sides of a pattern attached on opposite sides of a plate.

Model   Similar to pattern.

Molded plywood   Plywood shaped into a complex curve.

Mold side   The face of the component that is or was in direct contact with the mold.

Near net shape   A manufactured component that required little post-production work such as cutting, trimming, or milling.

Net shape   A manufactured component that requires no post-production work.

Non-ferrous alloy   An alloy without iron.

Open molds   Molds with large openings to access the mold cavity. Generally, open molds pro-duce components with one unfinished face.

Optic mold   See dip mold.

Over press   Extra medium that is squeezed out of the press because too much was put in the mold or the die. Over press should be removed with post-production finishing.

Parallel productions   Running two or more simul-taneous productions on the same or similar molds or tools. Parallel productions are done to speed up production schedules when mold costs are low.

Parison   Molten glass or plastic that is to be inflated inside of a mold.

Parting agent   A material or coating between the mold and the workpiece that allows the work-piece to be easily demolded.

Parting line   The seam between mold parts; typi-cally the seam is in a single plane.

Pattern   Tooling used to repeatedly form expend-able molds.

Pattern tree   In investment or lost-wax casting, a group of patterns organized around a central gating system. With a pattern tree, a single pour can manufacture multiple castings.

Permanent mold   A mold that is to last two or more cycles.

Plasticity   *Clay*: When the clay can be rolled and bent without cracking. *Metal*: When metal is stressed above its yield point but below is strain-harden-ing phase. Little to no additional loading is placed on the material, but it permanently deforms.

Post-production Secondary processes that are required after manufacturing.

Pot life The amount of time that a thermoset resin will remain viscous before setting.

Preform In blow molding, a semi-shaped pieced that has been formed in a mold, prior to being heated and blown into a mold cavity.

Production runs The number of components that are to be produced for a particular job.

Production times The amount of time it takes to produce all of the components for a particular job.

Production volume The number of components that are to be produced for a particular job or from a particular tool.

Progressive die In metal stamping, a die with multiple steps that work in sequence to shape sheet metal.

Pug mill A machine that is used to mix wet, or plastic, clay. Usually, modern pug mills have an auger that can both mix and feed the clay through an opening.

Punch die In metal stamping, the male part of the die.

Punching In metal stamping, cutting a piece of metal from a blank, in which the metal that is cut away is scrap (aka slug) and what is left of the blank is kept.

Quenching Quickly cooling a material after heating, typically done in water but can be done with oil.

Refractory Additive ingredients in a mold medium so that it withstands high temperatures.

Reglet A small notch in a material.

Restriking In metal stamping, after a workpiece is trimmed, it is placed into the die to be stamped again.

Reverse extrusion See indirect extrusion.

Rib A feature used to strengthen the component's wall without thickening the cross section.

Riser In metal casting, a riser stores molten metal so that as the metal shrinks during cooling, molten metal from the riser will flow into any resulting cavities.

Sacrificial tooling Tooling that is sacrificed during the manufacturing process and cannot be reused to produce additional components.

Sag The drooping of material over time.

Screed The straight-edged tool used to level concrete or the act of leveling concrete with a straight-edged tool.

Shot The amount of media that is placed in the mold for each cycle.

Sink marks A localized surface depression, due material shrinkage while cooling. They most often occur where the component walls are thickest, as those sections are the last to cool.

Slip block In concrete casting, a mold part that is sacrificed or removed as the component is demolded.

Slug *Stamping*: The scrap metal resulting from a punching operation. *Extruding stiff mud*: The mid-length section (approximately 8 ft long) that is rough cut as it comes out of the extruder.

Slump test A test to measure the workability of newly mixed concrete.

Spring back The elastic behavior in sheet metal forming, in which a newly deformed metal sheet partially returns to its original shape.

Stripping Removing concrete from a mold.

Stroke Each cycle of a metal stamping press.

Structural precast concrete Precast concrete in which its primary objective is to perform a structural function. Structural precast may be hidden or exposed.

Superplasticity Unusual capacity of a material to withstand large amounts of uniform strain without necking or rupturing.

Tempering glass Reheating annealed glass to a high temperature, and then quickly cooling the glass surface.

Thermoplastic Types of plastic with no bonds between their polymer chains. Thermoplastics are shaped by heating them to a prescribed temperature.

Thermoset plastic Types of plastic with chemical bonds between their polymer chains. Thermoset plastics are shaped before the bonds form; after bonds form, they can only be reshaped by breaking the bonds.

Tolerance The allowable variation of a component's dimensions.

Tooling *Manufacturing*: Tools, such patterns, molds, dies, and jigs, used to form the component. *Precast concrete*: Post-production finish tools, such as chisels, needle guns, and bush hammers, which change the cast surface of the concrete.

Tow A coarse or broken fiber used in forming composites.

Trim die A tool used to hold the workpiece during post-production trimming.

Tunnel kiln Large kiln that runs continuously. Room-temperature components are fed into one end of the kiln and emerging at the other end, having been elevated to their required temperature.

Undercuts A recess in the component's surface that make demolding difficult or impossible.

Veil For a composite, the outer layer that provides a smooth, finished surface.

Vents Small holes in tooling that allow for air flow.

Viscosity The resistance of a liquid to flow, due to internal friction.

Wandering parting line A seam between mold parts, not in a singular, geometric plane.

Warm working Working a material above room temperature but below hot working. For metal, this is below the metal's recrystallization temperature.

Warping Twisting of a member so that two corners do not fall in the same plane.

Wood products Manufactured materials made from wood or wood waste. These include oriented strand board (OSB), plywood, and particle board.

Workpiece The piece that is in the process of being manufactured. It is in an in-between state—neither the medium, blank, or final component.

Wythe A vertical layer of masonry unit or product.

Index